Fungal Biology

Series Editors

Fungal biology has an integral role to play in the development of the biotechnology and biomedical sectors. It has become a subject of increasing importance as new fungi and their associated biomolecules are identified. The interaction between fungi and their environment is central to many natural processes that occur in the biosphere. The hosts and habitats of these eukaryotic microorganisms are very diverse; fungi are present in every ecosystem on Earth. The fungal kingdom is equally diverse, consisting of seven different known phyla. Yet detailed knowledge is limited to relatively few species. The relationship between fungi and humans has been characterized by the juxtaposed viewpoints of fungi as infectious agents of much dread and their exploitation as highly versatile systems for a range of economically important biotechnological applications. Understanding the biology of different fungi in diverse ecosystems as well as their interactions with living and non-living is essential to underpin effective and innovative technological developments. This series will provide a detailed compendium of methods and information used to investigate different aspects of mycology, including fungal biology and biochemistry, genetics, phylogenetics, genomics, proteomics, molecular enzymology, and biotechnological applications in a manner that reflects the many recent developments of relevance to researchers and scientists investigating the Kingdom Fungi. Rapid screening techniques based on screening specific regions in the DNA of fungi have been used in species comparison and identification, and are now being extended across fungal phyla. The majorities of fungi are multicellular eukaryotic systems and therefore may be excellent model systems by which to answer fundamental biological questions. A greater understanding of the cell biology of these versatile eukaryotes will underpin efforts to engineer certain fungal species to provide novel cell factories for production of proteins for pharmaceutical applications. Renewed interest in all aspects of the biology and biotechnology of fungi may also enable the development of "one pot" microbial cell factories to meet consumer energy needs in the 21st century. To realize this potential and to truly understand the diversity and biology of these eukaryotes, continued development of scientific tools and techniques is essential. As a professional reference, this series will be very helpful to all people who work with fungi and should be useful both to academic institutions and research teams, as well as to teachers, and graduate and postgraduate students with its information on the continuous developments in fungal biology with the publication of each volume.

More information about this series at http://link.springer.com/series/11224

María Guadalupe Frías-De-León
Carolina Brunner-Mendoza
María del Rocío Reyes-Montes
Esperanza Duarte-Escalante
Editors

The Impact of Climate Change on Fungal Diseases

Springer

Editors
María Guadalupe Frías-De-León
Unit Research, Hospital Regional de Alta
Especialidad de Ixtapaluca
Mexico State, Mexico

María del Rocío Reyes-Montes
Microbiology and Parasitology Department
School of Medicine, Universidad Nacional
Autónoma de México
Mexico City, Mexico

Carolina Brunner-Mendoza
Microbiology and Parasitology Department
School of Medicine, Universidad Nacional
Autónoma de México
Mexico City, Mexico

Esperanza Duarte-Escalante
Microbiology and Parasitology Department
School of Medicine, Universidad Nacional
Autónoma de México
Mexico City, Mexico

ISSN 2198-7777 ISSN 2198-7785 (electronic)
Fungal Biology
ISBN 978-3-030-89663-8 ISBN 978-3-030-89664-5 (eBook)
https://doi.org/10.1007/978-3-030-89664-5

This Springer imprint is published by the registered company Springer Nature Switzerland AG
The registered company address is: Gewerbestrasse 11, 6330 Cham, Switzerland

To BChE María Eugenia Sánchez Conejo for her outstanding and tireless work in favor of the environment in Mexico City

Foreword

In 1969, it was announced that mankind had reached the Moon. Although this space event caused great astonishment among scientists and the general population until today, some believe it was a hoax.

Similarly, nowadays, there is much conversation about climate change and that various anthropogenic activities involving the use of fossil fuels have caused its acceleration. However, some do not believe that these events are interrelated.

The truth is that the world is experiencing climatic phenomena that have impacted human health. Hence, this book shows the impact that climate change has had or may have on the incidence and prevalence of diseases caused by fungi. Although relatively few fungi are capable of causing disease in humans, climate change may shift this picture. Therefore, it is crucial to expand our vision on the possible causal agents of mycoses as well as new perspectives on their diagnosis, treatment, and prevention.

Dirección General, Hospital Regional de Alta Especialidad de Ixtapaluca, Ixtapaluca
Estado de México, Mexico

Alma Rosa Sánchez-Conejo
María Eugenia Sánchez Conejo

Preface

The impending climate change is a cause for concern worldwide due to its negative influence on the incidence and prevalence of infectious and non-infectious diseases in humans, including fungal infections. However, mycoses are the ones that have received the least attention from health authorities, unlike those caused by other microorganisms. This book *The Impact of Climate Change on Fungal Diseases* evidences the changes in the ecology of fungi and the diseases they cause influenced by the temperature increase at a global level. Various aspects are reviewed, such as the emergence of pathogenic fungi, infections caused by fungi that were not considered pathogens, and the geographical expansion of endemic fungi, among others.

This book aims to draw the attention of health professionals to direct greater efforts on the prevention and control of fungal diseases. In addition, it exposes the urgent need for a unified approach in fungal identification to guide the treatment of those suffering from infections by these pathogens and therapeutic alternatives against fungal pathogens resistant to the main types of antifungals. Finally, given the need to prevent fungal infections, the current state of antifungal vaccines is presented, as well as vaccines with potential for commercialization in the coming years. The book contains 17 chapters, three of which explain the basics of climate change, the role of fungi in ecosystems, and the adaptation of fungi to high temperatures. Besides, nine chapters evidence the impact of climate change on infections caused by fungi; two chapters are focused on non-infectious diseases; and three chapters explain the challenges and new tools for the diagnosis, treatment and prevention of mycoses. The chapters also include tables, figures, and extensive references to guide readers.

The authors were invited to provide concise and up-to-date information on each topic based on their contributions and a particular interest in their subspecialty.

We thank all the authors for their valuable contributions and support in completing this project. We also thank Springer Nature for supporting the publication of this book, which is a pioneer in its field

Mexico State, Mexico María Guadalupe Frías-De-León
Mexico City, Mexico Carolina Brunner-Mendoza
Mexico City, Mexico María del Rocío Reyes-Montes
Mexico City, Mexico Esperanza Duarte-Escalante

Contents

Contributors

Mirta Álvarez Castelló University Hospital "General Calixto García", Havana, Cuba

Gustavo Acosta-Altamirano Unidad de Investigación, Hospital Regional de Alta Especialidad de Ixtapaluca, Ixtapaluca, Estado de México, Mexico

Yurixhi Paola Aguilera León Departamento de Microbiología y Parasitología, Facultad de Medicina, Universidad Nacional Autónoma de México, CDMX, Mexico City, Mexico

Michel Almaguer Chávez Faculty of Biology, University of Havana, Havana, Cuba

Eidi Victoria Alvarado-Ramírez Departamento de Microbiología y Parasitología, Facultad de Medicina, Universidad Nacional Autónoma de México, CDMX, Mexico City, Mexico

Raúl Ávila Sosa Departamento de Bioquímica-Alimentos, Facultad de Ciencias Químicas, Benemérita Universidad Autónoma de Puebla, Pue., Puebla, Mexico

Paola Betancourt-Cisneros Unidad de Investigación en Sistemática Vegetal y Suelo, Facultad de Estudios Superiores Zaragoza, Universidad Nacional Autónoma de México, CDMX, Mexico City, Mexico

Alexandro Bonifaz Dermatology Department, Hospital General de México "Dr. Eduardo Liceaga", CDMX, Mexico City, Mexico

Mycology Laboratory, Hospital General de México "Dr. Eduardo Liceaga", CDMX, Mexico City, Mexico

Carolina Brunner-Mendoza Microbiology and Parasitology Department, School of Medicine, Universidad Nacional Autónoma de México, Mexico City, Mexico

Laura Rosio Castañón Olivares Departamento de Microbiología y Parasitología, Facultad de Medicina, Universidad Nacional Autónoma de México, CDMX, Mexico City, Mexico

Elsa Castañeda Roldán Centro de Investigaciones en Ciencias Microbiológicas, Laboratorio de Micología, Posgrado en Ciencias Ambientales, Instituto de Ciencias-Benemérita Universidad Autónoma de Puebla, Pue., Puebla, Mexico

Esperanza Duarte-Escalante Microbiology and Parasitology Department, School of Medicine, Universidad Nacional Autónoma de México, Mexico City, Mexico

María Guadalupe Frías-De-León Unit Research, Hospital Regional de Alta Especialidad de Ixtapaluca, Mexico State, Mexico

Eduardo García-Salazar Unidad de Investigación, Hospital Regional de Alta Especialidad de Ixtapaluca, Estado de México, Mexico

Master of Health Sciences, Escuela Superior de Medicina, Instituto Politécnico Nacional, CDMX, Mexico City, Mexico

Gloria M. González School of Medicine, Autonomous University of Nuevo Leon, Monterrey, Nuevo Leon, Mexico

Rigoberto Hernández-Castro Departamento de Ecología de Agentes Patógenos, Hospital General Dr. Manuel Gea González, CDMX, Mexico City, Mexico

Francisca Hernández-Hernández Departamento de Microbiología y Parasitología, Facultad de Medicina, Unidad de Micología, Universidad Nacional Autónoma de México, Ciudad de México, CDMX, Mexico

Alejandro Jaramillo Centro de Ciencias de la Atmósfera, Universidad Nacional Autónoma de México, Ciudad de México, CDMX, Mexico

Faustino López Barrera Unidad de Investigación en Sistemática Vegetal y Suelo, Facultad de Estudios Superiores Zaragoza, Universidad Nacional Autónoma de México, CDMX, Mexico City, Mexico

Rosa Areli Martínez-Gamboa Laboratorio de Microbiología Clínica, Instituto Nacional de Ciencias Médicas y Nutrición "Salvador Zubirán", CDMX, Mexico City, Mexico

Luis Javier Méndez-Tovar Servicio de Dermatología y Micología Médica, Hospital de Especialidades, Centro Médico Nacional Siglo XXI, Instituto Mexicano del Seguro Social, Ciudad de México, CDMX, Mexico

Alma Mendoza-Ponce Centro de Ciencias de la Atmósfera, Universidad Nacional Autónoma de México, Ciudad de México, CDMX, Mexico

Beatriz Meraz-Ríos Microbiology and Parasitology Department, School of Medicine, Universidad Nacional Autónoma de México, Mexico City, Mexico

Soumya Moonjely Faculty of Mathematics and Science, Department of Biological Sciences, Brock University, St. Catharines, ON, Canada

Ricardo Munguía-Pérez Centro de Investigaciones en Ciencias Microbiológicas, Laboratorio de Micología, Posgrado en Ciencias Ambientales, Instituto de Ciencias-Benemérita Universidad Autónoma de Puebla, Pue., Puebla, Mexico

Laura Parra-Jaramillo Departamento de Microbiología y Parasitología, Facultad de Medicina, Universidad Nacional Autónoma de México, Ciudad de México, CDMX, Mexico

Angélica Pérez-Rodríguez Microbiology and Parasitology Department, School of Medicine, Universidad Nacional Autónoma de México, Mexico City, Mexico

Arturo Robles-Tenorio Instituto Dermatológico de Jalisco "Dr. José Barba Rubio", Zapopan, Mexico

Magnolia Lizbeth Martínez Aguilar Departamento de Microbiología y Parasitología, Facultad de Medicina, Universidad Nacional Autónoma de México, CDMX, Mexico City, Mexico

María del Rocío Reyes-Montes Microbiology and Parasitology Department, School of Medicine, Universidad Nacional Autónoma de México, Mexico City, Mexico

Alma Rosa Sánchez Conejo Dirección General, Hospital Regional de Alta Especialidad de Ixtapaluca, Ixtapaluca, Estado de México, Mexico

Karen Saez Gomez Centro de Investigaciones en Ciencias Microbiológicas, Laboratorio de Micología, Posgrado en Ciencias Ambientales, Instituto de Ciencias-Benemérita Universidad Autónoma de Puebla, Pue., Puebla, Mexico

Brenda Sáenz-Dávila Clinical Medicine Department, Instituto Mexicano del Seguro Social, CDMX, Mexico City, Mexico

Edith Sánchez Paredes Departamento de Microbiología y Parasitología, Facultad de Medicina, Universidad Nacional Autónoma de México, CDMX, Mexico City, Mexico

Andrés Tirado-Sánchez Dermatology Department, Hospital General de México "Dr. Eduardo Liceaga", CDMX, Mexico City, Mexico

Clinical Medicine Department, Instituto Mexicano del Seguro Social, CDMX, Mexico City, Mexico

Concepción Toriello Departamento de Microbiología y Parasitología, Facultad de Medicina, Universidad Nacional Autónoma de México, Ciudad de México, CDMX, Mexico

Denisse Vazquez-González Dermatology Department, Hospital General de México "Dr. Eduardo Liceaga", CDMX, Mexico City, Mexico

Juan Xicohtencatl-Cortes Laboratorio de Bacteriología Intestinal, Hospital Infantil de México Dr. Federico Gómez, CDMX, Mexico City, Mexico

Chapter 1
Climate Change Overview

Alejandro Jaramillo and Alma Mendoza-Ponce

Contents

Abbreviations

AR5	The fifth assessment report
CFCs	Chlorofluorocarbons
CH_4	Methane
CO	Carbon monoxide
CO_2	Carbon dioxide
GDP	Gross domestic product
GHGs	Greenhouse gases

A. Jaramillo (✉) · A. Mendoza-Ponce
Centro de Ciencias de la Atmósfera, Universidad Nacional Autónoma de México, Ciudad de México, CDMX, Mexico
e-mail: ajaramillo@atmosfera.unam.mx

M. G. Frías-De-León et al. (eds.), *The Impact of Climate Change on Fungal Diseases*, Fungal Biology, https://doi.org/10.1007/978-3-030-89664-5_1

IPCC	Intergovernmental panel on climate change
LUCC	Land-use and land cover change
N_2O	Nitrous oxide
Nox	Nitrogen oxides
O_3	Ozone
RCPs	Representative concentration pathways
RF	Radiative forcing
SO_2	Sulfur dioxide
SSPs	Socioeconomic pathways
VOC	Volatile organic compounds

1.1 Introduction

Climate change is a natural process that controls our planet's climate on timescales varying from decades to millennia. This process is mainly controlled by greenhouse gases (GHGs) that confine radiation in the atmosphere, where various feedback mechanisms occur until the Earth system reaches an energy balance. Natural forcings have controlled these processes at paleoclimatic scales, such as variations in solar luminosity, changes in the Earth's orbit, volcanic eruptions, and even continental drift (Hartmann 2016). However, anthropogenic emissions of gases such as carbon dioxide (CO_2), methane (CH_4), nitrous oxide (N_2O), and other GHGs have increased to unprecedented levels in at least the last 800,000 years, since the preindustrial era, driven by large economic and population growth (IPCC 2014a). In this chapter, we briefly describe an overview of the key concepts related to climate change.

1.2 Climate Variability

The weather has a large day-to-day variability. However, we are also aware of an underlying seasonal cycle; today may be hotter/colder than yesterday, but summer days are warmer than winter days on average. It is natural to think of the weather as a constant regular rhythm to which life has adapted. Nevertheless, if we explore past climates, we will find that the climate is not as invariable during the last thousands of years as it may seem (Hartmann 2016). Long time series of atmospheric variables (longer than 40 years) reveals that climate also has variability; for example, it is clear to us that July (the summer season) is warmer than January (the winter season) in the Northern Hemisphere, but July of this year seems hotter than July 50 years ago.

Atmospheric variables can be measured directly using instruments, which serve to create an instrumental record. However, we also have indirect observations that allow us to go further back in time, mainly using estimates from information stored in sediments of lakes and oceans, layers of ice, and tree rings, among other information sources. Additionally, new observational satellite-based systems have increased

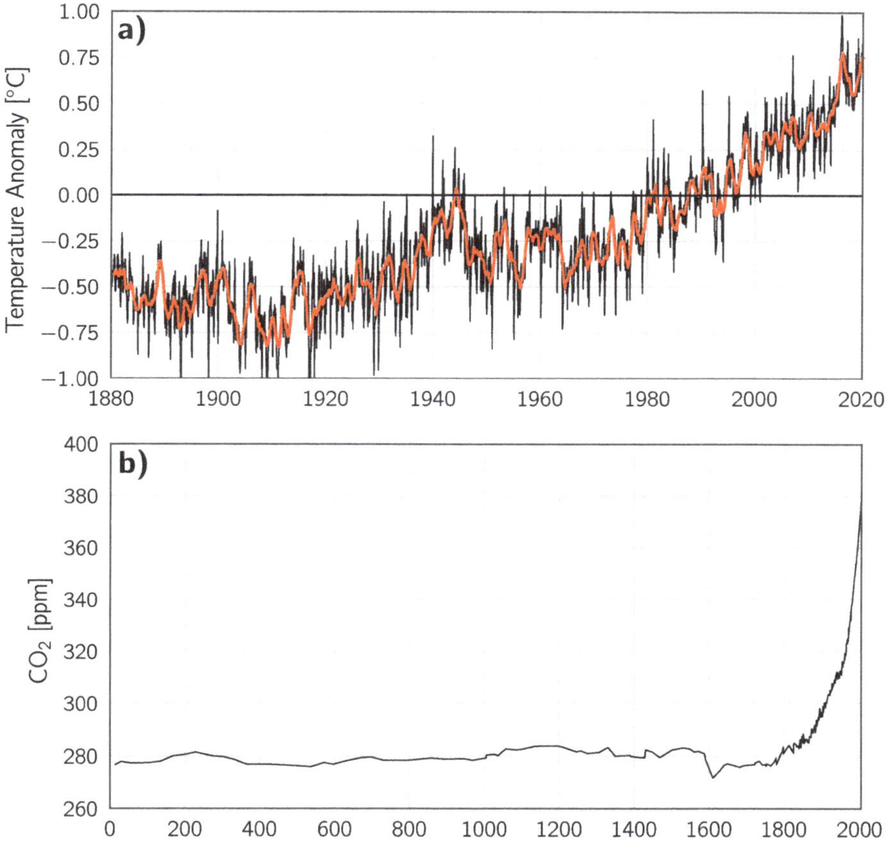

Fig. 1.1 Time series of (**a**) monthly global mean temperature anomalies for the period 1880 to 2020 for a reference period of 1971–2000 using data from Zhang et al. (2019, 2020). (**b**) Time series of estimated CO_2 concentrations of ice cores from Law Dome, Antarctica. Data from MacFarling et al. (2006)

the number of observations of the Earth's climate by orders of magnitude in the last decades (IPCC 2014b). Due to our interest in human-made global warming, considerable efforts have been made to estimate the global surface temperature based on the instrumental record and evaluate these estimated temporal trends over the past century (Hansen et al. 2010; Morice et al. 2012; Zhang et al. 2019). Figure 1.1a shows a clear increasing trend in temperature, especially since the 1980s, where the black line represents monthly global mean temperature anomalies from 1880 to 2020 with a reference period of 1971–2000 and the red line shows a 12-month moving average representing the underlying annual cycle.

Most of the climate variability is associated with natural processes. We know that the Earth has gone through freezing and cooling processes in the past, mainly associated with solar activity and the Earth's movements around the Sun. Paleoclimatic

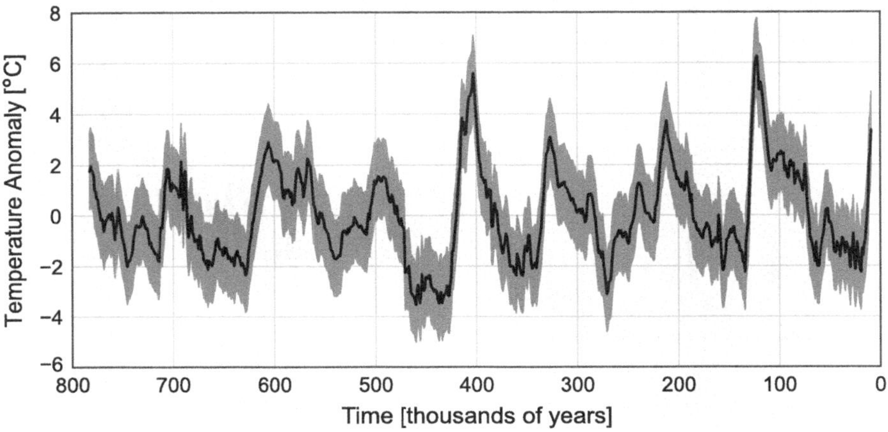

Fig. 1.2 Time series of global surface temperature anomalies from the last 784,000 years compared to the preindustrial era, reconstructed using 14 sea surface temperature paleo proxies using data from Friedrich et al. (2016)

reconstructions of key climate variables have provided insightful information about the Earth system's responses to various external forcings and its internal variability over a wide range of timescales (Hansen et al. 2010; Mann et al. 2008). Figure 1.2 shows a paleoclimatic reconstruction of a series of global surface temperature anomalies for the last 784,000 years. Natural forcings might explain these changes in the climate system, such as variations in solar luminosity, variations in the Earth's orbit around the Sun associated with Milankovitch cycles (e.g., Kerr 1987), volcanic eruptions, and even continental drift (Hartmann 2016).

Although much of climate variability is natural, human activities release large amounts of CO_2 and other GHGs to the atmosphere, which remained trapped for centuries and millennia in fossil fuels. Figure 1.1b shows a reconstruction of the estimated time series of CO_2 concentrations from ice cores in Law Dome, Antarctica. An increase in CO_2 concentrations can be observed from the mid-nineteenth century, while during the 1800 years before the industrial era, CO_2 levels remained stable at around 280 ppm. It seems that, since the industrial revolution, CO_2 levels are increasing at a faster rate than in the preindustrial period. This rise in CO_2 in the atmosphere has a clear human origin and plays a critical role in enhancing the greenhouse effect that we will explain in the next section.

1.3 The Greenhouse Effect and the Climatic Feedback Mechanisms

The greenhouse effect is a natural process where GHGs trap radiation in the atmosphere and modify the Earth's radiative balance with the Sun's incoming energy. The average temperature on the Earth's surface would be below the freezing point of water without it. Thus, Earth's natural greenhouse effect makes life as we know it possible (IPCC 2014b). Changes in the Earth system, both natural and anthropogenic, can perturb the Earth's energy balance, producing a radiative forcing (RF) that affects climate, which is a net change in the energy balance in response to an external perturbation (IPCC 2014b). When a forcing is applied, the climate system response will depend on the complex feedback mechanisms involved in the process. Therefore, the climate response will be in general different from a simple linear one depending on the kind of feedback mechanisms that can either amplify ("positive feedback") or diminish ("negative feedback") the effects of the RF (Hartmann 2016; IPCC 2014b).

An example of positive feedback is the interaction of an increase in temperature with the atmospheric water vapor. Water vapor is a potent GHG, and the capacity of air to contain water vapor increases with temperature. As the temperature increases, the amount of water vapor in the atmosphere rises, augmenting its effectiveness as a GHG. This mechanism, known as the water vapor feedback, is possibly the most significant positive feedback. Other examples of positive feedback are associated with the effects of the increase in temperature in the degradation of the permafrost (a ground layer that is completely frozen for at least 2 years) and the melting of ice and snow. The microbial decomposition in the permafrost releases methane and other GHG, contributing to even more permafrost degradation. The loss of ice and snow from the surface reduces the radiation reflected to space, warming up the surface and accelerating the melting.

Examples of negative feedback are the Planck and the lapse-rate feedbacks. The Planck feedback is related to the increase in radiation emitted by a body with increased temperature. As the body radiates more, the cooling also increases, reducing the temperature. Planck feedback is possibly the most important negative feedback mechanism. The lapse-rate feedback is related to the fact that temperature decreases with height in the Earth's troposphere. This lapse rate is critical for the greenhouse effect since the emission of infrared radiation from the Earth to space occurs mainly from the middle to high altitudes in the troposphere, where temperatures are lower than the surface. Therefore, a decrease in this rate will imply that the radiation emitted back to space will be closer to the surface temperature. Thus, the greenhouse effect will decrease as the surface cools. However, the positive water vapor feedback easily overcomes this lapse-rate feedback (Cess 1975).

Other feedback mechanisms that can be positive or negative in the energy balance are those associated with clouds and aerosols. Ice and clouds have a high albedo, reflecting large amounts of radiation into space, thereby cooling the Earth system. Besides, clouds can also retain the radiation emissions from the surface, trapping infrared radiation in the atmosphere. The behavior of clouds in cooling or warming the Earth's climate will depend on the cloud's physical properties and the condensation nuclei nature. This ambivalent effect has many uncertainties associated. Some aerosols increase atmospheric reflectivity, while others, like particulate black carbon, are strong GHGs. Aerosols also affect cloud albedo because many aerosols serve as cloud condensation nuclei or ice nuclei. Therefore, aerosol types and their distribution can result in small but important cloud albedo changes and lifetime (IPCC 2014b).

Moreover, biochemical activity can produce positive or negative feedback. Earth's climate has been shaped by life on Earth, and therefore, biology plays a vital role in climate sensitivity. Among the many ways that plants, animals, and other organisms can influence climate sensitivity, the most important is perhaps the ability to change the atmosphere's composition, such as the capture of CO_2 by plankton and plants or aerosols and the release of GHGs.

Furthermore, it is important to consider the timescale to understand the full impact of a feedback mechanism. Some feedback operates in short timescales (from hours to days), while others can operate in scales that go beyond (from years to centuries) (IPCC 2014b). Although water vapor is perhaps the most important greenhouse gas, its abundance in the atmosphere is not controlled by humans, and it responds freely to climatic conditions. Human activities have increased the emissions of GHGs, such as CO_2, CH_4, N_2O, and chlorofluorocarbons (CFCs). Besides, the production of pollutants (like carbon monoxide (CO), volatile organic compounds (VOC), nitrogen oxides (NOx), and sulfur dioxide (SO_2)) alters the concentration of GHGs like CH_4 and ozone (O_3) through atmospheric chemical reactions (IPCC 2014b). Most of these anthropogenic emissions have long lifetimes in the atmosphere, meaning that they can last for a long time in the atmosphere before being removed or circulated in other Earth system components. For instance, CH_4 can have lifetimes of about 10 years, and CO_2 lifetime can range from 50 to 200 years, while the lifetime of water vapor in the atmosphere is around 9 days (Hartmann 2016). Due to these lifetime differences, we can consider that long-lived gases provide forcing for the climate system. Moreover, changes in the abundance of water vapor will result from the feedback associated with this warming (IPCC 2014b).

Human activities also affect both the Earth's energy and water budgets by changing the land surface, including redistribution in the balance between latent and sensible heat fluxes. Land-use changes impact ecosystems, modifying the albedo, phenology, and its role in the carbon cycle (Foley 2005; Houghton 2003; IPCC 2014b). When climate feedback is considered, it is clear that human activities lead to an increase in temperatures due to an amplification of the greenhouse effect.

1.4 Current Impacts and the IPCC

The climate is shifting across our planet as a consequence of human activities (IPCC 2014a). The increase to unprecedented levels of GHGs caused by anthropogenic activities has affected the Earth system's energy balance due to an amplification of the greenhouse effect since the industrial era (Fig. 1.1b). Observations of the atmosphere, land, ocean, and cryosphere provide compelling evidence of climate change. The effects of GHG rise are observable nowadays and all over the world. There are different scientific groups investigating climate data, uncertainties, and processes through technologies and methodologies to offer a wide range of evidence on our planet's changing climate (IPCC 2014b). Moreover, there are multiple lines of evidence; some of the most well-known are mentioned below:

Global Temperature Rise: Instrumental observations show that land and sea surface temperatures have increased over the last 100 years. For the last 30 years, satellites have allowed a global and broad distribution of measurements. Observations indicate that the upper ocean temperature has increased at least since 1950 (Willis et al. 2010).

Sea-Level Rise: The historical tide gauge record shows that the average rate of global mean sea-level rise over the twentieth century was 1.7 ± 0.2 mm $year^{-1}$ and this rate increased to 3.2 ± 0.4 mm $year^{-1}$ since 1990, mostly because of increased thermal expansion and land ice contributions (Church and White 2011; IPCC 2014b).

Ocean Acidification: The increasing concentrations of CO_2 in the atmosphere and the rise in the ocean's uptake of CO_2 have a significant impact on seawater' chemistry, decreasing ocean pH. Long time series from several oceanic sites show this ongoing decline in pH, consistent with results from repeated measurements on ship transects spanning much of the globe (Byrne et al. 2010; Midorikawa et al. 2010).

Sea-Ice Loss: There has been a trend of decreasing the Northern Hemisphere sea-ice extent since 1978. The amount of multiyear sea ice has been thinning, and thus, the ice volume is reduced (Haas et al. 2008; Kwok et al. 2009).

Likelihood of the Occurrence of Extreme Events: Although the global temperature rise shows that winters are becoming warmer overall, an increase in extreme weather events becomes clear (Cohen et al. 2020). Climate change, driven by natural or human forcings, can lead to changes in the likelihood of the occurrence or strength of extreme weather and climate events such as extreme precipitation events or warm spells. It is possible to estimate the change in the probability distribution of extreme events using detection and attribution methodologies that use statistical tools to test the analyses from models relative to observations (IPCC 2014b; Stott et al. 2016). These methodologies have been used to show that human-induced climate change has changed the probability distribution of extreme events, like the July 2018 heat wave that affected Japan, which would not have happened without anthropogenic influence (Imada et al. 2019).

To compile all these observations and assess the most recent information published in the scientific, technical, and socioeconomic literature worldwide, the World Meteorological Organization and the United Nations Environment Program set up in 1988 the Intergovernmental Panel on Climate Change (IPCC). The IPCC prepares comprehensive assessment reports (IPCC 2014a) and special reports (IPCC 2014b; IPCC 2014c; IPCC 2014d), providing governments with a clear view of the current state of knowledge about the science of climate change, potential impacts, and options for adaptation and mitigation. IPCC reports are also a key input into international climate change negotiations (https://www.ipcc.ch/). The first IPCC assessment depended primarily on observed changes in surface temperature and climate model analyses; more recent assessments include multiple lines of evidence for climate change. The first line of evidence in assessing climate change focuses on the careful analysis of observational records of the atmosphere, land, ocean, and cryosphere systems (IPCC 2014a).

1.5 Future Impacts

1.5.1 The Representative Concentration Pathways and Socioeconomic Drivers of Climate Change

The Fifth Assessment Report (AR5) (IPCC 2014a) proposes three possible future climate change trajectories. These trajectories are known as the Representative Concentration Pathways (RCPs) based on diverse scenarios of anthropogenic GHG emissions. There are four RCPs, which include a stringent mitigation scenario (RCP2.6), two intermediate (RCP4.5 and RCP6.0), and one scenario with very high GHG emissions (RCP8.5). The RCP6.0 and the RCP8.5 refer to possible futures without additional efforts to constrain emissions; that is why they are called "baseline scenarios." The RCP2.6 assumes the aim to keep global warming below 2 °C above preindustrial temperatures.

The magnitude and extent of future impacts will depend not only on the dynamics of the Earth system but also on socioeconomic developments driven by population dynamics; economic development; technological change; social, cultural, and institutional changes; and policies (van Vuuren et al. 2014). Socioeconomic scenarios are used to understand the extent to which global societal developments can influence the nature and severity of climate change risks and response options (Frame et al. 2018). Mitigation and adaptation scenarios are useful to analyze such response strategies. A global climate scenario set was developed to visualize alternative futures and explore adaptation and mitigation challenges arising from alternative socioeconomic futures for different climate scenarios (O'Neill et al. 2017; Riahi et al. 2017). Many elements were integrated to develop climate change scenarios. The first element relates to the biophysical response of the RCPs. These alternatives depict different radiative forcing associated with varying levels of atmospheric concentrations of the GHG emissions (van Vuuren et al. 2014). Also, they integrate their impacts on temperature, precipitation, and sea level. The second element is based on the possible socioeconomic trends, best known as the shared socioeconomic pathways (SSPs), which describe future socioeconomic conditions, including their effects on the emissions of GHGs (O'Neill et al. 2014). The SSPs constitute the narrative and quantitative elements of socio-ecological systems such as demographic, political, social, cultural, institutional, lifestyle, economic, and technological trends (Frame et al. 2018).

Five reference pathways describe plausible alternative trends in the evolution of society and ecosystems over a century timescale, in the absence of climate change or climate policies (O'Neill et al. 2014). SSP1 depicts a future in which the development is consistent with the green growth (or sustainable development) paradigm (a more inclusive development respecting environmental boundaries). SSP2, known as the middle of the road, reflects an extension of the historical experience, particularly in terms of carbon and energy intensity improvements in its baseline. SSP3 is characterized by regional rivalry, and it faces high challenges both in mitigation and adaptation. This scenario results from the combination of high greenhouse gas emissions in the baseline (no climate mitigation policy) scenario and low mitigation capacity. SPP4, known as "Inequality" or "A Road Divided," assumes low challenges to mitigation and high challenges to adaptation (Calvin et al. 2017). The SSP4 points out high inequalities between and within countries and in terms of population growth is the second highest due to the high fertility rates in some areas (Samir and Lutz 2017). Finally, SSP5 is the pessimistic scenario in terms of emissions. It is characterized by rapid and fossil fuel development with high socioeconomic challenges to mitigation and low socioeconomic challenges to adaptation (Kriegler et al. 2017). Table 1.1 summarizes the main characteristics for each pathway.

The RCPs can be combined with the SSPs to obtain GHG emission scenarios based on the SSPs' socioeconomic assumptions and climate policies to reach the end of century radiative forcing levels as defined by the RCPs. Therefore, the SSPs and RCPs can be assembled into a matrix of climate forcing outcomes and socioeconomic development assumptions (van Vuuren et al. 2014). The original RCPs did not originate from the SSPs, as the SSPs were only developed a few years later. However, the emissions and forcing projections to drive the next round of climate change projections will use the new SSP-based scenarios (climatescenarios.org).

1.5.2 Projections of Changes in Temperature and Precipitation

The RCPs are very useful to project the likely increase (with uncertainty levels) of the global surface temperature over the twenty-first century. For the period 2016–2035 relative to 1986–2005, the RCPs are similar and range from 0.3 °C to 0.7 °C. By the mid-twenty-first century, the projected climate change magnitude is substantially affected by the emission scenario. Global surface temperature increases by the end of the twenty-first century (2081–2100) from 0.3 °C to 1.7 °C, according to the scenario RCP2.6. The projected rise in temperature is higher based on the scenarios RCP 4.5 and 6.0, which depict from 1.1 °C to 2.6 °C and 1.4 °C to 3.1 °C, respectively. The high emission scenario (RCP 8.5) projects a rise from 2.6 °C to 4.8 °C. In terms of geographical impacts, the Arctic region will continue to warm up more rapidly than the global mean. Moreover, more frequent hot and fewer cold temperature extremes are expected over most land areas on daily and seasonal timescales, while global mean surface temperature increases, i.e., heat waves will occur with a higher frequency and longer duration. Regarding precipitation, the AR5 indicates that changes will not be uniform either. The high latitudes and the equatorial Pacific will experience a growth in annual mean precipitation under the RCP8.5

Table 1.1 Socioeconomic pathway characteristics

	Scenario	SSP1	SSP2	SSP3	SSP4	SSP5
Driver		Sustainability	Middle of the road	Regional rivalry	Inequality	Fossil fuel development
Population	High[a]	Low	Medium	High	High	Low
	Low[a]	Low	Medium	High	Medium	Low
	Rich[a]	Medium	Medium	Low	Medium	High
Education	High	High	Medium	Low	Polarized[b]	High
	Low	High	Medium	Low	Polarized	High
	Rich	High	Medium	Low	Polarized	High
GDP		High	Medium	Very low	Low	Very high
Gini coefficients (equity)[c]	2050	0.41	0.48	0.52	0.58	0.38
	2100	0.33	0.41	0.73	0.72	0.25
Governance and institutions		Effective	Uneven	Weak	Institutions are only effective for the political and business elite	Increasingly effective
Fossil resources	Coal	Med/low	Medium	High	Med/low	High
	Conventional gas and oil	Medium	Medium	Medium	High/low	High
Fuel preference	Renewable	High	Medium	Medium	High	Medium
	Traditional biomass	Low	Low	High	Low	Low
Energy demand		Low	Medium	Low	Polarized	High
Electricity costs	Nuclear and renewables	High/low	Medium	High	Low	Medium
Agriculture and LUCC	Food demand	High	Medium	Low	Polarized	High
	Meat demand	Low	Medium	High	Medium	High
	Productivity	High	Medium	Low	Medium	High

Protected areas and environmental policies		Strong protected areas up to 30% of the terrestrial area	Medium, protected areas remain up to 17% of the terrestrial area	Low, protected areas remain at the current level	Contrasting with increase in rich countries and low/ medium in developing countries	Low/moderate
International trade		Connected markets, regional production	Moderate, semi-open globalization	Deglobalization and strongly constrained	Globally connected elites the rest poorly connected	High, strongly globalized
Pollutant emissions	Emission factors	Low	Medium	High	High	High
Marker	Model	IMAGE	MESSAGE-GLOBIOM	AIM	GCAM	REMID-MAGPIE
Challenges	Mitigation	Low	Medium	High	Low	High
	Adaptation	Low	Medium	High	High	Low
RCP 1.9		++++++[d]	++++	–	+	
RCP 2.6		++++++	++++++	++++	+++	+++
RCP 4.5		++++++	++++++	++++	+++	++++
RCP 6.0		++++++	++++++	++++	+++	++++
RCP 8.5						++++

GDP gross domestic product, *LUCC* land-use and land cover change, *RCPs* representative concentration pathways

[a]High fertility countries (TFR > 2.9) (mainly Asian and African countries); low fertility countries (TFR < 2.9) include North American countries except Mexico, European countries, Australia, Austria, Belgium, Canada, Czech, Republic, Denmark, Estonia, Finland, France, Germany, Greece, New Zealand, and Republic of Korea. Others refer to Latin American countries, Middle East countries, South African, Ukraine, and other Asian countries as Vietnam and Thailand. To see the full list, see Kc and Lutz (Samir and Lutz 2017)

[b]We are using polarized to refer to very contrasting conditions between developed and developing countries and inside the countries

[c]Gini coefficient allows quantifying the equality. This coefficient is weighted considering the population growth based on the SSPs for the projections. The coefficient goes from 0 to 1. Zero refers to the maximum equality in a society

[d]This number refers to the possible combination of SSP and RCP models. The number of + indicates the number of models that succeeded in making the scenario "work" out of the total number of models available for the SSP (maximum is 6)

Data were based on the following references: Bauer et al. (2017), Calvin et al. (2017), Crespo (2017), Dellink et al. (2017), Fricko et al. (2017), Fujimori et al. (2017), Samir and Lutz (2017), Kriegler et al. (2017), O'Neill et al. (2017), Popp et al. (2017), Riahi et al. (2017), and Rogelj et al. (2018)

scenario. However, in various midlatitude and dry subtropical regions, mean precipitation will decline, increasing in many midlatitude wet regions under the RCP8.5 scenario. Moreover, extreme precipitation events over most midlatitude land masses and wet tropical regions will become more intense and frequent. Besides, the ocean will continue to warm up during the twenty-first century, especially in the Northern Hemisphere and subtropical areas. Global mean sea-level rise will continue during the twenty-first century, faster than the observed from 1971 to 2010. For example, in the period 2081–2100, the rise will be between 0.26 and 0.55 m according to the RCP2.6 scenario compared to 1986–2005. According to the RCP 8.5, the increment projected reaches from 0.45 to 0.82 m f. It is important to highlight that, by the end of the twenty-first century, the sea level will not rise uniformly in more than 95% of the ocean. These augments will be especially experienced in 70% of the coastlines worldwide within ±20% of the global mean.

1.5.3 Climate Change Impacts on Ecosystems and Fungi

Climate change differentially affects ecosystems, biological communities, species, and populations. Climate impact modeling helps in understanding the potential consequences of the new climatic conditions at different levels. For example, (1) dynamic global vegetation models represent vegetation dynamics through photosynthesis, transpiration, and the interactions between land, atmosphere, and water. These models include ecological competition for resources, tissue turnover, population dynamics, soil organic matter, litter dynamics, and fire disturbance. Moreover, they include physiological, morphological, phenological, and bioclimatic responses, depending on the plant functional type. These vegetation processes feed the net ecosystem exchange and moisture of the global carbon exchange cycle (Sitch et al. 2003). (2) The species distribution models are numerical tools that combine observed data of species occurrence or abundance with environmental estimates. These models provide insights into ecological and evolutionary behaviors and predict distributions across landscapes, sometimes requiring extrapolation in space and time (Elith and Leathwick 2009). These numerical tools analyze how climate change will affect species distributions and species assemblages (Thuiller et al. 2005). The impacts on the distribution and viability of many taxa have already been reported. Climate change will heterogeneously affect the species and populations due to the diverse sensitivity, exposure to changes, and phenotypic plasticity (Fuller et al. 2016). Behavioral variations, such as a shift to nocturnal foraging or selecting a cool microclimate, could help free-living organisms face thermal and water stress (Hetem et al. 2014).

Assessing possible impacts of climate change on fungi is challenging because many species feed and live within substrates or underground, limiting our ability to predict their changes. Fungi's kingdom has between 2.2 and 3.8 million species (Hawksworth and Lücking 2017). This enormous diversity is controlled by climate and ecological interactions. On the one hand, temperature and precipitation that influence soil moisture and, on the other hand, the community of plants through the mycorrhizal interaction allow the survival and distribution of fungi (Bajpai and Johri 2019). Approximately 90% of plants are associated with mycorrhizal fungi. In this symbiosis, fungi release nitrogen, phosphorus, and other nutrients from soil to the plant while receiving carbohydrates, which improves plant performance. Some studies have found different effects of increasing temperature on the abundance of ectomycorrhizas. For example, some taxa like *Lactarius* and *Russula* decreased, while *Wilcoxina*, *Laccaria*, and *Clavulina* increased at higher temperatures (Mucha et al. 2018). Modification imposed by climate change on temperature and precipitation will affect fungal species and communities, and it might be predicted that future climate change may favor more thermophilic and drought-resistant microorganisms in many regions (Bajpai and Johri 2019). Regarding precipitation, it has been reported that after extreme rainfall, the taxa abundance changes, causing declines for groups like Basidiomycota and Zygomycota, while other taxa like *Plectopharella*, *Pilidium*, *Motierellaceae*, *Sporomiaceae*, *Cryptococcus*, and lichenized fungi like *Venturiaceae* and *Verrucaria andesiatica* increased (Barnes et al. 2018). These researches show that predicting the future distribution of fungi species is not straightforward due to processes like adaptation, migration, acclimatization, or extinction.

Evidence of individual species in laboratory conditions supports the unpredictability of many impacts of climate change on fungi. The growth of some fungi increases with temperature until reaching a maximum and then decreases. However, moisture levels cause a decrease in fungal growth when moisture is insufficient or excessive (Bidartondo et al. 2018). As a result, it is unclear how fungi species respond to changes in the communities and ecosystems. Climate change will modify the length of the reproductive season, the timing of the production of spore-bearing structures (Andrew et al. 2017; Büntgen et al. 2012; Yang et al. 2012), and the community structure (Mucha et al. 2018). Although scientists do not know with high certainty how the future distribution and abundance of fungi species will be under climate change, we know that we will face modifications on community structure and species spread. This is crucial when we consider infectious diseases.

The new climatic conditions can directly threaten human health because of the changing geographic ranges of vector-borne diseases caused by parasites, bacteria, fungi, and viruses transmitted to humans. For example, there is concern that global warming will bring new fungal diseases for mammals by reducing the gradient between the mammalian basal temperatures with the environment (Garcia-Solache and Casadevall 2010). Research on filamentous fungus has shown that these species can rapidly adapt to growth at higher temperatures (de Crecy et al. 2009). Therefore, higher temperatures due to global warming might lead to the selection of fungal lineages that can breach the mammalian thermal restriction zone (Casadevall et al. 2019; Garcia-Solache and Casadevall 2010). A possible example of this selection of thermally tolerant fungal lineages may be the rise of *Candida auris* as a human pathogen with the appearance of phylogenetically distant strains on three continents (Casadevall et al. 2019). Research is underway to reveal clues and predictive tools for determining where diseases might be located in the future (Ogden 2018).

1.6 Conclusion

Climate change is a natural process that controls our planet's climate on timescales varying from decades to millennia. This process is driven by the greenhouse effect, where atmospheric gases like water vapor, carbon dioxide, methane, nitrous oxide, and other GHGs retain radiation in the atmosphere. The greenhouse effect is a natural process, and the average temperature on the Earth's surface would be below the water's freezing point without it. Thus, Earth's natural greenhouse effect makes life as we know it possible. Nevertheless, human activities release large amounts of CO_2 and other GHGs into the atmosphere at unprecedented levels in at least the last 800,000 years. This release of GHGs enhances the greenhouse effect that is largely modified by human influence.

Changes in the Earth system, both natural and anthropogenic, perturb the Earth's energy balance. The climate system response will depend on the complex feedback mechanisms involved in the process, where positive and negative feedback mechanisms will work together to produce a climate's response that will be different from a simple linear. Examples of positive and negative feedback are the interaction of an increase in temperature with the atmosphere's water vapor, the Planck and the lapse-rate feedback, and the cloud and aerosol role in the energy balance. Moreover, Earth's climate has been shaped by life on Earth, and therefore, it is true that plants, animals, and other organisms can influence climate sensitivity; the most important is perhaps the ability to change the atmosphere composition, such as the capture of CO_2 by plankton and plants or the release of aerosols and GHGs. When climate feedback is considered, it is clear that human activities lead to an increase in temperatures due to an amplification of the greenhouse effect.

The effects of GHG increase in climate are observable nowadays and all over the world. Examples of such effects are evident from global temperature and sea-level rise, ocean acidification, sea-ice loss, and changes in the likelihood of extreme events. Different scientific groups worldwide are processing observations of the atmosphere, land, ocean, and cryosphere, examining climate data, uncertainties, and processes through technologies and methodologies to offer a wide range of compelling evidence on our planet's changing climate. The United Nations thought the IPCC provides assessments of the most recent information published in the scientific, technical, and socioeconomic literature worldwide. The IPCC goal is to provide governments with a clear view of the current state of knowledge about the science of climate change, potential impacts, and options for adaptation and mitigation.

Climate change affects differentially the hierarchical biological levels from genes to ecosystems. The impacts on the species distribution depend on the taxa's sensitivity, exposure to changes, and phenotypic plasticity. Computational modeling tools help visualize how climate change could affect the species distribution and their assemblages. Assessing the affectation of climate change IPCC on fungi is challenging, considering that they feed and live within substrates or underground, which impose many limitations. However, there is evidence of individual species in laboratory conditions that indicates that fungal growth could increase with an increase in temperature. There is concern that global warming will bring new fungal diseases for mammals by selecting fungal lineages that can breach the mammalian thermal restriction zone. Besides those findings and some reports on the sensitivity of fungi to moisture, it is unclear how fungi respond to climate changes in the real world, i.e., the length of the reproductive season and the timing of production or changes in geographic ranges. Finally, it is worth saying that the magnitude and extent of climate change will depend not only on the dynamics of the biophysical elements of the biosphere but also on socioeconomic development. Global climate change scenarios like the RCPs and the SSPs help visualize alternative futures and explore adaptation and mitigation challenges, which should be used to reinforce climate policies.

References

Andrew C, Heegaard E, Kirk PM, Bässler C, Heilmann-Clausen J, Krisai-Greilhuber I, et al. Big data integration: pan-European fungal species observations' assembly for addressing contemporary questions in ecology and global change biology. Fungal Biol Rev. 2017;31(2):88–98. https://doi.org/10.1016/j.fbr.2017.01.001.

Bajpai A, Johri BN. Climate change is real: fungal perspective! Kavaka. 2019;53:12. https://doi.org/10.36460/Kavaka/53/2019/12-21.

Barnes CJ, van der Gast CJ, McNamara NP, Rowe R, Bending GD. Extreme rainfall affects assembly of the root-associated fungal community. New Phytol. 2018;220(4):1172–84. https://doi.org/10.1111/nph.14990.

Bauer N, Calvin K, Emmerling J, Fricko O, Fujimori S, Hilaire J, et al. Shared socio-economic pathways of the energy sector – quantifying the narratives. Glob Environ Change. 2017;42:316–30. https://doi.org/10.1016/j.gloenvcha.2016.07.006.

Bidartondo MI, Ellis C, Kauserud H, Kennedy PG, Lilleskov E, Suz L, et al. Climate change: fungal responses and effects. In: Willis Katherine J, editor. State Worlds Fungi 2018. Kew: Royal Botanic Gardens; 2018. p. 62–9. stateoftheworldsfungi.org/2018/climate-change.html.

Büntgen U, Kauserud H, Egli S. Linking climate variability to mushroom productivity and phenology. Front Ecol Environ. 2012;10(1):14–9. https://doi.org/10.1890/110064.

Byrne RH, Mecking S, Feely RA, Liu X. Direct observations of basin-wide acidification of the North Pacific Ocean: pH changes in the North Pacific. Geophys Res Lett. 2010;37(2) https://doi.org/10.1029/2009GL040999.

Calvin K, Bond-Lamberty B, Clarke L, Edmonds J, Eom J, Hartin C, et al. The SSP4: a world of deepening inequality. Glob Environ Change. 2017;42:284–96. https://doi.org/10.1016/j.gloenvcha.2016.06.010.

Casadevall A, Kontoyiannis DP, Robert V. On the emergence of Candida auris: climate change, azoles, swamps, and birds. mBio. 2019;10(4). https://mbio.asm.org/content/10/4/e01397-19

Cess RD. Global climate change: an investigation of atmospheric feedback mechanisms. Tellus. 1975;27(3):193–8. https://doi.org/10.3402/tellusa.v27i3.9901.

Church JA, White NJ. Sea-level rise from the late 19th to the early 21st century. Surv Geophys. 2011;32(4–5):585–602.

Cohen J, Zhang X, Francis J, Jung T, Kwok R, Overland J, et al. Divergent consensuses on Arctic amplification influence on midlatitude severe winter weather. Nat Clim Chang. 2020;10(1):20–9.

Crespo CJ. Income projections for climate change research: a framework based on human capital dynamics. Glob Environ Change. 2017;42:226–36. https://doi.org/10.1016/j.gloenvcha.2015.02.012.

de Crecy E, Jaronski S, Lyons B, Lyons TJ, Keyhani NO. Directed evolution of a filamentous fungus for thermotolerance. BMC Biotechnol. 2009;9(1):74. https://doi.org/10.1186/1472-6750-9-74.

Dellink R, Chateau J, Lanzi E, Magné B. Long-term economic growth projections in the shared socioeconomic pathways. Glob Environ Change. 2017;42:200–14. https://doi.org/10.1016/j.gloenvcha.2015.06.004.

Elith J, Leathwick JR. Species distribution models: ecological explanation and prediction across space and time. Annu Rev Ecol Evol Syst. 2009;40(1):677–97. https://doi.org/10.1146/annurev.ecolsys.110308.120159.

Foley JA. Global consequences of land use. Science. 2005;309(5734):570–4. https://doi.org/10.1126/science.1111772.

Frame B, Lawrence J, Ausseil A-G, Reisinger A, Daigneault A. Adapting global shared socio-economic pathways for national and local scenarios. Clim Risk Manag. 2018;21:39–51. https://doi.org/10.1016/j.crm.2018.05.001.

Fricko O, Havlik P, Rogelj J, Klimont Z, Gusti M, Johnson N, et al. The marker quantification of the shared socioeconomic pathway 2: a middle-of-the-road scenario for the 21st century. Glob Environ Change. 2017;42:251–67. https://doi.org/10.1016/j.gloenvcha.2016.06.004.

Friedrich T, Timmermann A, Tigchelaar M, Elison Timm O, Ganopolski A. Nonlinear climate sensitivity and its implications for future greenhouse warming. Sci Adv. 2016;2(11):e1501923. https://doi.org/10.1126/sciadv.1501923.

Fujimori S, Hasegawa T, Masui T, Takahashi K, Herran DS, Dai H, et al. SSP3: AIM implementation of shared socioeconomic pathways. Glob Environ Change. 2017;42:268–83. https://doi.org/10.1016/j.gloenvcha.2016.06.009.

Fuller A, Mitchell D, Maloney SK, Hetem RS. Towards a mechanistic understanding of the responses of large terrestrial mammals to heat and aridity associated with climate change. Clim Change Responses. 2016;3(1):10. https://doi.org/10.1186/s40665-016-0024-1.

Garcia-Solache MA, Casadevall A. Global warming will bring new fungal diseases for mammals. MBio. 2010;1(1):e00061–10. https://doi.org/10.1128/mBio.00061-10.

Haas C, Pfaffling A, Hendricks S, Rabenstein L, Etienne J-L, Rigor I. Reduced ice thickness in Arctic transpolar drift favors rapid ice retreat. Geophys Res Lett. 2008;35(17):L17501. https://doi.org/10.1029/2008GL034457.

Hansen J, Ruedy R, Sato M, Lo K. Global surface temperature change. Rev Geophys. 2010;48:4. https://doi.org/10.1029/2010RG000345.

Hartmann DL. Global physical climatology (second edition). Boston, MA: Elsevier; 2016.

Hawksworth DL, Lücking R. Fungal diversity revisited: 2.2 to 3.8 million species. Microbiol Spectr. 2017;5(4) https://doi.org/10.1128/microbiolspec.FUNK-0052-2016.

Hetem RS, Fuller A, Maloney SK, Mitchell D. Responses of large mammals to climate change. Temperature. 2014;1(2):115–27. https://doi.org/10.4161/temp.29651.

Houghton RA. Revised estimates of the annual net flux of carbon to the atmosphere from changes in land use and land management 1850–2000. Tellus B. 2003;55(2):378–90. https://doi.org/10.1034/j.1600-0889.2003.01450.x.

Imada Y, Watanabe M, Kawase H, Shiogama H, Arai M. The July 2018 high temperature event in Japan could not have happened without human-induced global warming. SOLA. 2019;15A:8–12. https://doi.org/10.2151/sola.15A-002.

IPCC. In: Pachauri RK, Meyer LA, editors. Climate Change 2014: Synthesis Report. Contribution of Working Groups I, II and III to the Fifth Assessment Report of the Intergovernmental Panel on Climate Change. Geneva: IPCC; 2014a. p. 151.

IPCC. Climate Change 2013 - The Physical Science Basis: Working Group I Contribution to the Fifth Assessment Report of the Intergovernmental Panel on Climate Change [Internet]. Cambridge: Cambridge University Press; 2014b. [cited 2020 Oct 11]. http://ebooks.cambridge.org/ref/id/CBO9781107415324

IPCC. Climate Change 2014 Impacts, Adaptation, and Vulnerability: Working Group II Contribution to the Fifth Assessment Report of the Intergovernmental Panel on Climate Change [Internet]. Cambridge: Cambridge University Press; 2014c. [cited 2020 Oct 11]. http://ebooks.cambridge.org/ref/id/CBO9781107415379

IPCC. Climate Change 2014 Mitigation of Climate Change: Working Group III Contribution to the Fifth Assessment Report of the Intergovernmental Panel on Climate Change [Internet]. Cambridge: Cambridge University Press; 2014d. [cited 2020 Oct 11]. http://ebooks.cambridge.org/ref/id/CBO9781107415416

Kerr RA. Milankovitch climate cycles through the ages. Science. 1987;235(4792):973–4. https://doi.org/10.1126/science.235.4792.973.

Kriegler E, Bauer N, Popp A, Humpenöder F, Leimbach M, Strefler J, et al. Fossil-fueled development (SSP5): an energy and resource intensive scenario for the 21st century. Glob Environ Change. 2017;42:297–315. https://doi.org/10.1016/j.gloenvcha.2016.05.015.

Kwok R, Cunningham GF, Wensnahan M, Rigor I, Zwally HJ, Yi D. Thinning and volume loss of the Arctic Ocean sea ice cover: 2003–2008. J Geophys Res. 2009;114(C7):C07005. https://doi.org/10.1029/2009JC005312.

MacFarling M, Etheridge D, Trudinger C, Steele P, Langenfelds R, van Ommen T, et al. Law Dome CO_2, CH_4 and N_2O ice core records extended to 2000 years BP C. Geophys Res Lett. 2006;33:L14810. https://doi.org/10.1029/2006GL026152.

Mann ME, Zhang Z, Hughes MK, Bradley RS, Miller SK, Rutherford S, et al. Proxy-based reconstructions of hemispheric and global surface temperature variations over the past two millennia. Proc Natl Acad Sci. 2008;105(36):13252–7. https://doi.org/10.1073/pnas.0805721105.

Midorikawa T, Ishii M, Saito S, Sasano D, Kosugi N, Motoi T, et al. Decreasing pH trend estimated from 25-yr time series of carbonate parameters in the western North Pacific. Tellus Ser B Chem Phys Meteorol. 2010;62(5):649–59. https://doi.org/10.1111/j.1600-0889.2010.00474.x.

Morice CP, Kennedy JJ, Rayner NA, Jones PD. Quantifying uncertainties in global and regional temperature change using an ensemble of observational estimates: the HadCRUT4 data set. J Geophys Res Atmospheres 1984–2012. 2012;117(D8) https://doi.org/10.1029/2011JD017187.

Mucha J, Peay KG, Smith DP, Reich PB, Stefański A, Hobbie SE. Effect of simulated climate warming on the ectomycorrhizal fungal Community of Boreal and Temperate Host Species Growing near Their Shared Ecotonal Range Limits. Microb Ecol. 2018;75(2):348–63. https://doi.org/10.1007/s00248-017-1044-5.

O'Neill BC, Kriegler E, Riahi K, Ebi KL, Hallegatte S, Carter TR, et al. A new scenario framework for climate change research: the concept of shared socioeconomic pathways. Clim Chang. 2014;122(3):387–400. https://doi.org/10.1007/s10584-013-0905-2.

O'Neill BC, Kriegler E, Ebi KL, Kemp-Benedict E, Riahi K, Rothman DS, et al. The roads ahead: narratives for shared socioeconomic pathways describing world futures in the 21st century. Glob Environ Change. 2017;42:169–80. https://doi.org/10.1016/j.gloenvcha.2015.01.004.

Ogden LE. Climate change, pathogens, and people. Bioscience. 2018;68(10):733–9. https://doi.org/10.1093/biosci/biy101.

Popp A, Calvin K, Fujimori S, Havlik P, Humpenöder F, Stehfest E, et al. Land-use futures in the shared socio-economic pathways. Glob Environ Change. 2017;42:331–45. https://doi.org/10.1016/j.gloenvcha.2016.10.002.

Riahi K, van Vuuren DP, Kriegler E, Edmonds J, O'Neill BC, Fujimori S, et al. The shared socioeconomic pathways and their energy, land use, and greenhouse gas emissions implications: an overview. Glob Environ Change. 2017;42:153–68. https://doi.org/10.1016/j.gloenvcha.2016.05.009.

Rogelj J, Popp A, Calvin KV, Luderer G, Emmerling J, Gernaat D, et al. Scenarios towards limiting global mean temperature increase below 1.5 °C. Nat Clim Chang. 2018;8(4):325–32. https://doi.org/10.1038/s41558-018-0091-3.

Samir KC, Lutz W. The human core of the shared socioeconomic pathways: population scenarios by age, sex and level of education for all countries to 2100. Glob Environ Change. 2017;42:181–92. https://doi.org/10.1016/j.gloenvcha.2014.06.004.

Sitch S, Smith B, Prentice IC, Arneth A, Bondeau A, Cramer W, et al. Evaluation of ecosystem dynamics, plant geography and terrestrial carbon cycling in the LPJ dynamic global vegetation model. Glob Change Biol. 2003;9(2):161–85. https://doi.org/10.1046/j.1365-2486.2003.00569.x.

Stott PA, Christidis N, Otto FEL, Sun Y, Vanderlinden J-P, van Oldenborgh GJ, et al. Attribution of extreme weather and climate-related events. WIREs Clim Change. 2016;7:23–41. https://doi.org/10.1002/wcc.380.

Thuiller W, Lavorel S, Araújo MB, Sykes MT, Prentice IC. Climate change threats to plant diversity in Europe. Proc Natl Acad Sci U S A. 2005;102(23):8245–50. https://doi.org/10.1073/pnas.0409902102.

van Vuuren DP, Kriegler E, O'Neill BC, Ebi KL, Riahi K, Carter TR, et al. A new scenario framework for climate change research: scenario matrix architecture. Clim Chang. 2014;122(3):373–86. https://doi.org/10.1007/s10584-013-0906-1.

Willis J, Chambers D, Kuo C-Y, Shum CK. Global Sea level rise: recent progress and challenges for the decade to come. Oceanography. 2010;23(4):26–37. https://doi.org/10.5670/oceanog.2010.03.

Yang X, Luedeling E, Chen G, Hyde KD, Yang Y, Zhou D, et al. Climate change effects fruiting of the prize matsutake mushroom in China. Fungal Divers. 2012;56(1):189–98. https://doi.org/10.1007/s13225-012-0163-z.

Zhang H-M, Lawrimore J, Huang B, Menne M, Yin X, Sánchez-Lugo A, et al. Updated temperature data give a sharper view of climate trends. Eos [Internet]. 2019. [cited 2021 Apr 2];100. https://eos.org/science-updates/updated-temperature-data-give-a-sharper-view-of-climate-trends

Zhang H-M, Huang B, Lawrimore J, Menne M, Smith TM. NOAA global surface temperature dataset (NOAAGlobalTemp), version 4.0. NOAA National Centers for environmental Information; 2020. https://www.ncdc.noaa.gov/noaa-merged-land-ocean-global-surface-temperature-analysis-noaaglobaltemp-v4.

Chapter 2
Fungi: Essential Elements in the Ecosystems

Soumya Moonjely

Contents

Abbreviations

AM	Arbuscular mycorrhizal
ECM	Ectomycorrhizal
EIPF	Endophytic insect pathogenic fungi
NRPS	Non-ribosomal peptide synthetases

S. Moonjely (✉)
Faculty of Mathematics and Science, Department of Biological Sciences, Brock University, St. Catharines, ON, Canada

M. G. Frías-De-León et al. (eds.), *The Impact of Climate Change on Fungal Diseases*, Fungal Biology, https://doi.org/10.1007/978-3-030-89664-5_2

2.1 Introduction

The equilibrium of the ecosystem depends on the coexistence and interactions between organisms, including microorganisms, plants, and animals. Fungi play a functionally diverse role in different ecosystems and can be found in almost all ecosystems (Grossart et al. 2019). Fungi are essential players in our ecosystems as producers, consumers, and decomposers. In agroecosystems, fungal presence is essential in maintaining the fertility of the soil, cycling nutrients, and providing tolerance to plants against diseases and pests (Dighton 2003). In forest ecosystems, fungi are critical in decomposing plant and animal remains. Fungi play a significant role in several dimensions, including in global biogeochemistry by mobilizing carbon, nitrogen, and phosphorus, in medicine by producing pharmacologically important bioactive molecules, and as an important means of food. Fungi form the basis of many industries, including beverages, fermentation, enzymes, animal feed, bioinsecticide, plant fertilizer, and antimicrobial compounds (Fig. 2.1). Furthermore,

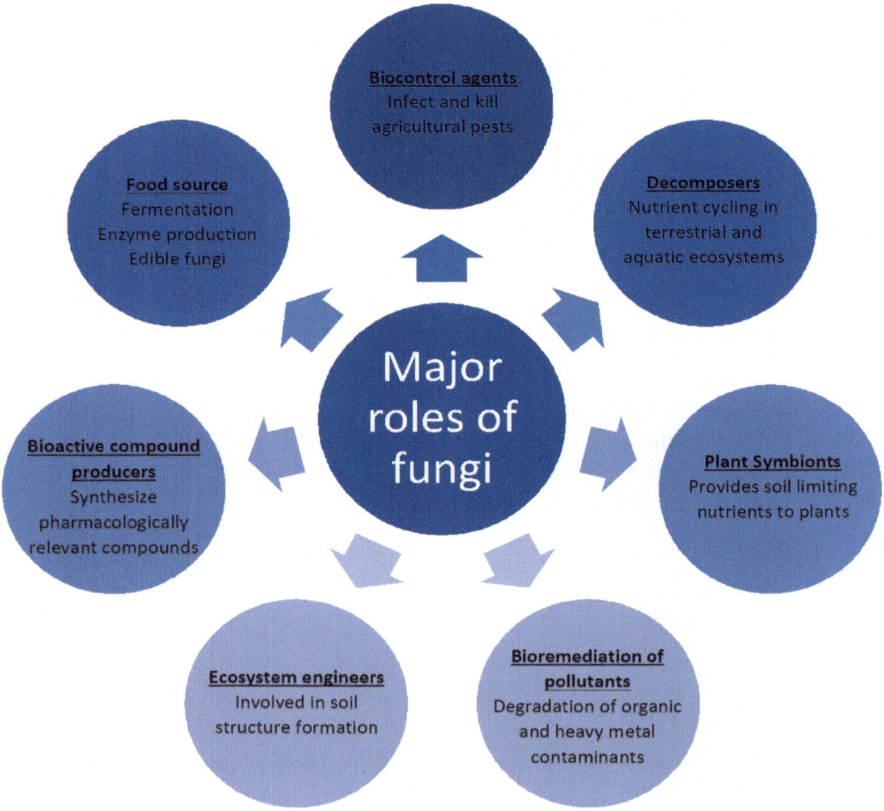

Fig. 2.1 Diagram showing the major contributions of fungi to the ecosystem and humanity

fungi have a significant role in addressing global challenges, including cleaning up pollutants and biodegrading toxic chemicals.

Currently, nine phylum-level clades have been recognized in the fungal kingdom after several taxonomic revisions: *Opisthosporidia*, *Chytridiomycota*, *Neocallimastigomycota*, *Blastocladiomycota*, *Zoopagomycota*, *Mucoromycota*, *Glomeromycota*, *Basidiomycota*, and *Ascomycota* (Naranjo-Ortiz and Gabaldón 2019). The fungal genera that belong to *Opisthosporidia* are known as intracellular parasites, and most of them are zoosporic. The members of *Chytridiomycota* are zoosporic, and they exist as either free-living saprobes or parasitoids. Parasitic members (e.g., *Dinomyces*) play a crucial role in the aquatic ecosystem by regulating algal bloom and recycling nutrients (Lepelletier et al. 2014). *Neocallimastigomycota* consists of obligate anaerobic and nonparasitic fungi. The members of this group are mainly inhabited in the gut of mammalian herbivores. Herbivorous mammals do not produce lignocellulolytic enzymes, and they rely on the symbiotic microbiota in their rumen to digest the ingested lignocellulosic plant materials (Gruninger et al. 2014). *Blastocladiomycota* consists of members that have a lifestyle as a saprobe in terrestrial or aquatic environments. Some fungi that belong to this clade are parasites of invertebrates or pathogens of algae or plants. The genus *Coelomomyces* can parasitize insects or crustacean hosts and has been explored as biocontrol agents (Powell 2017). The representatives from Zoopagomycota live either on decaying matter or soil as saprobes or parasitize amebae or other fungi (Naranjo-Ortiz and Gabaldón 2019). Many species of *Mucoromycota* grow as either plant parasites or mycoparasite. Some members are identified as rhizospheric fungi, endophytic fungi, or ectomycorrhizal fungi (Bonfante and Venice 2020). Certain species are reported to cause infections in humans or animals. However, species belonging to order *Mucorales* are industrially crucial for chitosan production, lipids, and carotenoids. All known representatives of *Glomeromycota* are obligate biotrophs, and almost 90% of the land plants form a symbiotic association with members of this group and are referred to as arbuscular mycorrhizal fungi. However, certain species (*Geosiphon pyriformis*) from this clade interacts with cyanobacteria (Schüßler 2006). The second most species-rich phylum of kingdom fungi is *Basidiomycota*. The members of this group are recognized by the presence of club-like reproductive structures referred to as basidia. More than 30,000 species have been identified in *Basidiomycota*. Species of this group are the main decomposers of wood and litter and thus play a critical role in the recycling of carbon and other nutrients in terrestrial ecosystems. They are also serving as an important source of food for both animals and humans. Many species from this group can form mutualistic, symbiotic, or pathogenic interactions with plant hosts (Zhao et al. 2017). Approximately two-thirds of all the identified belong to *Ascomycota*, therefore, the largest phylum in the fungal kingdom (Naranjo-Ortiz and Gabaldón 2019). Because of this reason, *Ascomycota* is the dominant species in the forest and agroecosystems. They play a significant role in soil stability, decomposition of plant and animal remains, and forming endophytic interactions with plant community. Many ascomycete species form pathogenic relationships with animal and plant hosts (Challacombe et al. 2019).

Fungi exist in all ecosystems, including freshwater, marine, and terrestrial ecosystems, and can survive in extreme environments. The proportion of the fungal community in aquatic ecosystems depends on the environmental settings, and there is a considerable variation within different aquatic habitats (Comeau et al. 2016). DNA sequencing studies revealed that, in freshwater habitats, the relative abundance of the fungal community was found to be >50% of all eukaryotic sequences. However, in saline aquatic ecosystems, the relative abundance was found to be low compared to freshwater systems (Grossart et al. 2019). In the terrestrial ecosystem, the diversity is mostly noticed in tropical ecosystems; however, the diversity of ectomycorrhizal fungi was found mostly in temperate or boreal ecosystems. Previous studies using structural equation models have shown that climatic variables directly affect fungal richness and diversity, the same as how plant and fungal community are positively correlated. The richness of the fungi is also related to the distance from the equator and annual mean precipitation. Approximately 0.8–5.1 million fungal species have been estimated globally (Wardle and Lindahl 2014). It is essential to study how the fungal population and diversity are affected due to anthropogenic activities and climate change. It is not easy to delve into different aspects that show why fungi are essential in our ecosystem. This chapter updates insight into how the fungal kingdom members make vital contributions to the ecosystem process and their beneficial services to humanity.

2.2 Fungi as Decomposers and Their Essential Role in Agriculture

Most plant species form a direct or indirect association with soil fungi. The fungal association with plants plays a substantial role in mobilizing nutrients, common pollutants, and radionuclides in agricultural and nonagricultural soils (Steiner et al. 2002). Plants form an indirect association with saprotrophic fungi in the rhizosphere or direct association with mycorrhizal or endophytic fungi. Plant-fungal associations impart several benefits to plants, including disease resistance, stress resistance, plant nutrition, efficient use of underground water, and increased photosynthetic ability. The two main groups of fungi in which plants form direct association are mycorrhizal and endophytic fungi (Behie and Bidochka 2014a, b). These fungi are found in most ecological niches. The term mycorrhiza is derived from the Greek words *mykes* and *rhiza*, meaning fungus and root, which is used to describe fungal association with plant roots. Over 200,000 plant species form an association with mycorrhizal fungi, and the 2 major groups of mycorrhizal fungi identified are ectomycorrhizal fungi and endomycorrhizal fungi (Bonfante and Anca 2009). Ectomycorrhizal (ECM) fungi belong to *Basidiomycota* and *Ascomycota*, forming a symbiotic association with roots of woody plants predominantly in forest ecosystems. ECM-root interactions are characterized by the presence of the mycelial sheath around the root tips called the mantle, from which the "Hartig net" originates by penetrating the plant root to form a mycelial network around the epidermis and/or outer cortical cells (Behie and Bidochka 2014a, b; Clasen et al. 2018). ECM is considered the functional extension of forest plants, which accounts for the major contributor of biomass in forest ecosystems. Furthermore, ECM contributes significantly to the decomposition of plant organic matter and can retrieve nitrogen and phosphorous from the decomposed plant litter. In forest ecosystems, ECM was a critical factor in regulating soil carbon dynamics (Lindahl and Tunlid 2015). Understanding the factors that control the stability and storage of nutrients in the soil by fungi is vital to mitigate the effects of natural or human-made environmental changes.

Endomycorrhizal fungi include arbuscular mycorrhizal, ericoid, and orchid mycorrhizal fungi. Arbuscular mycorrhizal (AM) fungi play a significant role in managing crop nutrition. Plants often face challenges in obtaining an adequate supply of soil limiting nutrients such as phosphorus and nitrogen. Many factors affect the acquisition of soil nutrients by plant roots. In particular soil types, the nutrients may be available in the form that plant roots cannot absorb. Plants compensate for this challenge by forming a mutualistic association with AM fungi. AM fungi are obligate biotrophs that belong to the phylum *Glomeromycota* that rely on host plants for carbon supply in exchange for mineral nutrients taken up from the surrounding environment. The partnership of AM fungi with the host plant is different from ECM fungi. Fungal spores germinate on the root surface, and the hyphae then penetrate the cortical cells of roots to establish an intracellular symbiosis by forming specialized structures called "arbuscules." Arbuscules are the main symbiotic interface to exchange nutrients between symbiotic partners. The fungi then produce extra-radicular mycelia that grow out to the plant roots to absorb mineral nutrients from the soil. This hyphal extension from the root has increased surface contact and penetration to the soil, enabling host plants to access nutrients that are otherwise inaccessible to roots (Behie and Bidochka 2014a, b).

Phosphate uptake is a major challenge for most plants. AM fungi benefit the host plant by facilitating the uptake of poor phosphate ions in the soil. Fungi absorb phosphate through extra-radicular mycelium via high-affinity phosphate transporters. Besides, AM fungi facilitate the transport of another soil limiting nutrient, nitrogen. Nitrogen is a major requirement for the growth process in plants, as it is needed for the synthesis of several compounds and macromolecules, including enzymes, other proteins, nucleotides, and certain carbohydrates. Studies have shown that AM fungi benefit host plants by assimilating organic and inorganic nitrogen from the soil. AM fungus *Glomus intraradices* encodes multiple ammonium transporter proteins involved in the uptake of ammonia from the soil and subsequent transfer to plant hosts (López-Pedrosa et al. 2006; Pérez-Tienda et al. 2011).

Endophytic fungi live within the plant tissues without any disease symptoms. Endophytic fungi exhibit a mutualistic or symbiotic relationship with their plant counterparts. More than one million fungal endophytes have been estimated, and a significant proportion of these fungi belong to *Ascomycota*. Based on evolutionary relatedness, taxonomy, and type of plant hosts, two major types of fungal endophytes have been described: the clavicipitaceous endophytes (class 1) and non-clavicipitaceous endophytes. The classic example of clavicipitaceous endophytes is *Epichole* sp. predominantly associated with plants belonging to the *Poaceae* family (grasses). Non-clavicipitaceous fungi are primarily isolated from both vascular and nonvascular plant species. The non-clavicipitaceous endophytes are further divided into three categories: class 2, 3, and 4. Class 2 endophytes can colonize both above- and belowground plant parts, whereas class 3 and 4 endophytes are primarily colonized in aboveground and belowground tissues, respectively. The association with endophytic fungi implies several beneficial effects to their host plants, including growth promotion and resistance to biotic and abiotic stresses. The root colonizing facultative endophyte *Piriformospora indica* forms a symbiotic association with several crops. Studies have shown that the symbiotic association of *P. indica* improved the tolerance of host plants to phytopathogens. Moreover, plants colonized with *P. indica* exhibited systemic resistance to foliar pathogens. A study conducted on the endophytic colonization by *Fusarium equiseti* and *Pochonia chlamydosporia* on barley roots showed increased resistance of barley to phytopathogen *Gaeumannomyces graminis* var. *tritici*. Furthermore, the colonization of nematode egg parasitic and endophytic fungus, *P. chlamydosporia*, enhanced the growth of barley under greenhouse conditions (Maciá-Vicente et al. 2009). *P. chlamydosporia* demonstrated the ability to colonize a wide variety of plants, including monocots and dicots (Moonjely and Bidochka 2019). Endophytic insect pathogenic fungus (*Metarhizium* sp.) can form a symbiotic association with several species of monocots and dicots. These fungi also form a parasitic association with insect hosts. Studies using stable isotopes (N^{15}) have shown that *Metarhizium* sp. were able to transfer nitrogen from the infected insect to the plant host in exchange for carbon (Behie and Bidochka 2014a, b; Behie et al. 2017). Besides, the colonization of *Metarhizium* imparts resistance to host plants against phytopathogen, *Fusarium solani* (Sasan and Bidochka 2013). These studies suggest that the interaction of fungi with the plant hosts and the environment is dynamic and diverse than anticipated. These studies also signify the essential role of fungi in the ecological cycling of nutrients to plant communities.

Fungi have a significant role in the formation of soil and mineral cycling in soil. Soil is created due to the dissociation of parent rock into mineral particles. Weathering of parent rock to mineral particles occurs as a result of abiotic and biotic factors. As a biotic factor, fungi play an important role in creating the mineral particles and in modifying the chemical composition of the soil. Both mycorrhizal fungi and saprotrophic fungi act synergistically with algae and bacteria in the dissolution of parent rock to release mineral nutrients (Dighton 2003). Lichens are the symbiotic association with algae and fungi. The fungal counterparts of lichens are mainly the species from *Ascomycota* and *Deuteromycota*, occasionally with *Basidiomycota*. Lichens contribute to the biogeochemical process of weathering of rock to form mineral particles by providing water and organic acids. The lichens act as a point of condensation of water, especially in dry areas, thus acting as a center of "water-related rock-weathering process." Moreover, the presence of lichens increases the carbon dioxide concentration due to its respiration which subsequently combines with water to form carbonic acid. Carbonic acid is a weak organic acid that contributes to the dissolution of calcium-rich rocks. Lichens also contribute to the chemical weathering of rocks by producing organic acids such as oxalic, citric, lichenic, and tartaric acid. In addition, fungus alone has the ability to solubilize rocks such as *Aspergillus niger* (solubilization of metamorphic rocks), *Penicillium* (solubilization of calcium-rich rocks), and *Resinicium bicolor* (solubilization of strontianite) (Dighton 2003).

2.3 Role of Fungi in the Food and Pharmaceutical Industry

Many fungal species form a significant part of the human diet. In boreal ecosystems, mushrooms are the staple food for reindeers and moose. Moreover, fungal hyphae, spores, or truffles form the food for soil-inhabiting insects and small rodents (Dighton 2003). More than 2500 fungal species are reported to be edible mushrooms. However, the harvest of many edible mushrooms has been found to decline due to the change in their natural environment (Yun and Hall 2004). Certain edible mushrooms are in high demand in the world market, with an annual value of US$ 2 billion. The commercially available edible mushrooms include *Volvariella volvacea* (straw mushroom), *Pleurotus ostreatus* (oyster mushroom), *Agaricus bisporus* (small button or large portobello mushroom), and lentinula edodes (Shiitake mushroom). Edible mushrooms are an excellent source of protein, essential amino acids, β-carotene, and dietary fiber, and most edible mushrooms can be easily cultivated at a low cost. Certain edible mushrooms have medicinal properties as well. Species of *Pleurotus* sp. can be easily grown on straw or other lignocellulose waste from agro biomass (Ghorai et al. 2009).

Many ingredients used for food processing are derived from fungi through industrial fermentation. Fungi are used as biotransformation agents to produce organic acids, food colorants, vitamins, fatty acids, and enzymes. Fungi can produce pigments and colorants for the food industry, including carotenoids, melanins, flavins, phenazins, quinones, monascins, violacin, and indigo. Some strains of *Monascus* sp. produce red and yellow pigments. *Aspergillus* sp. found to synthesize red pigment were found to produce yellow (physcion and emodin) and red (erythroglaucin and catenarin) compounds (Caro et al. 2012). Similarly, species of *Eurotium* sp. and *Dermocybe* sp. have also been identified to produce yellow and red pigments (Bechtold and Mussak 2009; Gessler et al. 2013). However, the pigment produced by many of these fungi has not been approved as a food colorant in many countries due to safety concerns. Many of these pigments were found to be contaminated by toxic metabolites (e.g., citrinin) (Dufossé et al. 2014).

More than 95% of the global citric acid production is through industrial fermentation using fungi. Citric acid is an important component in jams, jellies, beverages, and frozen foods. It is mainly included in food as an additive for tartness or pH regulator. The most common fungal species used in the production of citric acid is *A. niger*. The other species that are reported to produce organic acids are *A. wentii*, *P. citrinum*, and *Candida lipolytica*. Enzymes have a wide variety of uses in food processing as a clarifier or as a hydrolysis agent. Filamentous fungi are the primary source of many enzymes such as amylases and proteases by *Aspergillus* sp., invertase by *Saccharomyces* sp., lactase by *Aspergillus* sp. and *Kluyveromyces* sp., and galactosidase by *Mortierella vinaceae* (Copetti 2019).

Saccharomyces cerevisiae, commonly known as baker's yeast, is the most extensively used microorganism in the food industry and household for food purposes (Ghorai et al. 2009). Yeast mediated biochemical transformations are widely used for brewing, preparing beverages, and fermenting baking goods. There are several properties of *S. cerevisiae* that make it excellent for industrial use, particularly in the brewing industry, including efficient growth properties and tolerance to environmental stress. Dried *S. cerevisiae* has been used as animal feed because of its high nutritive value and is an excellent source of protein for feeding livestock, including cattle and other ruminants (Ferreira et al. 2010).

2.4 Fungi as Biocontrol Agents

2.4.1 Entomopathogenic Fungi

Several fungi have been identified as biological control agents against plant pathogens and insect pests. It is an economically feasible and environment-friendly approach to control pest damages. A biological control refers to the use of living organisms to control the population density of a pest organism. One of the significant causes of crop loss is insect pests in an agricultural setting, and the potential of entomopathogenic fungi to control the insect pests in agricultural settings has been in use in several countries. Approximately 700 species of entomopathogenic fungi have been identified; however, the species from *Isaria*, *Metarhizium*, *Beauveria*, and *Lecanicillium* have been commercially produced and used as mycoinsecticides. The entomopathogenic lifecycle starts with the attachment of fungal conidia on the insect cuticle. The initial adherence of fungal conidia on the insect cuticle is facilitated by hydrophobic interactions or via the expression of specific proteins. For example, in *Metarhizium* sp., the specific protein, *Metarhizium* adhesin 1 (Mad 1), is involved in the initial attachment of fungal spores to insect cuticle (Wang and St. Leger 2007). The adhered fungal conidia subsequently germinate and form specialized infection structures called appressoria. Appressoria penetrates the insect exoskeleton through a combination of mechanical pressure and enzymatic activity. The fungi employ enzymes including proteases, chitinases, esterases, and lipases for the degrading insect cuticle. Once inside the insect hemocoel, the fungi differentiate into yeastlike bodies called blastospores. These fungi are capable of exhibiting different adaptive mechanisms to evade the insect immune system. The fungus multiplies in the insect hemocoel, and once the nutrient is depleted, the fungi emerge out of the insect cadaver and mummify the insect (Moonjely et al. 2016). Entomopathogenic fungi-based biocontrol formulations are an important alternative to chemical pesticides and may limit the need for environmentally damaging nitrogen-rich fertilizers, contribute to improved plant productivity, and improve agricultural soil quality.

2.4.2 Nematophagous Fungi

Another biologically important class is the nematophagous or nematicidal fungi, which can digest or trap nematodes. More than 700 species of nematophagous fungi to have been identified belong to different phyla, including *Ascomycota*, *Basidiomycota*, *Chytridiomycota*, and *Zygomycota*, that have the ability to trap or consume nematodes, springtails, or amebae (Li et al. 2015). These fungi exist in natural and agricultural soil as saprotrophs and play a critical role in recycling carbon, nitrogen, and other elements that originated from the nematode's degradation. Nematicidal or nematophagous fungi are divided into four groups: endoparasitic, predator, ovicidal fungi, and toxin-producing fungi (Braga and De Araújo 2014).

Endoparasites use conidia to infect nematodes. The majority of the endoparasitic fungi are obligate parasites, and therefore they have no or limited saprophytic phase. The infection process of endoparasitic fungi is similar to entomopathogenic fungi, in which the infection process initiates when the fungal conidia adhere to nematode cuticle or the nematode sensory structures. After initial attachment, the fungi develop the infection vesicles in the cuticle layers and then produce hyphae responsible for absorbing the nutrients from the nematode. Generally, the fungi kill the nematodes within 3 days, and new conidia emerge out from the nematode cadaver. Endoparasitic fungi have a limited market as a biological control agent. Because of its obligate parasitic lifestyle, limited or no saprophytic phase, and restricted host range, its in vitro production is difficult and expensive on an industrial scale. Nematophagous predators are referred to as nematode-trapping fungi that belong to a monophyletic group in the order of *Orbiliales*. The main species identified as predator-type nematophagous fungi belong to the genera *Arthrobotrys*, *Dactylellina*, *Drechslerella*, *Duddigtonia*, and *Monacosporium*. They live saprophytically in the terrestrial as well as aquatic ecosystems and produce specialized hyphae to trap nematodes. These fungi create specialized trapping devices, and the morphology of the trapping structure varies among different species, including constricting rings, adhesive nodules, adhesive networks, adhesive columns, or non-constricting rings (Li et al. 2015). The process of trapping of nematodes has been described in *Arthrobotrys oligospora*, one of the best-studied nematode-trapping fungi that capture nematodes through adhesive network traps. *A. oligospora* exhibits a dual lifestyle, lives in soil or decaying matter as a saprophyte, colonize plant roots, and can parasitize other fungi as well. *A. oligospora* recognizes the presence of nematode and initiates a multistep nematode infection process including attraction, recognition, trap formation, adhesion, penetration, and immobilization. The presence of the nematodes and other environmental factors such as low nutrients induce the saprotrophic mycelium to switch to parasitic phase (Yang et al. 2011). The adhesive network trap of *Arthrobotrys* consists of several loops connected to each other in a three-dimensional form, whereas the genus *Monacosporium* produces adhesive knobs and branches to capture nematodes. Genus *Dactylellina* is characterized by stalked adhesive knobs or non-constricting rings. Constricting ring nematode trap is the characteristic feature of the genus *Dreschslerella* (Li et al. 2015; Vidal-Diez de Ulzurrun and Hsueh 2018). Subtilisin-like serine proteases have been identified as the main virulence factor in nematode-trapping fungi to breach the collagenous rich nematode cuticle (Yang et al. 2011).

Helminth infections are one of the major problems in livestock management. Effective biological control of gastrointestinal nematodes in horses using nematode-trapping fungi has been described. The sodium alginate pellets containing fungal mycelia formulations of *Duddingtonia flagrans* were orally given to horses at 15-day intervals for 6 months. The analysis of the fecal samples has shown a significant reduction of nematode egg count in *D. flagrans* treated horses (Braga et al. 2009). Similarly, another species of the nematode-trapping fungi, *Monacosporium thaumasium*, was found effective in controlling cyathostomin larvae in horses in Brazil (de Oliveira Tavela et al. 2011). Cyathostomin infections have been described as a major cause of colic and diarrhea in horses. These studies show the potential of nematophagous fungi as a promising biological control agent in parasite disease management in domestic animals. However, the duration of treatment, the correct interval between the treatment, and the dose administered to the animals are crucial to the effectiveness of any biological control strategy (Buzatti et al. 2015). Ovicidal fungi target immobile stages of nematodes and use appressoria to parasitize nematode eggs and cysts. These fungi have great potential in controlling the density of viable nematode eggs in the soil that are harmful to animal or plant health. Several species from *Ascomycota* have been identified as ovicidal fungi, including *Paecilomyces lilacinus*, *Clonostachys rosea*, *Dactyella ovoparasitica*, *Lecanicillium psalliotae*, and *Pochonia chlamydosporia*. Some of the ovicidal fungi identified are soil-inhabiting fungus as well. Therefore, these fungi can be easily grown in the laboratory for biological control applications. The eggshells of nematodes are made up of protein and chitin. Egg parasitic fungi produce hydrolytic enzymes, such as protease and chitinases, to disintegrate the eggshell layers. Ovicidal effects have been classified into three types. In type 1 effect, the hyphae do not colonize internally in the eggs but result in lytic effect. The infection of fungi does not cause any morphological damage to the eggshell. Type 2 effect is similar to type 1 effect; however, the fungi cause alters the morphology of the eggshell and the embryo. In type 3 effect, the fungal hyphae penetrate and colonize internally and cause subsequent morphological alterations to the nematode eggshell and embryo (Braga and De Araújo 2014).

Phytoparasitic nematodes are one of the major causes of global annual crop loss, which accounts for an average of 12.3%. Approximately 4100 plant parasites have been reported, which cause symptoms such as wilting, nutrient deficiency, root lesions, reduced flowering, poor yield, or death of the plant (Degenkolb and Vilcinskas 2016). Several members of the *Pleurotus* genus produce nematotoxic microdroplets that paralyze nematodes. Approximately 280 species from

Ascomycota and *Basidiomycota* have been identified to produce nematicidal metabolites. Approximately 179 compounds belonging to different chemical classes were isolated. The chemical class includes alkaloids, peptides, terpenoids, macrolides, oxygen heterocycle, and benzo compounds, quinones, and sterols (Li et al. 2015). These metabolites have nematicidal properties and therefore are an environmentally safe approach to be used to fight against phytoparasitic nematodes.

2.5 Fungi as a Source of Bioactive Compounds

The role of fungi in producing pharmaceutically important compounds has been known since the discovery of the antibiotic penicillin by Alexander Fleming in 1929. Penicillin was first identified in a fungus *Penicillium* sp. which belongs to *Ascomycota*. Bioactive compounds produced by fungi have yielded metabolites of therapeutic applications, including immunosuppressants, cholesterol-lowering agents, prebiotics, antitumor agents, antimicrobials, and antiparasitic drugs (Demain and Martens 2017). Cyclosporin is widely used as an immunosuppressant drug after organ transplant surgery and is a fungal metabolite produced by *Tolypocladium nivenum* (Liu et al. 2015). Cyclosporins have antiviral properties and were found active against coronavirus. The biosynthesis of another immunosuppressant, antiviral, and antitumor drug has been reported in *P. stoloniferum*. Several fungal genera (*Penicillium*, *Aspergillus*, *Monascus*, *Doratomyces*, *Eupenicillium*, *Gymnoascus*, *Hypomyces*, *Paecilomyces*, *Trichoderma*, *Pleurotus*) have the ability to produce statins, the hypolipidemic drug (cholesterol-lowering drug). The synthetic production of Taxol, a steroidal antitumor drug, is unfeasible due to its complex chemical structure. Fungal species such as *Taxomyces*, *Pestaltipsis*, *Tubercularia*, *Phyllosticta*, *Metarhizium*, *Colletotrichum*, *Fusarium*, and *Pestalotiopsis* have the ability to produce Taxol (Demain and Martens 2017).

The intimate association of fungal partners suggests a continuous metabolic interaction with the host plant (Brader et al. 2014). Research on these complex interactions, and the subsequently secreted metabolome, have led to the discovery of many novel bioactive molecules of pharmaceutical importance, including paclitaxel, camptothecin, podophyllotoxin, hypericin, and emodin (Kusari and Spiteller 2011). Several of the pharmacologically important metabolites were originally thought to be derived from plants; however, it has later been revealed that many fungal endophytes residing in plant roots are able to synthesize the same metabolites that occur in plants. Previous studies have revealed the similarity of biosynthetic pathways of plant-associated metabolites in both plants and endophytic fungi. This suggests the potential for the independent production of these secondary metabolites by endophytic fungi. It has been suggested that the genetic recombination of an endophyte with its plant partner has led to the integration of these biosynthetic genes. It has also been hypothesized that endophytes possess gene clusters that have remained silent during coevolution and might be activated by suitable ecological associations (Aly et al. 2011; Kusari and Spiteller 2011). A deeper understanding of the expression of biosynthetic genes during plant-fungal association will help to optimize metabolite production under laboratory conditions. The microbial diversity found in the rhizosphere of medicinal plants and their potential role in eliciting plant secondary metabolites are well documented (Kaul et al. 2012). The relationship of plant hosts with fungal root colonizers is often a balanced symbiotic interaction, characterized by reciprocal nutrient exchange, whereby fungus received plant-derived carbon sources in exchange for limited soil nutrients (Aly et al. 2011). Fungal endophytes usually perform compatible interactions with the host plant, which not only contribute to the growth promotion but simultaneously contribute to the secondary metabolite accumulation in the plant during abiotic and biotic stress (Kusari and Spiteller 2011). A study performed on mycorrhizal (*Glomus* sp.) colonization on annual wormwood, *Artemisia annua* L plants, revealed enhanced yield in artemisinin (antimalarial drug) along with other terpenes (Rapparini et al. 2008).

Entomopathogenic and nematophagous fungi have the potential for the synthesis of bioactive compounds or secondary metabolites. Many of these secondary metabolites have pharmaceutical, nematicidal, or insecticidal properties. In endophytic insect pathogenic fungi (EIPF), *Metarhizium robertsii*, 85 genes that have been involved in the biosynthesis of secondary metabolites have been identified, encoding non-ribosomal peptide synthetases (NRPS), NRPS-like, polyketide synthases (PKS), PKS-like, and hybrid NRPS-PKS (Gibson et al. 2014). *Metarhizium* sp. is known for producing cyclic depsipeptides, destruxins. Destruxins are well-known for their insecticidal property and have been used as an alternative for chemical insecticides in controlling pests in agroecosystems. Destruxins are also known for their immunosuppressive (Pal et al. 2007), phytotoxic (Pedras et al. 2000), antiproliferative (Yeh et al. 2012), and cytotoxic effects (Dornetshuber-Fleiss et al. 2013). Recently, the role of destruxins in osteoporosis and cancer therapy was reported due to their activity as vacuolar H^+-ATPase inhibitor (Liu and Tzeng 2012). In some cell types, vacuolar H^+-ATPases are linked to disease progression, including osteoporosis and cancer (Holliday 2017), and therefore vacuolar H^+-ATPase inhibitors like destruction is a potential target for such diseases.

Several secondary metabolites have been identified in EIPF *Beauveria bassiana* as well. Of these, the well-studied metabolites are beauvericin and bassionolide. Beauvericin is a cyclic hexadepsipeptide, known for its insecticidal, antimicrobial, antiviral, and antitumor activities (Wang and Xu 2012). Studies have also shown its potential to reverse the multidrug resistance in yeast. The biosynthesis of beauvericin is reported in multiple fungal species, including nematophagous and plant colonizing fungi *Paecilomyces* and *Fusarium*, respectively (Gibson et al. 2014). Brassinolide, a octacyclo depsipeptide isolated from fungal species including *Beauveria* sp. (Xu et al. 2009), *Lecanicillium* sp. (Ravindran et al. 2018), and *Xylaria* sp. (Jirakkakul et al. 2008), is reported as cytotoxic, and antiplasmodial and antimicrobial activities were assessed in vitro. Apart from these, bassianolide also demonstrated inhibitory activity on acetylcholine-induced smooth muscle contraction. Moreover, 48 secondary metabolite biosynthetic gene clusters are predicted in EIPF *B. bassiana*, including 13 NRPS, 12 PCS, 7 NRPS-like, 1 PKS-like, 3 hybrid NRPS-PKS, and 12 genes

Table 2.1 Some pharmaceutical compounds produced by fungi

Product	Function/application	Source	References
Asterriquinone	Anticancer	*Aspergillus terreus* *Humicola grisea*	Li (2010)
Camptothecin	Anticancer	*Entrophospora infrequens*	Tansuwan et al. (2007)
Cephalosporin	Antimicrobial	*Cephalosporium acremonium*	Cruz et al. (2004)
Cyclosporin	Immunosuppressant	*Tolypocladium inflatum*	Bushley et al. (2013)
Ergot alkaloids	Oxytocic effect Migraine relief	*Claviceps purpurea*	Schardl et al. (2006)
Fumosorinone	Metabolic syndrome drug	*Isaria fumosorosea*	Liu et al. (2015)
Mevalonin	Antibiotic	*A. terreus*	
Mycophenolic acid	Immunosuppressant Antiviral Antitumor	*Penicillium stoloniferum* *P. viridicatum* *P. Brevicompactum*	Min et al. (2019)
Penicillin	Antimicrobial	*Penicillium chrysogenum*	Müller et al. (1991)
Statins	Hypocholesterolemic drug	*Aspergillus terreus*, *Penicillium* sp. *Trichoderma*	Subhan et al. (2016)
Sclerotiorin	Apoptotic properties	*Cephalotheca faveolata*	Giridharan et al. (2012)
Taxol	Anticancer	*Metarhizium*, *Taxomyces*, *Pestaltipsis*, *Tubercularia*, *Phyllosticta*, *fusarium*	Demain and Martens (2017)
Ustilagic acid	Antimicrobial	*Pseudozyma fusiformata* *Ustilago maydis*	Kulakovskaya et al. (2005)
Xylariaquinone A	Antimalarial	*Xylaria* sp.	Tansuwan et al. (2007)

related to the biosynthesis of terpene/steroid (Gibson et al. 2014). The biosynthesis of fumosorinone, a potential drug target for the treatment of metabolic syndromes like type 2 diabetes, has been identified in entomopathogenic fungi *Isaria fumosorosea* (Liu et al. 2015). These studies suggest the potential of many members in the fungal kingdom to contribute to the production of bioactive metabolites with pharmaceutical and other industrial applications. Fungi have provided significant contribution to the field of pharmaceutical industry (Table 2.1).

2.6 Conclusion

Fungi have made significant contributions to several fields that are beneficial to the environment and human health. The fungal kingdom evolved a close interaction with the environment, plant, and animals as a part of the evolution process. Previous studies have established the beneficial role of fungi plant colonizer to improve plant health and a sustainable source for producing bioactive molecules and other lifesaving drugs. However, little is known about how fungal interactions affect vital ecosystem functions due to climate or anthropogenic changes, which is an important topic to consider for research. More studies are needed on the genetic aspects of the persistence of different fungal populations in our ecosystems and their interactions with different hosts due to climate change.

References

Aly AH, Debbab A, Proksch P. Fungal endophytes: unique plant inhabitants with great promises. Appl Microbiol Biotechnol. 2011;90(6):1829–45. https://doi.org/10.1007/s00253-011-3270-y.

Bechtold T, Mussak R. Handbook of natural colorants; 2009. p. 151–82. https://doi.org/10.1002/9780470744970.

Behie SW, Bidochka MJ. Nutrient transfer in plant-fungal symbioses. Trends Plant Sci. 2014a;19(11):734–40. https://doi.org/10.1016/j.tplants.2014.06.007.

Behie SW, Bidochka MJ. Ubiquity of insect-derived nitrogen transfer to plants by endophytic insect-pathogenic fungi: an additional branch of the soil nitrogen cycle. Appl Environ Microbiol. 2014b;80(5):1553–60. https://doi.org/10.1128/AEM.03338-13.

Behie SW, Moreira CC, Sementchoukova I, Barelli L, Zelisko PM, Bidochka MJ. Carbon translocation from a plant to an insect-pathogenic endophytic fungus. Nat Commun. 2017;8:14245. https://doi.org/10.1038/ncomms14245.

Bonfante P, Anca IA. Plants, mycorrhizal fungi, and bacteria: a network of interactions. Annu Rev Microbiol. 2009;63:363–83. https://doi.org/10.1146/annurev.micro.091208.073504.

Bonfante P, Venice F. Mucoromycota: going to the roots of plant-interacting fungi. Fungal Biol Rev. 2020;34(2):100–13. https://doi.org/10.1016/j.fbr.2019.12.003.

Brader G, Compant S, Mitter B, Trognitz F, Sessitsch A. Metabolic potential of endophytic bacteria. Curr Opin Biotechnol. 2014;27(100):30–7. https://doi.org/10.1016/j.copbio.2013.09.012.

Braga FR, de Araújo JV. Nematophagous fungi for biological control of gastrointestinal nematodes in domestic animals. Appl Microbiol Biotechnol. 2014;98(1):71–82. https://doi.org/10.1007/s00253-013-5366-z.

Braga FR, Araújo JV, Silva AR, Araujo JM, Carvalho RO, Tavela AO, et al. Biological control of horse cyathostomin (Nematoda: Cyathostominae) using the nematophagous fungus Duddingtonia flagrans in tropical southeastern Brazil. Vet Parasitol. 2009;163(4):335–40. https://doi.org/10.1016/j.vetpar.2009.05.003.

Bushley KE, Raja R, Jaiswal P, Cumbie JS, Nonogaki M, Boyd AE, et al. The genome of tolypocladium inflatum: evolution, organization, and expression of the cyclosporin biosynthetic gene cluster. PLoS Genet. 2013;9(6):e1003496. https://doi.org/10.1371/journal.pgen.1003496.

Buzatti A, de Paula SC, Fernandes MA, Yoshitani UY, Sprenger LK, dos Santos CD, et al. Duddingtonia flagrans in the control of gastrointestinal nematodes of horses. Exp Parasitol. 2015;159:1–4. https://doi.org/10.1016/j.exppara.2015.07.006.

Caro Y, Anamale L, Fouillaud M, Laurent P, Petit T, Dufosse L. Natural hydroxyanthraquinoid pigments as potent food grade colorants: an overview. Nat Prod and Bioprospect. 2012;2(5):174–93. https://doi.org/10.1007/s13659-012-0086-0.

Challacombe JF, Hesse CN, Bramer LM, McCue LA, Lipton M, Purvine, et al. Genomes and secretomes of Ascomycota fungi reveal diverse functions in plant biomass decomposition and pathogenesis. BMC Genomics. 2019;20(1):1–27. https://doi.org/10.1186/s12864-019-6358-x.

Clasen BE, Silveira ADO, Baldoni DB, Montagner DF, Jacques RJS, Antoniolli ZI. Characterization of ectomycorrhizal species through molecular biology tools and morphotyping. Sci Agric. 2018;75(3):246–54. https://doi.org/10.1590/1678-992x-2016-0419.

Comeau AM, Vincent WF, Bernier L, Lovejoy C. Novel chytrid lineages dominate fungal sequences in diverse marine and freshwater habitats. Sci Rep. 2016;6(1):1–6. https://doi.org/10.1038/srep30120.

Copetti MV. Fungi as industrial producers of food ingredients. Curr Opinion Food Sci. 2019;25:52–6. https://doi.org/10.1016/j.cofs.2019.02.006.

Cruz AJG, Pan T, Giordano RC, Araujo MLGC, Hokka CO. Cephalosporin C production by immobilized Cephalosporium acremonium cells in a repeated batch tower bioreactor. Biotechnol Bioeng. 2004;85(1):96–102. https://doi.org/10.1002/bit.10877.

de Oliveira Tavela A, Araújo JV, Braga FR, Silva AR, Carvalho RO, Araujo JM, et al. Biological control of cyathostomin (Nematoda: Cyathostominae) with nematophagous fungus Monacrosporium thaumasium in tropical southeastern Brazil. Vet Parasitol. 2011;175(1–2):92–6. https://doi.org/10.1016/j.vetpar.2010.09.035.

Degenkolb T, Vilcinskas A. Metabolites from nematophagous fungi and nematicidal natural products from fungi as an alternative for biological control. Part I: metabolites from nematophagous ascomycetes. App Microbiol Biotechnol. 2016;100(9):3799–812. https://doi.org/10.1007/s00253-015-7233-6.

Demain AL, Martens E. Production of valuable compounds by molds and yeasts. J Antibiot. 2017;70(4):347–60. https://doi.org/10.1038/ja.2016.121.

Dighton J. Fungi in ecosystem processes. New York: Marcel Dekker; 2003. https://doi.org/10.1201/9780203911440.

Dornetshuber-Fleiss R, Heffeter P, Mohr T, Hazemi P, Kryeziu K, Seger C, et al. Destruxins: fungal-derived cyclohexadepsipeptides with multifaceted anticancer and antiangiogenic activities. Biochem Pharmacol. 2013;86(3):361–77. https://doi.org/10.1016/j.bcp.2013.05.022.

Dufossé L, Fouillaud M, Caro Y, Mapari SA, Sutthiwong N. Filamentous fungi are large-scale producers of pigments and colorants for the food industry. Curr Opin Biotechnol. 2014;26:56–61. https://doi.org/10.1016/j.copbio.2013.09.007.

Ferreira IMPLVO, Pinho O, Vieira E, Tavarela JG. Brewer's saccharomyces yeast biomass: characteristics and potential applications. Trends Food Sci Technol. 2010;21(2):77–84. https://doi.org/10.1016/j.tifs.2009.10.008.

Gessler NN, Egorova AS, Belozerskaya TA. Fungal anthraquinones. Appl Biochem Microbiol. 2013;49(2):85–99. https://doi.org/10.1134/S000368381302004X.

Ghorai S, Banik SP, Verma D, Chowdhury S, Mukherjee S, Khowala S. Fungal biotechnology in food and feed processing. Food Res Int. 2009;42(5–6):577–87. https://doi.org/10.1016/j.foodres.2009.02.019.

Gibson DM, Donzelli BGG, Krasnoff SB, Keyhani NO. Discovering the secondary metabolite potential encoded within entomopathogenic fungi. Nat Prod Rep. 2014;31(10):1287–305. https://doi.org/10.1039/c4np00054d.

Giridharan P, Verekar SA, Khanna A, Mishra PD, Deshmukh SK. Anticancer activity of sclerotiorin, isolated from an endophytic fungus Cephalotheca faveolata Yaguchi, Nishim & Udagawa. Indian J Exp Biol. 2012;50(7):464–8.

Grossart HP, Van den Wyngaert S, Kagami M, Wurzbacher C, Cunliffe M, Rojas-Jimenez K. Fungi in aquatic ecosystems. Nat Rev Microbiol. 2019;17(6):339–54. https://doi.org/10.1038/s41579-019-0175-8.

Gruninger RJ, Puniya AK, Callaghan TM, Edwards JE, Youssef N, Dagar SS, et al. Anaerobic fungi (phylum Neocallimastigomycota): advances in understanding their taxonomy, life cycle, ecology, role and biotechnological potential. FEMS Microbiol Ecol. 2014;90:1–17. https://doi.org/10.1111/1574-6941.12383.

Holliday LS. Vacuolar H+-ATPases (V-ATPases) as therapeutic targets: a brief review and recent developments. Biotarget. 2017;1(7):18. https://doi.org/10.21037/biotarget.2017.12.01.

Jirakkakul J, Punya J, Pongpattanakitshote S, Paungmoung P, Vorapreeda N, Tachaleat A, et al. Identification of the nonribosomal peptide synthetase gene responsible for Bassianolide synthesis in wood-decaying fungus Xylaria sp. BCC1067. Microbiology. 2008;154(4):995–1006. https://doi.org/10.1099/mic.0.2007/013995-0.

Kaul S, Gupta S, Ahmed M, Dhar MK. Endophytic fungi from medicinal plants: a treasure hunt for bioactive metabolites. Phytochem Rev. 2012;11(4):487–505. https://doi.org/10.1007/s11101-012-9260-6.

Kulakovskaya TV, Shashkov AS, Kulakovskaya EV, Golubev WI. Ustilagic acid secretion by Pseudozyma fusiformata strains. FEMS Yeast Res. 2005;5(10):919–23. https://doi.org/10.1016/j.femsyr.2005.04.006.

Kusari S, Spiteller M. Are we ready for industrial production of bioactive plant secondary metabolites utilizing endophytes? Nat Prod Rep. 2011;28(7):1203–7. https://doi.org/10.1039/c1np00030f.

Lepelletier F, Karpov SA, Alacid E, Le Panse S, Bigeard E, Garcés E, et al. Dinomyces arenysensis gen. Et sp. nov. (Rhizophydiales, Dinomycetaceae fam. Nov.), a chytrid infecting marine dinoflagellates. Protist. 2014;165(2):230–44. https://doi.org/10.1016/j.protis.2014.02.004.

Li SM. Prenylated indole derivatives from fungi: structure diversity, biological activities, biosynthesis and chemoenzymatic synthesis. Nat Prod Rep. 2010;27(1):57–78. https://doi.org/10.1039/b909987p.

Li J, Zou C, Xu J, Ji X, Niu X, Yang J, et al. Molecular mechanisms of nematode-nematophagous microbe interactions: basis for biological control of plant-parasitic nematodes. Annu Rev Phytopathol. 2015;53:67–95. https://doi.org/10.1146/annurev-phyto-080614-120336.

Lindahl BD, Tunlid A. Ectomycorrhizal fungi - potential organic matter decomposers, yet not saprotrophs. New Phytol. 2015;205(4):1443–7. https://doi.org/10.1111/nph.13201.

Liu BL, Tzeng YM. Development and applications of destruxins: a review. Biotechnol Adv. 2012;30(6):1242–54. https://doi.org/10.1016/j.biotechadv.2011.10.006.

Liu L, Zhang J, Chen C, Teng J, Wang C, Luo D. Structure and biosynthesis of fumosorinone, a new protein tyrosine phosphatase 1B inhibitor firstly isolated from the entomogenous fungus Isaria fumosorosea. Fungal Genet Biol. 2015;81:191–200. https://doi.org/10.1016/j.fgb.2015.03.009.

López-Pedrosa A, González-Guerrero M, Valderas A, Azcón-Aguilar C, Ferrol N. GintAMT1 encodes a functional high-affinity ammonium transporter that is expressed in the extraradical mycelium of Glomus intraradices. Fungal Genet. Biol. 2006;43;102–110. https://doi.org/10.1016/j.fgb.2005.10.005.

Maciá-Vicente JG, Rosso LC, Ciancio A, Jansson HB, Lopez-Llorca LV. Colonisation of barley roots by endophytic fusarium Equiseti and Pochonia chlamydosporia: effects on plant growth and disease. Ann Appl Biol. 2009;155(3):391–401. https://doi.org/10.1111/j.1744-7348.2009.00352.x.

Min C, Dong H, Liu X, Zhang Z. Screening and identification of a Penicillium brevicompactum strain isolated from the fruiting body of Inonotus obliquus and the fermentation production of mycophenolic acid. Ann Microbiol. 2019;69(13):1351–60. https://doi.org/10.1007/s13213-019-01517-z.

Moonjely S, Bidochka MJ. Generalist and specialist Metarhizium insect pathogens retain ancestral ability to colonize plant roots. Fungal Ecol. 2019;41:209–17. https://doi.org/10.1016/j.funeco.2019.06.004.

Moonjely S, Barelli L, Bidochka MJ. Chapter 4. Insect pathogenic fungi as endophytes. In: Lovett B, St. Leger RJ, editors. Advances in genetics. Academic Press Inc; 2016. p. 107–35.

Müller WH, van der Krift TP, Krouwer AJ, Wösten HA, van der Voort LH, Smaal EB, et al. Localization of the pathway of the penicillin biosynthesis in Penicillium chrysogenum. EMBO J. 1991;10(2):489–95. https://doi.org/10.1002/j.1460-2075.1991.tb07971.x.

Naranjo-Ortiz MA, Gabaldón T. Fungal evolution: diversity, taxonomy and phylogeny of the fungi. Biol Rev. 2019;94(6):2101–37. https://doi.org/10.1111/brv.12550.

Pal S, St. Leger RJ, Wu LP. Fungal peptide destruxin a plays a specific role in suppressing the innate immune response in Drosophila melanogaster. J Biol Chem. 2007;282(12):8969–77. https://doi.org/10.1074/jbc.M605927200.

Pedras MSC, Biesenthal CJ, Zaharia IL. Comparison of the phytotoxic activity of the phytotoxin destruxin B and four natural analogs. Plant Sci. 2000;156(2):185–92. https://doi.org/10.1016/S0168-9452(00)00253-3.

Pérez-Tienda J, Testillano PS, Balestrini R, Fiorilli V, Azcón-Aguilar C, Ferrol N. GintAMT 2, a new member of the ammonium transporter family in the arbuscular mycorrhizal fungus Glomus intraradices. Fungal Genet Biol. 2011;48:1044–55. https://doi.org/10.1016/j.fgb.2011.08.003.
Powell MJ. In: Archibald JM, Simpson AGB, editors. Handbook of protists; 2017. p. 1497–521.
Rapparini F, Llusià J, Peñuelas J. Effect of arbuscular mycorrhizal (AM) colonization on terpene emission and content of Artemisia annua L. Plant Biol. 2008;10(1):108–22. https://doi.org/10.1055/s-2007-964963.
Ravindran K, Sivaramakrishnan S, Hussain M, Dash CK, Bamisile BS, Qasim M, et al. Investigation and molecular docking studies of bassianolide from Lecanicillium lecanii against Plutella xylostella (Lepidoptera: Plutellidae). Comp Biochem Physiol C Toxicol Pharmacol. 2018;206–207:65–72. https://doi.org/10.1016/j.cbpc.2018.03.004.
Sasan RK, Bidochka MJ. Antagonism of the endophytic insect pathogenic fungus Metarhizium robertsii against the bean plant pathogen Fusarium solani f. sp. phaseoli. Can J Plant Pathol. 2013;35(3):288–93. https://doi.org/10.1080/07060661.2013.823114.
Schardl CL, Panaccione DG, Tudzynski P. Ergot alkaloids—biology and molecular biology. Alkaloids Chem Biol. 2006;63:45–86. https://doi.org/10.1016/s1099-4831(06)63002-2.
Schüßler A. The Geosiphon-nostoc endosymbiosis and its role as a model for arbuscular mycorrhiza research. In: Esser K, editor. Fungal associations, the Mycota. Berlin, Heidelberg; 2006. p. 77–91.
Steiner M, Linkov I, Yoshida S. The role of fungi in the transfer and cycling of radionuclides in forest ecosystems. J Environ Radioactivity. 2002;58(2–3):217–41. https://doi.org/10.1016/S0265-931X(01)00067-4.
Subhan M, Faryal R, Macreadie I. Exploitation of aspergillus terreus for the production of natural statins. J Fungi. 2016;2(13):1–13. https://doi.org/10.3390/jof2020013.
Tansuwan S, Pornpakakul S, Roengsumran S, Petsom A, Muangsin N, Sihanonta P, et al. Antimalarial benzoquinones from an endophytic fungus. Xylaria sp J Natural Prod. 2007;70(10):1620–3. https://doi.org/10.1021/np0701069.
Vidal-Diez de Ulzurrun G, Hsueh YP. Predator-prey interactions of nematode-trapping fungi and nematodes: both sides of the coin. Appl Microbiol Biotechnol. 2018;102(9):3939–49. https://doi.org/10.1007/s00253-018-8897-5.
Wang C, St. Leger RJ. The MAD1 adhesin of Metarhizium anisopliae links adhesion with blastospore production and virulence to insects, and the MAD2 adhesin enables attachment to plants. Eukaryot Cell. 2007;6(5):808–16. https://doi.org/10.1128/EC.00409-06.
Wang Q, Xu L. Molecules Beauvericin, a bioactive compound poroduced by fungi: a short review. Molecules. 2012;17:2367–77. https://doi.org/10.3390/molecules17032367.
Wardle DA, Lindahl BD. Ecology. Disentangling global soil fungal diversity. Science. 2014;346(6213):1052–3. https://doi.org/10.1126/science.aaa1185.
Xu Y, Orozco R, Wijeratne EK, Espinosa-Artiles P, Leslie Gunatilaka AA, Patricia Stock S, et al. Biosynthesis of the cyclooligomer depsipeptide bassianolide, an insecticidal virulence factor of Beauveria bassiana. Fungal Genet Biol. 2009;46(5):353–64. https://doi.org/10.1016/j.fgb.2009.03.001.
Yang J, Wang L, Ji X, Feng Y, Li X, Zou C, et al. Genomic and proteomic analyses of the fungus arthrobotrys oligospora provide insights into nematode-trap formation. PLoS Pathog. 2011;7(9):e1002179. https://doi.org/10.1371/journal.ppat.1002179.
Yeh CT, Rao YK, Ye M, Wu WS, Chang TC, Wang LS, et al. Preclinical evaluation of destruxin B as a novel Wnt signaling target suppressing proliferation and metastasis of colorectal cancer using non-invasive bioluminescence imaging. Toxicol Appl Pharmacol. 2012;261(1):31–41. https://doi.org/10.1016/j.taap.2012.03.007.
Yun W, Hall IR. Edible ectomycorrhizal mushrooms: challenges and achievements. Can J Bot. 2004;82(8):1063–73. https://doi.org/10.1139/B04-051.
Zhao RL, Li GJ, Sánchez-Ramírez S, Stata M, Yang ZL, Wu G, Dai YC, et al. A six-gene phylogenetic overview of Basidiomycota and allied phyla with estimated divergence times of higher taxa and a phyloproteomics perspective. Fungal Divers. 2017;84:43–74. https://doi.org/10.1007/s13225-017-0381-5.

Chapter 3
Thermotolerance and Adaptation to Climate Change

Michel Almaguer Chávez

Contents

Abbreviations

CYA Czapek Yeast Autolysate Agar
HSP Heat shock proteins

M. Almaguer Chávez (✉)
Faculty of Biology, University of Havana, Havana, Cuba
e-mail: michelalm@fbio.uh.cu

M. G. Frías-De-León et al. (eds.), *The Impact of Climate Change on Fungal Diseases*, Fungal Biology, https://doi.org/10.1007/978-3-030-89664-5_3

3.1 Introduction

Temperature is closely related to fungal biology, affecting various processes such as growth, reproduction, spore formation, germination, and indeed all activities of the organism (Paterson and Lima 2019). The values of environmental temperature allow fungi with limits outside a certain range of variation to be excluded. This occurs because the tolerable conditions do not allow it to develop or to complete its life cycle. This exclusion effect is immediate and determines the fungal populations that are present in certain habitats. There may also be a less immediate effect whereby this environmental factor may have a selective effect among fungi whose development ranges are within the ambient temperature range. In this way, fungi whose optimum temperatures are close to the prevailing temperature will be the most favored (de Menezes et al. 2019). In this way, this environmental factor affects the ecology of fungi.

Each fungal species has a growth rate to a certain temperature, and hence cardinal temperatures have been identified. These are the minimum temperature where growth occurs, the optimum temperature where optimal growth occurs, and the maximum temperature at which it can grow (Fig. 3.1). In the temperature-growth rate curve, there is characteristically a linear portion in which growth increases directly with temperature, an optimum range that may be narrow or rather broad, and a descending limb as the temperature becomes too high for growth (Cochrane 1958).

According to the response to this environmental factor, fungi have been classified as psychrophilic, mesophilic, thermotolerant, and thermophilic (Fig. 3.2). There are divergences in the grouping and classification of some species because it is

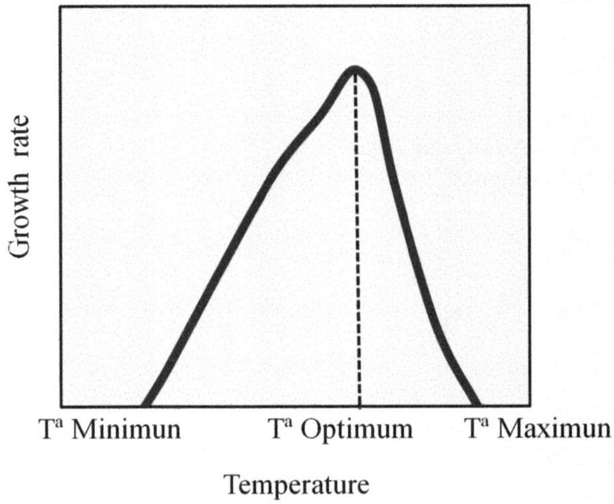

Fig. 3.1 Typical curve of growth as a function of temperature. Minimum, optimum, and maximum temperature where growth occurs

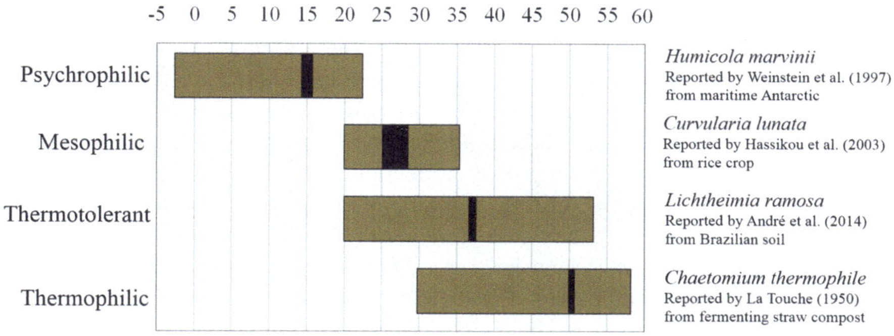

Fig. 3.2 Temperature ranges for the growth of some psychrophilic, mesophilic, thermotolerant, and thermophilic fungal species. ■ Optimum temperature range, ■ temperature range of fungal growth

difficult to specify precisely the limits in each group. However, this classification helps to understand the relationship of fungal growth with this environmental factor. There is evidence that in the same species there may be a variable range of response, which shows certain diversity (Rangel et al. 2005).

Few fungi are psychrophilic or psychrotolerant, with the ability to survive but not to grow at temperatures below 5 °C (Fig. 3.2). Such is the case of *Humicola marvinii* reported by Weinstein et al. (1997) on Signy Island, Antarctic. In this group, the term psychrophilic has also been proposed, for those organisms that have an optimum growth at temperatures not higher than 16 °C and a maximum at around 20 °C (Morris and Clarke 1987). Most of the known fungi are mesophilic and can develop between 10 and 35 °C, with optimum growth temperature between 20 and 30 °C (Cooney and Emerson 1964). Hassikou et al. (2003) reported the mesophilic growth rates of a *Curvularia lunata* strain isolated from a rice crop. Relatively few fungi can grow at temperatures above 37 °C, and it is suggested that the upper limit for the growth of any fungus is approximately 62 °C. This group of fungi that can grow at high temperatures is thermophilic, with minimum growth of about 20 °C, optimal development in the range of 40 to 50 °C, and a maximum of up to 50 °C or more. For example, some hyphomycetes isolated from hot springs by Tansey (1973) had a maximum growth temperature of 61.5 °C (Brown and Fritz 2019). Such is the case of *Chaetomium thermophile* reported by La Touche (1950) from fermenting straw compost. Fungi, being eukaryotic organisms, are much more sensitive than other domains of life (*Archaea* and *Bacteria*), and at temperatures above 65 °C, their membranes become more sensitive or can be irreparably damaged. However, since the investigations of Cooney and Emerson (1964), several fungi that can develop at high temperatures have been described. The information on the taxonomy, physiological characteristics, and ecology of these fungi capable of surviving under high temperatures has been provided (Maheshwari et al. 2000).

There are mesophilic fungi that may have the ability to survive or maintain viability at high temperatures and are classified as thermotolerant. Such is the case of

Lichtheimia ramosa reported by André et al. (2014) from Brazilian soil (Fig. 3.2). This group is distinguished from those with the true ability to actively grow as a thermophilic at high temperatures (Magan 2007). Thermophilic and thermotolerant fungal species are distinguished based on their minimum and maximum growth temperature (Fig. 3.2). Thermophilic fungi have a minimum growth temperature of 20 °C or higher and a maximum growth temperature of 50 °C or higher. However, thermotolerant have a wider growth temperature range, ranging from less than 20 °C to about 55 °C (Cooney and Emerson 1964; Maheshwari et al. 2000). Thermotolerant species can often be part of fungal communities that colonize a variety of humid organic substrates and where temperatures can be high. These substrates include hay, straw or cereal-based fertilizers, bird nests, and tropical soils (Magan 2007).

Evans (1971) studied more than 30 species isolated from coal spoil tips and delineated the differences between thermophilic and thermotolerant, according to their growth at various temperatures. The aforementioned study argued that the definition of Cooney and Emerson (1964) was becoming artificial and obscure, delimiting heat-tolerant fungi into three groups. Group 1 (strong thermophilic) constitutes the obligate thermophilic, while Group 2 (weak thermophilic) and Group 3 (strong thermotolerant) species form a transition stage between true thermophilia and general thermotolerance. This last term covers the species included in Group 4 (thermotolerant in general). Then it was suggested that the strains of certain fungi were transitional between the thermophilic and thermotolerant groups (de Oliveira and Rodrigues 2019). In this sense, a thermotolerant fungus could grow at 45 °C or higher, usually with a maximum close to 50 °C, but it can also grow below 20 °C. Another important contribution from Evans (1971) is the identification of two broad groupings within thermotolerant. In one group were strong thermotolerant fungi that grew well at 50 °C and often up to 55 °C. However, they grew little at temperatures below 20 °C, with a minimum growth temperature close to this temperature. Due to this behavior, he did not consider them to be true thermophilic (Fig. 3.3). The other group was generally termed thermotolerant, as they grew well over a wide temperature range. Species in this group had their optimum at around 40 °C, were able to grow at temperatures well below 20 °C, and mostly grew very slowly or not at all above 50 °C.

The thermotolerance of some fungal species may also allow a less immediate exclusion effect in the face of an increase in this environmental factor, with ecological implications. In nature, the temperature fluctuates intraday (i.e., day and night), between seasons (i.e., spring, summer, autumn, winter, dry period-rainy period), and even between long-time series (Roy and Bhattacharya 2020). This contributes to the fact that habitat conditions are generally not always optimal for the same fungal population and is related to cyclical patterns of species succession in communities.

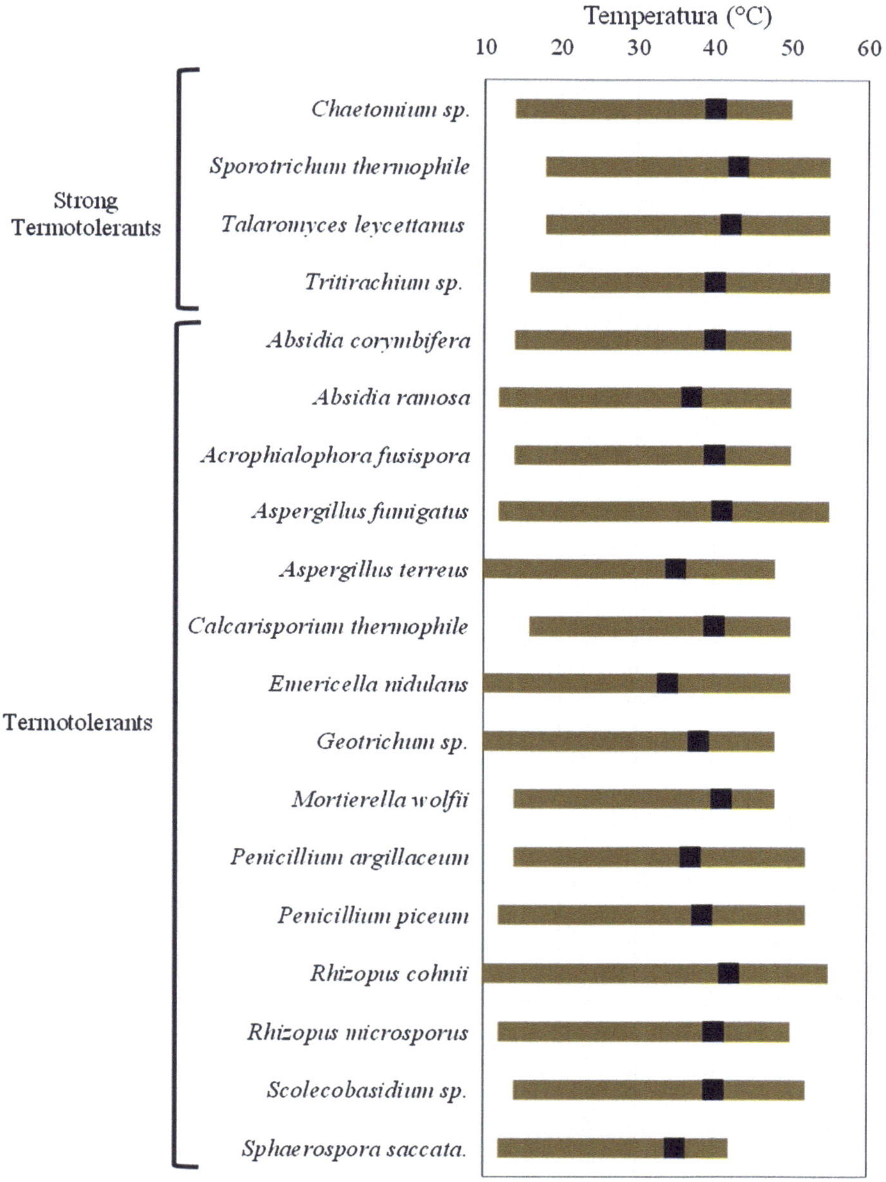

Fig. 3.3 Optimal, minimum, and maximum temperatures for the growth of some thermotolerant fungi. ■ Optimal temperatures, ■ minimum and maximum temperatures (Data from Evans 1971)

3.2 Mechanisms of Fungal Thermotolerance

Temperature can exert a marked influence on all aspects of fungal development, and various cellular processes contribute to the tolerance of different fungi to wide ranges of this abiotic factor. Joshi et al. (2020) based on what was argued by Grime (1977) suggest that fungi use strategies to survive and thrive in different environments. This is based on evidence for the existence of three primary strategies and their relevance to ecological and evolutionary theory. These three primary strategies can be competitive, stressful, and rude, but there may even be secondary strategies in combination. Competitive strategies are used when conditions are not stressful and maximize the exploitation of resources. Competitive characteristics are evident in fungi such as *Armillaria mellea* and *Serpula lacrymans*, which are responsible for long-term infections of timber and produce a consolidated mycelium which may extend rapidly through the production of rhizomorphs. Stress strategies involve adaptation to ongoing stress. Examples of stress-tolerant fungi appear to include the slow-growing basidiomycetes, which form the terminal stages of fungal succession on decaying matter, and the various fungi that occur in lichens and ectotrophic mycorrhizas. All of the stress-tolerant fungi are characterized by slow-growing, relatively persistent mycelia and low reproductive effort. Ruderal strategies are used under conditions rich in nutrients drastically disturbed by reducing their availability and increasing the reproductive potential of the fungus. Ruderal lifestyles are particularly characteristic of the mucorales, in which most species are ephemeral colonists of organic substrates. These fungi grow exceedingly rapidly and exploit the initial abundance of sugars, but as the supply of soluble carbohydrate declines, they cease mycelial growth and sporulate profusely.

The optimum temperature for growth is usually defined as the temperature at which the growth rate is maximum. This phenotypic character is related to the occurrence and velocity of various cellular and biochemical processes involved in growth. The involvement of these mechanisms has also been hypothesized to explain the basis for thermophilia and thermotolerance (Crisan 1973; Magan 2007). These cellular and biochemical processes include lipid solubilization, rapid resynthesis of essential metabolites, molecular thermostability, and ultrastructural thermostability.

Structural thermostability is of vital importance since there is the possibility that solubilization of cellular lipids occurs at high temperatures, with the danger that the cells lose their integrity. An increase in temperature can result in cellular lipids that contain more saturated fatty acids that have a higher melting point than those present in most mesophilic that do not show thermotolerance and that contain markedly less saturated lipids. Therefore, this characteristic contributes to maintain cell integrity at higher temperatures than in the case of mesophilic. It has also been suggested that increased fluidity of saturated lipids at high temperatures may enable metabolic activity and cell function to allow active growth at temperatures above 40 °C (Magan 2007).

Several investigations have focused on the ability of various fungi to produce heat shock proteins (HSP) when exposed to extreme temperatures (Tiwari et al. 2015; Wang et al. 2020). In heat shock, HSPs perform the equally important but different function of molecular chaperones. They participate in the stabilization of proteins and the adaptive modification of the composition of cellular proteins. This includes the identification of defective proteins, their partial proteolysis and refolding, as well as the control of the folding of newly synthesized polypeptides. However, the work on fungal HSPs is less than on other organisms. The studies carried out in *Aspergillus nidulans*, *Neurospora crassa*, *Saccharomyces cerevisiae*, and *Schizophyllum commune* stand out (Ewaze et al. 2007; Higgins and Lilly 1993; Newbury and Peberdy 1996). In most of the referred cases, the studies have involved the exposure of strains of these species to temperatures between 45 and 55 °C. Such conditions have been found to reduce the growth but not the development of aerial hyphae and often result in the appearance of various proteins that are rapidly synthesized or increased. In the basidiomycete *Schizophyllum commune*, proteolytic processes were also found to be significantly affected by such exposure (Higgins and Lilly 1993; Magan 2007). Different fugal HSPs related to thermotolerance were reviewed by Tiwari et al. (2015) based on their molecular weight, cellular localization, and functional characterization. Mainly expression of Hsp10, Hsp30, Hsp60, Hsp90, and Hsp104 increased in fungi exposed to heat stress (Table 3.1).

Fungi exposed to nonlethal heat shock can acquire resistance to lethal heat shock and then turn on acquired thermotolerant (Tereshina 2005). Piper (1993) reported molecular events associated with the acquisition of heat tolerance by the yeast *Saccharomyces cerevisiae*. Cells of this yeast that grow optimally at 25 °C acquire resistance to a lethal temperature of 50 °C when exposed to a temperature of 37 °C.

Maheshwari et al. (2000) reported acquired thermotolerance in mesophilic species, related to a greater survival at a lethal temperature, after a brief exposure to sublethal temperatures. The nonlethal heat shock could be associated with some mechanisms: synthesis of HSPs, synthesis of trehalose (up to 0.5 M), maintenance of intracellular pH through regulation of membrane ATPase activity, a decrease in the water activity and/or its redistribution in cytosol compartments, and detoxification of reactive oxygen species (Piper 1993; Tereshina 2005; Wang et al. 2020).

Attention has been more frequently focused on HSPs, which can be synthesized at high levels in response to brief exposure (i.e., 1 h) to high temperatures, such as 45–55 °C (Deacon 2013). HSPs are ubiquitous, present both under normal conditions and at elevated temperatures, and act as chaperones, ensuring that the cellular proteins fold correctly and that damaged proteins are destroyed. Maheshwari et al. (2000) stated that the synthesis of HSPs is an adaptive response to increased thermotolerance and survival in stressful conditions. Trent et al. (1994) observed that conidia of *Thermomyces lanuginosus*, germinated at 50 °C and subjected to heat shock at 55 °C for 60 min before exposure to 58 °C, showed greater survival compared to non-shocked conidia thermal. The experimental evidence they obtained allowed them to detect an increase in the synthesis of eight HSPs at 55 °C (Maheshwari et al. 2000; Magan 2007). Guimaraes et al. (2011), Iyalla (2017), and Zhang and Xu (2018) reported that a transient synthesis of HSP60 was involved in

Table 3.1 Some fungal heat shock proteins related to thermotolerance

HSPs	Cellular location	Fungi	References
HSP10	Mitochondria	*Saccharomyces cerevisiae* *Aureobasidium melanogenum*	Ricke and Bielinsky (2006), Tiwari et al. (2015), Jia et al. (2018)
HSP12	Cytoplasm, cell wall, plasma membrane	*Saccharomyces cerevisiae*	Varela et al. (1995), Welker et al. (2010), Tiwari et al. (2015)
HSP30	Plasma membrane	*Saccharomyces cerevisiae*	Amoros and Estruch (2001), Tiwari et al. (2015)
HSP40	Cytosol, mitochondria, endoplasmic reticulum	*Lentinula edodes*	Hartl and Hayer-Hartl (2002), Fan et al. (2003), Tiwari et al. (2015)
HSP42	Cytosol	*Aspergillus fumigatus*	Haslbeck et al. (2004), Ghazaei (2017), Tiwari et al. (2015)
HSP60	Mitochondria, cytosol	*Penicillium marneffei*	Chandler et al. (2008), Cappello et al. (2009), Tiwari et al. (2015), Bakar et al. (2020)
HSP70	Cytosol, nucleus, endoplasmic reticulum, ribosomes, mitochondria	*Vairimorpha necatrix* *Trichoderma* sp. *Candida albicans*	Hirt et al. (1997), Montero-Barrientos et al. (2008), Leach et al. (2012), Tiwari et al. (2015)
HSP78	Mitochondria	*Vairimorpha necatrix*	Hirt et al. (1997), Tiwari et al. (2015)
HSP90	Cytosol, endoplasmic reticulum, nucleus	*Saccharomyces cerevisiae* *Candida albicans*	Sorger (1991), Leach et al. (2012), Tiwari et al. (2015), O'Meara et al. (2017)
HSP104	Cytosol	*Saccharomyces cerevisiae* *Cryptococcus neoformans*	Boreham and Mitchel (1994), Glover and Lindquist (1998), Tiwari et al. (2015), Yang et al. (2017)

the development of *Histoplasma capsulatum* and orchestrates the adaptation to high-temperature stress. Advances in the sequences of the complete genome of thermotolerant fungi will make it possible to examine the physiological pathways involved and the number of genes that regulate their production, which will allow a more complete understanding of the groups of genes involved in this process and therefore in thermophilia or thermotolerance.

In addition to the function of HSPs at elevated temperatures, the concomitant production of polyhydric alcohols (polyols) has also been detected (Magan 2007; Maheshwari et al. 2000). Polyols (i.e., glycerol and mannitol) also tend to accumulate in response to stress conditions. Heat stabilizing factors in the cytosol can also contribute to the thermostability of enzymes (Deacon 2013).

It is also suggested that trehalose acts as a general protector against stress in the cytosol and stabilizes the membranes during dehydration. Tereshina (2005) stated that the parallel synthesis of HSPs and trehalose occurs in response to heat shock and is related to the acquisition of thermotolerance by fungal cells.

Among the roles of trehalose are to reserve carbohydrate used during germination and in the storage process of fungal spores, membrane protector under various types of stress (thermal, oxidative, and osmotic stresses and stresses caused by heavy metals, drugs, and metabolic inhibitors), regulator of the glycolysis process and the cellular concentrations of glucose and ATP, and chemical chaperone involved in the stabilization and folding of proteins.

Trehalose synthetase and trehalase, the enzymes involved in trehalose metabolism, may well be attributed to HSPs. The roles of HSPs and trehalose in the acquisition of thermotolerance can vary, depending on the stage of development and the rate of protein synthesis related to it (Liu et al. 2019). In the spores, which are more thermotolerant latent structures, HSPs were not revealed since protein synthesis is absent. However, accumulation of trehalose is observed. The thermoprotective function of trehalose is related to the characteristic features of its structure and the ability to form intermolecular hydrogen bonds, which contribute to stabilizing proteins and membranes. In addition, it performs protective and chaperone functions, as well as a regulatory function by controlling the evolution of glucose in glycolysis and the activity of membrane-associated ATPase (Wang et al. 2020). The other aspect of the thermoprotective function of trehalose is related to the specific characteristics of the regulation of trehalose synthesis by the induction of the expression of the trehalose synthetase gene under heat shock, as well as by the direct activation of this enzyme by the rise in temperature. Thus, trehalose acts both in emergencies and in the long-term protection of subcellular structures during exposure to stressors, participating, together with molecular companions, in a unified process of protection against stress and quality control of proteins (Paixão et al. 2019; Tereshina 2005).

High temperatures are a problem in many biological systems, particularly those involved in complex macromolecular assemblages such as ribosomes. This relationship between thermotolerance, the virulence of a pathogenic fungus, and ribosome biogenesis have been observed in *A. fumigatus*, which may offer new avenues for the development of new antifungals that disrupt thermotolerant ribosome assemblage (Bhabhra and Askew 2005).

Fungi that can grow in desertification conditions are exposed to a wide temperature range that can have an action of thermostable enzymes (Joshi et al. 2020). For example, *Thermomyces lanuginosus* produces a large number of chitinases with greater stability against high temperatures (Khan et al. 2015). Ahirwar et al. (2017) isolated thermotolerant fungi producing hemicellulases from heated environments.

Melanin deposition on the cell wall can also influence thermotolerance and could be related to fungal geographical distribution and ecology (Cordero and Casadevall 2017). The ubiquitous presence of Ascomycota in soil samples worldwide is attributed to its highly competitive capacities together with the presence of melanin (Egidi et al. 2019; Lin and Xu 2020). Environmental factors have been shown to influence the deposition of melanin content in the membrane of strain XJ5-1 of the yeast *Aureobasidium melanogenum*, isolated from the desert (Jiang et al. 2016). In this particular example, the presence of melanin confers higher tolerance to heat treatment (40 °C), besides UV radiation, oxidation, salt shock, desiccation, and strong acid hydrolysis at high temperature, for example, 80 °C (Lin and Xu 2020). Melanization increased tolerance to heat stress in *C. neoformans* and *Wangiella dermatitidis* (Cordero and Casadevall 2017). Joshi et al. (2020) compile several investigations that refer that melanin is also related to the adaptability of fungi in other hostile environments. Some current studies with melanotic fungi have revealed a wide spectrum of functions for this kind of pigment, ranging from drought and radiation resistance to increased virulence in fungal pathogens (Lin and Xu 2020). In *Monilinia fructicola*, melanin functions to provide the conidia with resistance to a variety of environmental stresses such as high temperatures (Rehnstrom and Free 1996; Cordero and Casadevall 2017). The role of melanin in fungal thermotolerance is scarce, that is why it is necessary to explore the various functions ascribed to this biological pigment in fungi and its relationship with climate change.

3.3 Spores, Resistance, and Survival

Temperature also influences the survival of various fungi, and this is related to the ability to form various structures that can be resistant to high temperatures. Most fungal propagules die on exposure to 80 °C for approximately 1 min. However, the ascospores of some food spoilage fungi such as *Neosartorya fischeri* can survive even for 1 h of exposure to temperatures around 85 °C (Baggerman and Samson 1988; Rajashekhara et al. 1998). In the investigation of Jesenska et al. (1993) with soil fungi, it was detected that the most heat-tolerant survived exposure to 90 °C for 1 h. For their part, the ascospores of *Byssochlamys fulva* and *Talaromyces macrospora* are resistant to high temperatures and require heat activation for germination (Dijksterhuis et al. 2002).

Suryanarayanan et al. (2011) reported thermotolerance of mesophilic fungal spores that were common saprotrophs. They detected six ascomycetes, isolated from a semiarid tropical habitat subject to fire in India, which were able to grow after incubating their spores at 100 °C for 2 h in dry heat, in which two survived at 110 °C and one at 115 °C. Related to these results, they suggested the possibility that the form of constitutive heat tolerance expressed by fungi could represent a specific adaptation to prolonged periods of drought and the prevalence of fires in their dry habitat. This physiological trait may be widespread, putting a focus on the evolution of opportunistic pathogens on a warmer planet.

Some strains of entomopathogenic fungi such as *Beauveria* spp., *Isaria* spp. and *Metarhizium* spp. show an important survival of the conidia at high temperatures, with a relationship between the thermotolerance of the conidia and their origin from geographic isolation. Most entomopathogenic fungi are mesophilic and can develop between 10 and 40 °C, but optimal growth is between 25 and 35 °C. The temperature has the greatest impact on conidial stress that inhibits metabolic processes, such as decreased morphogenesis during germination, protein denaturation, and membrane disorganization. Thermal stress directly impacts the ability of fungal strains to attack arthropods and their performance of an environmental activity (Paixão et al. 2019).

In spores, which are more thermotolerant latent structures, HSPs are not reported since protein synthesis is absent. However, accumulation of trehalose is observed with an important thermoprotective function previously analyzed (Paixão et al. 2019; Tereshina 2005).

3.4 Thermotolerance and Its Relationship with Environmental Changes

Ever since the basis of the thermotolerance studies, fungi in natural environments have provided abundant evidence that the active fungal community depends on thermal conditions (Carreiro and Koske 1992a, b, c; Widden 1986). Cochrane (1958) suggested that subtropical strains of a particular fungus may have higher optimal growth temperatures than strains from temperate regions. Therefore, isolates of the same species from different geographical areas may differ in their temperature-growth relationships. Later Evans (1971) concluded that the strains of certain fungi are transitional between thermotolerant and thermophilic, depending on environmental pressures. Thermotolerant fungal strains that grow in a habitat subjected to constant high temperatures can become dependent on these high temperatures for continued growth, and this could be a natural step in the evolution toward thermophilia. Evans (1971) also suggested that some thermotolerant species such as *Aspergillus fumigatus* can grow at lower temperatures and, therefore, will be able to compete with mesophilic populations. For their part, Craveri et al. (1964) stated that weak thermophilic species and strong thermotolerant species form a transition group between true thermophilia and general thermotolerance. Thermotolerant species grow in a wide temperature range, that is, they grow in the normal mesophilic range (10–35 °C) but can also grow up to 48–50 °C and occasionally more. These high growth temperatures can be important in determining the habitats in which thermotolerant fungi occur, and a versatile temperature range for growth will be especially important in environments subject to wide temperature fluctuations.

Maheshwari et al. (2000) analyzed the evolutionary pathways toward fungal thermotolerance. They argued that any new mutation in a fungal population would spread, ultimately replacing the wild type. The replacement rate will depend on your growth advantage, which in turn depends on selective pressure. De Crecy et al. (2009) observed that favorable variants take longer to prevail during certain cycles, which seems to indicate that evolution is occurring in discrete steps and that a different evolutionary path may encounter different thermal barriers.

For most organisms, growth occurs best in a relatively narrow temperature range and lags well outside this range (Huey and Stevenson 1979; McLean et al. 2005). Temperatures that allow for relatively rapid growth can vary between genotypes, and natural selection favor particular variants given spatial and temporal patterns of environmental temperature. The theory of evolution of thermal reaction norms for yield, such as growth rate, is based on the assumptions that yield contributes additively to fitness and there is a trade-off between yields at different temperatures. Based on these assumptions, this theory predicts that thermal specialists will outperform thermal generalists in most patterns of environmental variation. A fundamental trade-off exists between specialists and generalists that results from specialization for environmental condition (Moor et al. 2020). The generalist species can survive in a wide variety of temperatures with different resources, while specialist species have narrower requirements and tolerances. According to Angilletta et al. (2010), generalists can perform over a wider range of temperatures but perform poorer than specialists at the optimum temperature. Therefore, organisms must tolerate a wide range of environmental conditions but restrict growth under specific environmental conditions. In fact, generalists are only expected to outperform specialists when the environmental temperature is relatively stable within generations but varies between generations (Gilchrist 1995).

It has been argued that the urbanization of our planet has created rapid changes in the climate. Air and ground temperatures have increased in urban areas compared to their surroundings, and this phenomenon is known as the urban heat island effect. Causes include both increased heat absorption by pavement and buildings and increased heat production by air-conditioners, traffic, and factories in urban areas. In general, urban heat islands are characterized by higher minimum, average, and maximum temperatures. Urban heat islands have been exerting evolutionary pressure on organisms for at least 50 years and for more than 100 years in some cities (Brazel et al. 2000; McLean et al. 2005; Tereschenko and Filonov 2001).

Given the gradual increase in temperatures that has been associated with urbanization, fungi have begun to adapt to the warmer conditions of urban soils. McLean et al. (2005) studied the thermal reaction norms (18–26 °C) for the growth rate of four species of chitinolytic fungi (*Chrysosporium pannorum*, *Penicillium bilaii*, *Torulomyces lagena*, and *Trichoderma koningii*) compared between urban and rural sites near Louisville, Kentucky. Urban isolates of *C. pannorum* and *T. koningii* grew faster than rural isolates at 26 °C but grew more slowly than rural isolates at 18 °C. This difference in the thermal reaction norms between urban and rural isolates allowed them to suggest that urbanization caused thermal specialization. In this process, natural selection was able to produce urban genotypes that grow faster than rural genotypes at high temperatures but slower than rural genotypes at low temperatures. On the other hand, urban isolates of *T. lagena* and *P. bilaii* grew as fast as or faster than rural isolates at all temperatures, suggesting that countergradient variation could also occur. In such variation against the gradient, natural selection could produce urban genotypes that grow faster than rural genotypes at all temperatures. Thermal adaptation of the fungi analyzed was similar to that observed in other ectothermic organisms where the fungal thallus temperature varies according to the environmental temperature where it is found (Angilletta et al. 2002, 2003; Conover and Schultz 1995; Huey and Kingsolver 1993; McLean et al. 2005).

The emergence of these patterns may be because the genetic structure of populations changes at the beginning of urbanization, causing different responses to selection in each species. If there are no specialized genotypes for high-temperature growth within a population, selection for more optimal growth at high temperatures could produce a counter-gradient variation. Another possibility is that the specific biology of each species makes one form of adaptation more favorable than another (McLean et al. 2005). Thermal specialization and counter-gradient variation in growth rate involve different proximate mechanisms, which in turn lead to different evolutionary trade-offs (Angilletta et al. 2003). If the fitness consequences of these trade-offs vary between species, natural selection might favor thermal specialization in some species and counter-gradient variation in others. Unlike the first possible explanation, the latter does not imply genetic restrictions as the cause of the different patterns of intraspecific variation (McLean et al. 2005). Different responses to urbanization can disrupt competitive relationships between fungi, which can affect organic matter decomposition and nutrient cycling and trophic networks (Bonkowski et al. 2000; Carreiro and Koske 1992b; Marín et al. 1998; McLean et al. 2005; Naár and Kecskés 1998).

3.5 Habitats of Thermotolerant Fungi

Changes in climate can have impacts on fungal biodiversity patterns. Variation in thermal reaction norms could have broader impacts on the dynamics of communities in urban settings and the processes of the ecosystems that contain those communities (McLean et al. 2005).

Thermotolerant fungi represent a heterogeneous group made up of a large number of morphological and physiological forms with variability in their adaptation to high temperatures and therefore to climatic variations (Narayan and Prakash 2019). The habitat of thermotolerant fungi ranges from the hot natural environments to artificial environments (Fig. 3.4). Several investigations reported a wide range of these organisms causing spoilage of bagasse, grains, groundnut, hay, palm kernels, peat, wood chips, and other agricultural crops. These fungi are also detected in animal and industrial wastes and processes related to the biodegradation of lignocellulose.

Several investigations reported the detection of thermotolerant fungi in soils with high temperatures. The results of Murgia et al. (2019) showed high biodiversity in the desert sand of the Middle East, consisting predominantly of fungi that have adapted to survive in the extreme environmental conditions that prevail in deserts, indicating the resistance of fungi of the Dothideomycetes classes, Pezizomycetes, and Sordariomycetes, such as *Albifimbria terrestris*, *Fusarium equiseti*, *F. redolens*, *F. solani*, and *Chaetomium madrasense*. The presence of these fungi in the soil may be related to the solar heating they receive (Saroj et al. 2017), and this may determine that the fungi present in that habitat can grow at temperatures above 50 °C or even have viability at 80 °C (Joshi et al. 2020). Rajavaram et al. (2010) made soil isolates from mines, animal manures, cattle dung, composts, industrial wastes, poultry manure, and vermicompost and found *Aspergillus fumigatus* in all

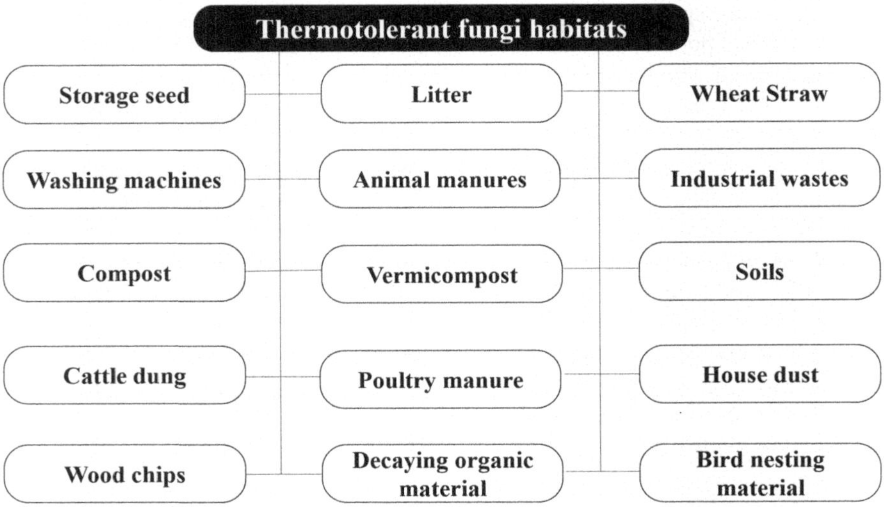

Fig. 3.4 Wide range habitats of thermotolerant fungi ranging from hot natural environment to artificial environments

substrates. Salar and Aneja (2006) found six species of thermotolerant fungi in temperate soils of northern India (*Aspergillus fumigatus*, *Chaetomium senegalense*, *Chrysosporium tropicum*, *Emericella nidulans*, *Penicillium chrysogenum*, and *Rhizopus microsporus*). Singh and Sandhu (1986) isolated thermotolerant strains of *A. aculeatus*, *A. carneus*, *A. fischeri*, *A. flavus*, and *A. nidulans* in soils of Port Blair.

Industrialization and the development of technologies have made possible the design, commercialization, and distribution of equipment that can reach temperatures higher than ambient temperatures, and consequently, new habitats have emerged for fungal populations. In this sense, washing machines and dishwashers can be considered extreme environments due to the high temperatures from which thermotolerant fungi have also been isolated (Zalar et al. 2011). Tischner et al. (2019) reported that 32% of the sampled washers in Hungary were highly contaminated with various species of fungi and 46% by human opportunistic pathogenic species. None of the fungi examined could grow at temperatures above 50 °C, so they can be considered thermotolerant species. Babic et al. (2015) hypothesized that washing machines could act as reservoirs for thermotolerant contaminating or pathogenic fungi.

Changes in climate and those expected in the coming decades are expected to have an impact on fungal biodiversity patterns, which, in turn, may also have implications for public health. Therefore, it is of vital importance to evaluate the ecological responses of fungi (Damialis et al. 2019).

Climate change is positive for all biomes but its effect increases moderately with latitude (Seidl et al. 2017). The studies conducted by Hoegh-Guldberg et al. (2019) show that climate warming of 1.5–2 °C is predicted to reduce permafrost in cold regions by 28–53%, resulting in large deposits of carbon available for microbial respiration and the emission of greenhouse gases that could lead to local extinction of cold-adapted fungi (Joshi et al. 2020). Gomes et al. (2018) reported diversity in Antarctic soils by identifying 218 fungi belonging to 22 taxa of 15 different genera belonging to the *Zygomycota*, *Ascomycota*, and *Basidiomycota* divisions. The genera *Antarctomyces*, *Mortierella*, *Penicillium*, and *Pseudogymnoascus* were the most frequently isolated, and *P. tardochrysogenum*, *P. verrucosus*, *Goffeauzyma gilvescens*, and *Mortierella* sp. were the most abundant taxa. Common yeasts in Antarctic soil belong to the genera *Candida*, *Cryptococcus*, *Debaryomyces*, *Guehomyces*, *Leucosporidiella*, *Metschnikowia*, and *Rhodotorula* (Martinez et al. 2016). Among these, there are genera with a wide distribution also in tropical areas. Further research directed toward the change in the stoichiometry (carbon/nitrogen ratio) of the soil of the polar region related to the changing climatic conditions would be a key step to predict the adverse effects on fungal communities (Alves et al. 2019).

Gange et al. (2007) reported rapid and recent changes in fungal fruiting patterns analyzing information on over 300 autumnal fruiting species, from more than a thousand zones, located in Southern England during 45 years (1950–2005). This study concludes that climate warming appears to have caused significant numbers of species to begin fruiting in spring as well as autumn. Until the time of the aforementioned study, several investigations had found recent phenological responses to climate change in higher organisms, such as plants, insects, or birds, and were restricted to events in spring (Parmesan and Yohe 2003; Toljander et al. 2006). Later Gange et al. (2011) informed about the relationship between climate change and host shifts in fungi and concluded that changing temperature causes altered growth rates, germination rates, and combative ability of one fungal species relative to another. Specifically, the case of *Auricularia auricula-judae* may have resulted in the ability to colonize a wider host range. The mean autumn fruiting period is nowadays later for saprotrophic and mycorrhizal fungi in Norway (Kauserud et al. 2008) and Switzerland (Büntgen et al. 2011). Moore et al. (2008) reported that since 1975, many species of fleshy fungi that previously fruited principally in autumn in the United Kingdom now also fruit during spring. Meanwhile, Kauserud et al. (2010) described spring fruiting nowadays occurs earlier than it did in the past century. According to Kauserud et al. (2008), these recent changes in fruiting are dependent on fungus life-history strategy, with variances between early fruiting species and late fruiting species. Kauserud et al. (2012) reported the temporal changes in the timing of fruit body formation to occur in all these European countries but are particularly pronounced in the United Kingdom, and it is related to geographical and bioclimatological features.

3.6 Thermotolerance within the Environmental Fungi

Thermotolerant fungi compose the mycobiota that grows in piled masses of plant material, heaps of agricultural and forestry products, stored food grains, and other accumulations of organic matter. In these ecosystems, the warm, humid, and aerobic environment provide the basic conditions for their development, and the fungal spores could become airborne from these native sources. For example, Narayan and Prakash (2019) reported 27 species of airborne thermotolerant fungi in India, from some of the aforementioned substrates, and highlighted *Aspergillus versicolor* and *Paecilomyces variotii*.

Studies on the thermotolerant airborne fungi are not as much of those carried out in other ecosystems. However, the relationship to air temperature and climate has been reported. Earlier and longer fungal sporulation seasons have been reported in the United Kingdom (Gange et al. 2007), Norway (Kauserud et al. 2012), and other European countries (Kasprzyk et al. 2016). Damialis et al. (2015a, b) demonstrated the beginning of a long-term trend in airborne spore concentrations in Thessaloniki during 1987–2005. Helfer (2014) showed that some fungal populations could be vulnerable to global environmental change. The sporulation and the formation of reproductive bodies of fungi are influenced by meteorological factors, such as rainfall, air temperature, and relative humidity (Carlile et al. 2001; Senanayake et al. 2020). Therefore, the variability in different stages of reproduction could cause modifications in the diversity of fungi and the temporal dynamics of the dispersion of their propagules. There is evidence that changing climate might cause alterations in phenology (Corden et al. 2003; Gange et al. 2007; Kauserud et al. 2012) and the dynamics of fungal communities (Gange et al. 2011).

Martínez-Herrera et al. (2016) reported that the ability to grow at temperatures of 55 °C and survive at 75 °C distinguished species in environmental strains of *Aspergillus* section *Fumigati* and allows the organisms to grow in decomposing organic matter and infect host mammals. *Aspergillus fumigatus* is particularly common, found in a wide range of environments, growing between 12 and 55 °C. However, it is considered more thermotolerant than thermophilic, because it can grow at temperatures as low as 12 °C and its optimal temperature is below 40 °C. This fungus is extraordinary for its wide growth temperature range and its equally varied range of habitats. It commonly grows in compost, substrates in indoor environments, moldy grains, and other decaying organic matter. It can also develop on the hydrocarbons of aviation fuel. In recent years, *A. fumigatus* has become a major problem in operating rooms, especially transplant units, because it can colonize wounds and grow within the tissues of transplant patients. However, it is an essentially saprotrophic species (Deacon 2013).

In *Aspergillus* section *Nigri*, the growth and conidia maturation occur more rapidly at temperatures between 28 and 30 °C (Marín et al. 1998). *Aspergillus niger* is a mesophilic species (Belli et al. 2004). However, Rojas et al. (2010) found that some environmental strains, isolated from Cuba, grew from 18 to 40 °C in Czapek Yeast Autolysate Agar (CYA) medium. However, Rojas et al. (2010) surprisingly

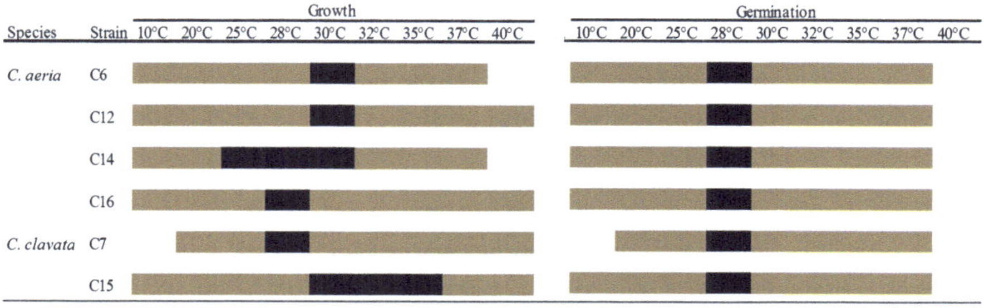

Fig. 3.5 Growth and conidia germination of *C. aeria* and *C. clavata* strains at different temperatures. Fungal strains were isolated from the atmosphere of a rice agroecosystem in Cuba

found the presence of several strains with optimal growth between 28 and 37 °C. In general, black aspergilli are less thermotolerant than *Aspergillus* section *Fumigati*, which perhaps leads to the fact that the majority of nosocomial cases of aspergillosis are caused mainly by *A. fumigatus* and not by *A. niger* (Scazzocchio 2009).

Other thermotolerant fungi have been detected in some investigations concerning the physiology of environmental fungal strains. A growth study, from 10 to 40 °C with mesophilic strains of *Curvularia* isolated from the atmosphere of a rice agroecosystem in Cuba, determined that two strains of *C. aeria* and two of *C. clavata* grew at 40 °C (Almaguer et al. 2013). Besides, conidia germination was observed between 10 and 37 °C (Fig. 3.5). In the aforementioned study, the differences between some strains show that the physiological behavior of each strain of the same species can differ considerably as a function of the moment of collection, the characteristics of the ecosystem, and the source of the inoculum.

Higher concentrations of airborne spores could be expected in tropical climates, where these temperature values are common throughout the year. Garcia-Solache and Casadevall (2010) warned about the potential risks of the adaptation of fungi to high temperatures and the importance of studying the possible effects of global warming on fungal populations since a warmer climate could change the distribution of species and favor those that are more thermotolerant. Additionally, under strong selection pressure, the prevalence of species adapted to heat tolerance may increase. In this sense, it has been shown that in urban areas, some fungi grow faster at high temperatures than in rural environments, and an increase in temperature can influence higher growth and an increase in the production of spores in *C. cladosporioides* (Damialis et al. 2015a, b). Sánchez and Almaguer (2018) confirmed the formation of conidia within the 10–34 °C interval and optimal growth in 25–30 °C, in environmental isolates of *Cladosporium cladosporioides*, collected from the atmosphere of Havana, Cuba. It was also observed that the isolates could be divided into three groups according to their physiological response to temperatures. One group had lower growth, sporulation, and germination values, and they were related to the lower temperatures around the isolation date. The species that formed the other two groups were collected on days when the average temperature was higher.

In addition, the variability of the response at different temperatures among the airborne propagules in Cuba was also found for several isolates of *Pyricularia oryzae* from a rice agroecosystem (Almaguer-Chávez et al. 2019). In this investigation, the formation of four groups between strains of *P. oryzae* isolated at various times during a year was observed. All these findings in airborne thermotolerant fungi may be related to genetic plasticity and could influence the ability to adapt to different environmental conditions. In this sense, Huo et al. (2012) stated that warmer conditions could alter the distribution of species and the incidence of fungal diseases in crops of economic interest. Moreover, climate change brings the specter of new infectious diseases (Casadevall 2020).

3.7 Perspectives of Fungal Thermotolerance Studies

Climate change can affect the diversity of the structure of fungal communities, since the survival of species sensitive to high temperatures may be affected, while resistant species will grow excessively creating an overpopulation of habitats (Joshi et al. 2020; Moreno et al. 2013). The specialized functions carried out by fungi in the ecosystem could be affected, such as decomposition, pathogenesis, and symbiosis. These changes in community structure and fungal diversity affect the ecosystem (Joshi et al. 2020; Setälä and McLean 2004). Climate change affects the diversity, geographic distribution, and composition of fungal communities around the world, and virulent fungal species capable of withstanding climatic disturbances will replace weakened species (Joshi et al. 2020; Moreno et al. 2013). For example, Newsham et al. (2016) recorded an increase of approximately five fungal taxa per °C of temperature increase. The diversity of fungi is greater in warmer habitats, but the evolutionary and environmental factors that shape the biogeography of fungi are not fully understood. However, according to the meta-analysis by Vetrovsky et al. (2019), high altitudes support concentrated fungal diversity and indicate a narrow climate tolerance of mycorrhizal fungi compared to pathogenic fungi. The aforementioned authors assembled a large set of previously generated mycobiome data linked to specific geographic locations around the world and used this dataset to describe the distribution of fungi and look for correlations with different environmental factors, including the weather. That meta-study identified climate as an important driver of different aspects of fungal biogeography, including the global distribution of common fungi, as well as the composition and diversity of fungal communities. The diversity of fungi was concentrated at high latitudes, in contrast to the opposite pattern shown previously for plants and other organisms. Mycorrhizal fungi appear to have closer climatic tolerances than pathogenic fungi. Then Vetrovsky et al. (2019) speculate that climate change could affect the functioning of the ecosystem due to the close climatic tolerances of the main fungal taxa.

The thermotolerance that is evidenced in several groups of environmental fungi has been related to the ability to grow in mammals and higher temperatures, and eventually global warming can exert pressure for the emergence of new diseases in mammals. For this reason, there is a concern about new diseases caused by fungal agents (Garcia-Solache and Casadevall 2010).

According to Casadevall (2020), climate change brings the specter of new infectious diseases. The pillar of endothermy is robust in humans (37 °C), but it can be broken if fungi adapt to higher temperatures. Although most species of fungi cannot tolerate the temperatures of mammals, some species can adapt to survive higher temperatures by gradually being exposed to warmer temperatures. Thus, global warming will lead many fungal species to adapt to higher temperatures, and some with pathogenic potential for humans will break the thermal defensive barrier provided by endothermy (Robert and Casadevall 2009).

Fungi that are pathogenic to insects can be experimentally adapted to replicate at mammalian temperatures through progressive heating cycles, demonstrating that fungi can rapidly adapt to higher temperatures (Casadevall 2020; de Crecy et al. 2009). Rangel et al. (2005) reported variability in the conidial thermotolerance of *Metarhizium anisopliae*, isolated from different geographical origins. In general, isolates from higher latitudes demonstrated greater susceptibility to heat than isolates closer to the equator.

It has been suggested that in tropical and subtropical climates, thermotolerant fungi should be more abundant. Many fungal pathogens that are not found in temperate regions can be found in tropical regions, such as *Paracoccidioides brasiliensis*, *Penicillium marneffei*, and *Cryptococcus gattii* (Casadevall 2020). The incidence of fungal infections in these locations tends to be higher, possibly due to increased numbers of fungi that can manage with human body temperatures. That is why the potential effects of global warming on fungi must be considered. Global warming could have a significant effect on fungal populations. First, a warmer climate could change the distribution of heat-tolerant and susceptible species by favoring those that are more thermotolerant and by creating the conditions for more environmental fungi to spread and come into closer contact with human populations. Second, under strong selective pressure, the prevalence of heat-tolerant species may increase (Raimalani and Panchamia 2020). Global warming means the narrowing of the thermal gradient between ambient temperature and that of mammals. The current gradient is approximately 22 °C, and consequently each degree of increase in global mean temperature reduces the gradient by approximately 5%. Casadevall et al. (2019) suggest that with current global warming, the prevalence of fungal diseases will increase. The risk of newly emerging fungal pathogens could be magnified by the fact that there are few antifungal medications available and there are no licensed vaccines.

The environment is likely to contain a large number of potentially pathogenic fungal species that are currently nonpathogenic to humans because they lack the ability to grow at the temperatures of mammals. The direct and indirect effects of climate changes induced by an exponentially growing human population as drivers of fungal evolution should be an area of intense investigation in the coming decades. The widening of the geographic range of innately pathogenic thermotolerant fungi and the acquisition of virulence traits in thermotolerant nonpathogenic environmental fungi may shape the twenty-first century as an era of expanding fungal diseases

(Casadevall et al. 2019). The recent appearance of *Candida auris* on three continents has been proposed to be related to global warming and the adaptations in certain microorganisms that make them pathogens for humans. This emergence evidences the need for increased surveillance and continuous monitoring (Casadevall 2020; Casadevall et al. 2019; Lockhart et al. 2017).

Global warming, strong climatic fluctuations, changes in rainfall patterns and drought, temperature changes, and forest fires are very evident in different parts of the world and affect fungal communities. Franco-Manchon et al. (2019) studied burned and unburned Scots pine forests, concluding that the dynamics of fungal communities were affected by forest fires. For example, mycorrhizal taxa were more affected than saprotrophic taxa by forest fires, as they were not observed the following year after forest fires. Nsa et al. (2020) isolated *Lichtheimia ramosa*, in a freshly burned field in Ikorodu, Nigeria. The genus *Lichtheimia* belongs to the class *Zygomycete* and includes saprotrophic species that can be isolated from decaying soil and plant material (Nagao et al. 2005). Its thermotolerance has been linked to its virulence (Alastruey-Izquierdo et al. 2010a; Schwartze et al. 2012; Schwartze and Jacobsen 2014). Attention to the thermotolerant genus *Lichtheimia* increased remarkably since its separation from the mesophilic genus *Absidia* and its taxonomic revision (Alastruey-Izquierdo et al. 2010b; André et al. 2014; Hoffmann et al. 2007; Hoffmann 2010). Ecologically, *L. ramosa* is abundant in soil, decaying plant debris, and food and is one of the causative agents of mucormycosis in humans (Nsa et al. 2020).

3.8 Climate Change and Aeromycological Dynamics

Fungi are adapting at a rapid rate in the face of environmental change. Joshi et al. (2020) reported that climate change caused an earlier and longer period of sporulation. Temperature is the environmental factor that systematically changes for decades and can significantly affect the growth and development of fungi, including the abundance of their sporulation (Kasprzyk et al. 2016). The concentration of airborne spores can increase in a wide range of temperatures as has been documented in countries around the world. In addition, the annual amounts of spores and their relationship with temperature were different between cities with climatic differences (Table 3.2).

Although the above examples explain the influence of temperature on the atmospheric dynamics of fungal propagules, it is not yet clear how medium- or long-term changes in climate, and particularly in temperature, affect the abundance of fungi in the air. Some investigations reported the probable impacts of climate change on airborne allergenic fungi and spore production (Barnes et al. 2013; Sindt et al. 2016). However, there is only limited evidence of trends in fungal development (Cecchi et al. 2010). Corden and Millington (2001) in Derby, UK, suggested a trend toward increasing annual counts of *Alternaria*, especially after 1992. Corden et al. (2003) reported that the abundance of *Alternaria* spores

Table 3.2 Some studies about relationships between temperature and airborne fungal spore concentration

Sampling period	Location	Fungi	Findings	References
2014–2015	Salamanca, España	*Alternaria*	Significant and positive correlation coefficients with temperature and total daily insolation	Fuentes et al. (2019)
2014–2016	Lisboa, Portugal	*Cladosporiumcladosporioides*, *Coprinus*, *Leptosphaeria*, *Agaricus*, *Cladosporium herbarum*, *Ustilago*, *Alternaria*	The mean temperature was the factor that exerted the greatest influence on the spore levels	Ferro et al. (2019)
2002–2014	Bratislava, Slovakia	*Alternaria*, *Epicoccum*, and *Stemphylium*	Considering that climate predictions for Central Europe indicate an increase in temperature events, shifts toward earlier onset, later end, longer duration, and increasing intensity of spore seasons of some fungal taxa in future years are expected, potentially exacerbating allergic respiratory disorders in humans	Ščevková et al. (2016)
1987–1996	Tulsa, Oklahoma	*Ganoderma*	Cumulative season total was significantly related to temperature	Craig and Levetin (2000)
2011–2014	SW of Iberian Peninsula	*Alternaria*	For a predicted increase in temperature in a climate change scenario, *Alternaria* spore production is foreseen to increase temperatures reaching optimal growing conditions in the two seasonal growing periods in Mediterranean climate	Maya-Manzano et al. (2016)

(continued)

Table 3.2 (continued)

Sampling period	Location	Fungi	Findings	References
2013	Szczecin, Poland	21 types of spores	The largest amount of the total variance in the spore composition was explained by the air temperature	Grinn-Gofrón et al. (2016)
2011–2012	Poland	*Cladosporium*	Warm and dry weather accelerate and elongate *Cladosporium* spore seasons in Poland	Kasprzyk et al. (2016)
2012–2013	Mexico City, Mexico	*Aspergillus*	Thermotolerance among *aspergillus* isolates from hospital environments	Martínez-Herrera et al. (2016)
2004–2013	Szczecin and Cracow, Poland	*Alternaria* and *Cladosporium*	As regards temperature, it was higher in Cracow and was still increasing in the studied years. If it increases, the peak value, total number of spores, and their average concentrations in a season will also increase	Grinn-Gofrón et al. (2016)
2003–2008	Madeira, Portugal	14 fungal genera	Temperature was the meteorological parameter that favored the most release and dispersal of the conidia	Sousa et al. (2016)
2003–2004	Taipei, China	Viable airborne fungi	Ambient temperature was the most consistent environmental factor positively correlated with fungal concentrations	Wu et al. (2007)

(continued)

Table 3.2 (continued)

Sampling period	Location	Fungi	Findings	References
1993–1996	Hualien, Taiwan	*Cladosporium*, *Arthrinium/Papularia*, *aspergillus/Penicillium*, *Ganoderma*, *Curvularia*, *fusarium*, *Ulocladium/Stemphylium*, *Cercospora*, *Alternaria*, *Drechslera*, *Torula*, *botrytis*, *Nigrospora*, *Pithomyces*, *Periconia*, ascospores, and other fungi	Among the meteorological factors, temperature was the most consistent predictor of fungal concentrations. Except for *aspergillus/Penicillium* and other fungi, higher temperature was positively associated with higher fungal levels	Ho et al. (2005)
2005	San Juan, Puerto Rico	Fungal spores	There were positive correlations between airborne fungal spore concentrations and temperature	Quintero and Rivera-Mariani (2010)
2001	Barbados	Fungal spores	Fungal spore concentration rose earlier in mid-July and remained so until October	Blades (2002)
2002	Medellín, Colombia	Fungal spores	Fungal spore concentration pollen levels in the atmosphere were associated with an increase or decrease in temperature	Angilletta et al. (2002)
1992–1995	Sydney, Australia	*Alternaria*	Fungal spore concentration pollen levels in the atmosphere were associated with an increase or decrease in temperature	Stennett and Beggs (2004)

(continued)

Table 3.2 (continued)

Sampling period	Location	Fungi	Findings	References
1980–1989	Stockholm, Sweden	*Cladosporium* and *Alternaria*	Daily mean temperatures of more than 15 °C when combined with a sufficient amount of precipitation seem to optimize the sporulation conditions for *Cladosporium*	Hjelmroos (1993)
1992	Waterloo, Canada	*Cladosporium*, *Epicoccum*, *Polythrincium*, *aspergillus/Penicillium*, *Leptosphaeria*, *Coprinus*, *Ganoderma*	The meteorological factors were, in descending order of importance, mean temperature, minimum temperature, and maximum temperature	Li and Kendrick (1995)
1991	Mexico City, Mexico	*Conidia*, *Basidiospores*, *Ascospores*	Daily mean spore concentrations were significantly correlated in southern and central areas with maximum temperature	Calderón et al. (1997)
1991–1993	Yokohama, Japan	*Cladosporium*, *Penicillium*, *Pestalotia*, *fusarium*, *aspergillus*, and others	The concentrations of fungi in outdoor air ($n = 288$) were significantly correlated with the maximum temperature of the day, minimum temperature of the day, average temperature of the day	Takahashi (1997)

(continued)

Table 3.2 (continued)

Sampling period	Location	Fungi	Findings	References
1997	Turin, Italy	*Cladosporium, Penicillium, Epicoccum, aspergillus*, and many others	Climatic and seasonal characteristics are prevalently determined by temperature; summer samplings are further forward on temperature gradient vector	Marchisio and Airaudi (2001)
1999–2000	Edirne, Turkey	*Cladosporium, aspergillus, Penicillium, Curvularia, Drechslera, Scytalidium, fusarium*	Temperature was the meteorological parameter correlated with spore levels	Asan et al. (2002)

varies greatly from one place to another depending on the regional vegetation and other local factors, among which temperature does not always play a determining role. Sindt et al. (2016) analyzed the temporal variations in a wide spectrum of characteristics of the atmospheric dynamics of *Cladosporium* in several French localities with different climatic conditions, with the aim of investigating the relationship between the levels of conidia and the evolution of the climate. On an annual basis and for 10 years, they demonstrated the climatic sensitivity of this fungal genus, the main component of the airborne fungi, and stated that the general increase in annual temperatures observed in recent years in several stations was consistent with those reported on warming. Regarding the potential impact of climate change, significant trends are identified by Hollins et al. (2004) when analyzing the relationships between airborne fungal spore concentration of *Cladosporium* and summer climate in Britain.

Hjelmroos (1993) suggested an upward linear trend in mean spore concentrations over time, in parallel with a positive trend in temperature. In Thessaloniki, Greece, a significant increase in air temperature has been observed during the period 1987–2005, and the annual mean concentrations of *Cladosporium* have decreased considerably and significantly between the years 1987 and 1992 and the years 1999 and 2005 (Damialis et al. 2015a), simulating different environmental regimes. Damialis et al. (2015b) have shown several fungi showed dramatic growth responses, with faster mycelium growth and lower spore production at higher temperatures. Several Spanish studies (Rodríguez-Rajo et al. 2005; Sabariego et al. 2000) demonstrated that average temperatures of 23–29 °C correspond to optimal conditions for high *Cladosporium* spore counts. Average temperature has been considered the most adequate predictor of its spore concentrations (Aira et al. 2012; Grinn-Gofroń et al. 2019; Hollins et al. 2004). For *Alternaria* (Thibaudon and Lachasse 2006), they stated that latitude had an undeniable impact on the concentration of spores. The majority of these works was conducted in temperate climates; however, the situation can be quite different in hot climates.

3.9 Conclusion and Future Prospects

Thermotolerance is present among fungi and could be essential for adaptation to climate change. Thermotolerance in fungi can be acquired in mesophilic species where environmental pressures exert selection on environmental mesophilic populations. Therefore, the gradual increase in temperatures, a consequence of climate change, proves to cause fungi to begin to adapt to warmer conditions and to change their ecological patterns. This incidence of changes in climate on fungal diversity may also have public health implications since fungal adaptation to survive at higher temperatures can undermine the pillar of endothermy. Therefore, it is of vital importance to evaluate the ecological responses of fungi to temperature.

Acknowledgments The author is thankful to Teresa I. Rojas Flores for helpful suggestions and proofreading and the Faculty of Biology, University of Havana, Cuba, for providing infrastructural facilities and continuous encouragement.

References

Ahirwar S, Soni H, Prajapati BP, Kango N. Isolation and screening of thermophilic and thermotolerant fungi for production of hemicellulases from heated environments. Mycology. 2017;8(3):125–34. https://doi.org/10.1080/21501203.2017.1337657.

Aira MJ, Rodríguez-Rajo FJ, Fernández-González M, Seijo C, Elvira-Rendueles B, Gutiérrez-Bustillo M, et al. *Cladosporium* airborne spore incidence in the environmental quality of the Iberian Peninsula. Grana. 2012;51(4):293–304. https://doi.org/10.1080/00173134.2012.717636.

Alastruey-Izquierdo A, Hoffmann K, de Hoog GS, Rodriguez-Tudela JL, Voigt K, Bibashi E, et al. Species recognition and clinical relevance of the zygomycetous genus *Lichtheimia* (syn. *Absidia* pro parte, Mycocladus). J Clin Microbiol. 2010a;48:2154–70. https://doi.org/10.1128/JCM.01744-09.

Alastruey-Izquierdo A, Cuesta I, Walther G, CuencaEstrella M, Rodríguez-Tudela JL. Antifungal susceptibility profile of human-pathogenic species of *Lichtheimia*. Antimicrob Agents Chemother. 2010b;54(7):3058–60. https://doi.org/10.1128/AAC.01270-09.

Almaguer M, Rojas TI, Dobal V, Batista A, Aira MJ. Effect of temperature on growth and germination of conidia in *Curvularia* and *Bipolaris* species isolated from the air. Aerobiologia. 2013;29(1):13–20. https://doi.org/10.1007/s10453-012-9257-z.

Almaguer-Chávez M, Rojas TI, Dobal V. Caracterización morfo-fisiológica de aislados ambientales de *Pyricularia oryzae*. Nova Acta Ci Compostelana Biol. 2019;26:125–31.

Alves IMS, Goncalves VN, Oliveira FS, Schaefer CEGR, Rosa CA, Rosa LH. The diversity, distribution, and pathogenic potential of cultivable fungi present in rocks from the south Shetlands archipelago, Maritime Antarctica. Extremophiles. 2019;23:327–36. https://doi.org/10.1007/s00792-019-01086-8.

Amoros M, Estruch F. Hsf1p and Msn2/4p cooperate in the expression of Saccharomyces cerevisiae genes HSP26 and HSP104 in a gene- and stress type-dependent manner. Mol Microbiol. 2001;39(6):1523–32. https://doi.org/10.1046/j.1365-2958.2001.02339.x.

André AL, Hoffmann K, Lima DX, de Oliveira RJ, Vieira HE, Malosso E, et al. A new species of *Lichtheimia* (Mucoromycotina, Mucorales) isolated from Brazilian soil. Mycol Prog. 2014;13:343–52. https://doi.org/10.1007/s11557-013-0920-8.
Angilletta MJ, Niewiarowski PH, Navas CA. The evolution of thermal physiology in ectotherms. J Thermal Biol. 2002;27:249–68. https://doi.org/10.1016/S0306-4565(01)00094-8.
Angilletta MJ, Wilson RS, Navas CA, James RS. Tradeoffs and the evolution of thermal reaction norms. Trends Ecol Evol. 2003;18:234–40. https://doi.org/10.1016/S0169-5347(03)00087-9.
Angilletta MJ, Cooper BS, Schuler MS, Boyles JG. The evolution of thermal physiology in endotherms. Front Biosci. 2010;E2:861–81. https://doi.org/10.2741/e148.
Asan A, Sen B, Sarica S. Airborne fungi in urban air of Edirne city (Turkey). Sect Bot. 2002;1:23–4. https://doi.org/10.1007/s10661-007-0130-4.
Babic MN, Zalar P, Zenko B, Schroers HJ, Dzeroski S, Gunde-Cimerman N. *Candida* and *Fusarium* species known as opportunistic human pathogens from customer-accessible parts of residential washing machines. Fungal Biol. 2015;119:95–113. https://doi.org/10.1016/j.funbio.2014.10.007.
Baggerman WI, Samson RA. Heat resistance of fungal spores. In: Samson RA, van Reenen-Hoekstra ES, editors. Introduction to food-borne fungi. Baarn: Centraalbureau vor Schimmelcultures; 1988. p. 262–7.
Bakar NA, Karsani SA, Alias SA. Fungal survival under temperature stress: a proteomic perspective. PeerJ. 2020;8:e10423. https://doi.org/10.7717/peerj.10423.
Barnes CS, Alexis NE, Bernstein JA, Demain CJR, JG, Horner E. Climate change and our environment: the effect on respiratory and allergic disease. J Allergy Clin Immunol Pract. 2013;1(2):137–41. https://doi.org/10.1016/j.jaip.2012.07.002.
Belli N, Marin S, Sanchis V, Ramos AJ. Influence of water activity and temperature on growth of isolates of *aspergillus* section Nigri obtained from grapes. Int Food Microbiol. 2004;96(1):19–27. https://doi.org/10.1016/j.ijfoodmicro.2004.03.004.
Bhabhra R, Askew DS. Thermotolerance and virulence of *aspergillus fumigatus*: role of the fungal nucleolus. Med Mycol. 2005;43(1):S87–93. https://doi.org/10.1080/13693780400029486.
Blades ED. The characterization of pollen and fungal spore aeroallergens in the atmosphere over Barbados and their association with asthma in this Caribbean island. Ams Confex. 2002;1:1–10.
Bonkowski M, Griffiths BS, Ritz K. Food preferences of earthworms for soil fungi. Pedobiologia. 2000;44:666–75. https://doi.org/10.1078/S0031-4056(04)70080-3.
Boreham DR, Mitchel REJ. Regulation of heat and radiation stress responses in yeast by hsp-104. Radiat Res. 1994;137(2):190–5.
Brazel A, Selover N, Vose R. Heisler G the tale of two climates—Baltimore and Phoenix urban LTER sites. Clim Res. 2000;15:123–35. https://doi.org/10.3354/cr015123.
Brown SR, Fritz SC. Eukaryotic organisms of continental hydrothermal systems. Extremophiles. 2019;23:367–76. https://doi.org/10.1007/s00792-019-01101-y.
Büntgen U, Kauserud H, Egli S. Linking climate variability to mushroom productivity and phenology. Front Ecol Environ. 2011;10:14–9. https://doi.org/10.1890/110064.
Calderón C, Lacey J, Mccartney A, Rosas I. Influence of urban climate upon distribution of airborne Deuteromycete spore concentrations in Mexico City. Int J Biometeorol. 1997;40(2):71–80. https://doi.org/10.1007/s004840050021.
Cappello F, Bellafiore M, Palma A. 60KDa chaperonin (HSP60) is over-expressed during colorectal carcinogenesis. Eur J Histochem. 2009;47(2):105–10. https://doi.org/10.4081/814.
Carlile M, Watkinson S, Gooday G. The fungi. Londres: Academic Press; 2001. p. 588.
Carreiro MM, Koske RE. Effect of temperature on decomposition and development of microfungal communities in leaf litter microcosms. Can J Bot. 1992a;70:2177–83. https://doi.org/10.1139/b92-269.
Carreiro MM, Koske RE. The effect of temperature and substratum on competition among three species of forest litter microfungi. Mycol Res. 1992b;96:19–24. https://doi.org/10.1016/S0953-7562(09)80991-X.

Carreiro MM, Koske RE. Room temperature isolations can bias against selection of low temperature microfungi in temperate forest soils. Mycologia. 1992c;84:886–900. https://doi.org/10.1080/00275514.1992.12026221.

Casadevall A. Climate change brings the specter of new infectious diseases. J Clin Invest. 2020;130(2):553–5. https://doi.org/10.1172/JCI135003.

Casadevall A, Kontoyiannis DP, Robert V. On the emergence of *Candida auris*: climate change, azoles, swamps, and birds. mBio. 2019;10:e01397–19. https://doi.org/10.1128/mBio.01397-19.

Cecchi L, D'Amato G, Ayres JG, Galan C, Forastiere F, Forsberg B. Projections of the effects of climate change on allergic asthma: the contribution of aerobiology. Allergy. 2010;65(9):1073–81. https://doi.org/10.1111/j.1398-9995.2010.02423.x.

Chandler JM, Treece ER, Trenary HR, Brenneman JL, Flickner TJ, Frommelt JL, et al. Protein profiling of the dimorphic, pathogenic fungus, *Penicillium marneffei*. Proteome Sci. 2008;6:17. https://doi.org/10.1186/1477-5956-6-17.

Cochrane VW. Physiology of fungi. New York: J. Wiley; 1958. p. 508.

Conover DO, Schultz ET. Phenotypic similarity and the evolutionary significance of counter gradient variation. Trends Ecol Evol. 1995;10:248–52. https://doi.org/10.1016/S0169-5347(00)89081-3.

Cooney DG, Emerson R. Thermophilic fungi. An account of their biology, activities and classification. San Fran-cisco, CA: W H Freeman & Co.; 1964. p. 256.

Corden JM, Millington WM. The long-term trends and seasonal variation of the aeroallergen Alternaria in Derby, UK. Aerobiologia. 2001;17:127–36. https://doi.org/10.1023/A:1010876917512.

Corden JM, Millington WM, Mullins J. Long-term trends and regional variation in the aeroallergen Alternaria in Cardiff and Derby UK – are differences in climate and cereal production having an effect? Aerobiologia. 2003;19:191–9. https://doi.org/10.1023/B:AERO.0000006529.51252.2f.

Cordero RJ, Casadevall A. Functions of fungal melanin beyond virulence. Fungal Biol Rev. 2017;31(2):99–112. https://doi.org/10.1016/j.fbr.2016.12.003.

Craig R, Levetin E. Multi-year study of *Ganoderma* aerobiology. Aerobiologia. 2000;16(1):75–81. https://doi.org/10.1023/A:1007682600175.

Craveri R, Manachini PL, Craveri A. Eumiceti termofili presenti nel suolo. Annal Microbiol Enzimol. 1964;14:13–26.

Crisan EV. Current concepts of thermophilism and the thermophilic fungi. Mycologia. 1973;65:1171–98.

Damialis A, Mohammad AB, Halley JM, Gange AC. Fungi in a changing world: growth rates will be elevated, but spore production may decrease in future climates. Int J Biometeorol. 2015a;59:1157–67. https://doi.org/10.1007/s00484-014-0927-0.

Damialis A, Vokou A, Gioulekas D, Halley JM. Long-term trends in airborne fungal-spore concentrations: a comparison with pollen. Fungal Ecol. 2015b;13:150–6. https://doi.org/10.1016/j.funeco.2014.09.010.

Damialis A, Traidl-Hoffmann C, Treudler R. Climate change and pollen allergies. In: Marselle MR, Stadler J, Korn H, Irvine KN, Bonn A, editors. Biodiversity and health in the face of climate change. Cham: Springer; 2019. p. 47–66.

de Crecy E, Jaronski E, Lyons B, Lyons TJ, Keyhani NO. Directed evolution of a filamentous fungus for thermotolerance. BMC Biotechnol. 2009;1:74. https://doi.org/10.1186/1472-6750-9-74.

de Menezes GCA, Porto BA, Simões JC, Rosa CA, Rosa LH. Fungi in snow and glacial ice of Antarctica. In: Rowlan., editor. Fungi of Antarctica. Cham: Springer; 2019. p. 127–46.

de Oliveira TB, Rodrigues A. Ecology of thermophilic fungi. In: Tiquia-Arashiro SM, Grube M, editors. Fungi in extreme environments: ecological role and biotechnological significance. Cham: Springer; 2019. p. 39–57. https://doi.org/10.1007/978-3-030-19030-9_3.

Deacon JW. Fungal biology. 4th ed. Australia: Blackwell Publishing Ltd.; 2013.

Dijksterhuis J, van Driel KGA, Sanders MG, Molenaar D, Houbraken JAMP, Samson RA, et al. Trehalose degradation and glucose efflux precede cell ejection during germination of heat-resistant ascospores of *Talaromyces macrosporus*. Arch Microbiol. 2002;178:1–7. https://doi.org/10.1007/s00203-002-0410-x.

Egidi E, Delgado-Baqueriz M, Plett JM, Wang J, Eldridge DJ, Bardget R, et al. A few Ascomycota taxa dominate soil fungal communities worldwide. Nat Commun. 2019;10 https://doi.org/10.1038/s41467-019-10373-z.

Evans HC. Thermophilous fungi of coal spoil tips. II. Occurrence, distribution and temperature relationships. Trans Br Mycol Soc. 1971;57(2):255–66. https://doi.org/10.1016/S0007-1536(71)80007-4.

Ewaze JO, Summerbell RC, Scott JA. Physiological studies of the warehouse staining fungus, *Baudoinia compniacensis*. Mycol Res. 2007;111(12):1422–30. https://doi.org/10.1016/j.mycres.2007.09.010.

Fan CY, Lee S, Cyr DM. Mechanisms for regulation of Hsp70 function by Hsp40. Cell Stress Chaperon. 2003;8(4):309–16. https://doi.org/10.1379/1466-1268(2003)008<0309:mfrohf>2.0.co;2.

Ferro R, Nunes C, Camacho I, Paiva M, Morais-Almeida M. Monitorização de esporos de fungos em Lisboa, 2014–2016. Rev Port Imumoalergol. 2019;27(1):29–39.

Franco-Manchon I, Oria-De-Rueda JA, Martin-Pinto P, Salo K, Bonet JA, Bonet JA. Are wildfires a threat to fungi in European Pinus forests? A case study of boreal and Mediterranean forests. Forests. 2019;10(4):309. https://doi.org/10.3390/f10040309.

Fuentes AS, Rodríguez de la Cruz D, Sánchez Sánchez J, Sánchez RE. Estudio preliminar del contenido atmosférico de esporas de *Alternaria* en la ciudad de Salamanca. Rev Salud Ambient. 2019;19:350–66.

Gange AC, Gange EG, Sparks TH, Boddy L. Rapid and recent changes in fungal fruiting patterns. Science. 2007;316:71. https://doi.org/10.1126/science.1137489.

Gange AC, Gange EG, Mohammad AB, Boddy L. Host shifts in fungi caused by climate change? Fungal Ecol. 2011;4:154–90. https://doi.org/10.1016/j.funeco.2010.09.004.

Garcia-Solache MA, Casadevall A. Hypothesis: global warming will bring new fungal diseases for mammals. MBio. 2010;1(1):e00061–10. https://doi.org/10.1128/mBio.00061-10.

Ghazaei C. Molecular insights into pathogenesis and infection with *Aspergillus fumigatus*. Malays J Med Sci. 2017;24(1):10. https://doi.org/10.21315/mjms2017.24.1.2.

Gilchrist GW. Specialists and generalists in changing environments. Fitness landscapes of thermal sensitivity. Am Nat. 1995;146:252–70.

Glover JR, Lindquist S. Hsp104, Hsp70, and Hsp40: a novel chaperone system that rescues previously aggregated proteins. Cell. 1998;94(1):73–82. https://doi.org/10.1016/s0092-8674(00)81223-4.

Gomes ECQ, Godinho VRM, Silva DBAS, De Paula MTR, Vitoreli GA, Zani CL. Cultivable fungi present in Antarctic soils: taxonomy, phylogeny, diversity, and bioprospecting of antiparasitic and herbicidal metabolites. Extremophiles. 2018;22:381–93. https://doi.org/10.1007/s00792-018-1003-1.

Grime JP. Evidence for the existence of three primary strategies in plants and its relevance to ecological and evolutionary theory. Am Nat. 1977;111:1169–94. https://doi.org/10.1086/283244.

Grinn-Gofrón A, Strzelczak A, Danuta S, Myszkowska D. A 10-year study of *Alternaria* and *Cladosporium* in two polish cities (Szczecin and Cracow) and relationship with the meteorological parameters. Aerobiologia. 2016;32:83–94. https://doi.org/10.1007/s10453-015-9411-5.

Grinn-Gofroń A, Nowosad J, Bosiacka B, Camacho I, Pashley C, Belmonte J, Skjøth C. Airborne *Alternaria* and *Cladosporium* fungal spores in Europe: forecasting possibilities and relationships with meteorological parameters. Sci Total Environ. 2019;653:938–46. https://doi.org/10.1016/j.scitotenv.2018.10.419.

Guimaraes AJ, Nakayasu ES, Sobreira TJP, Cordero RJB, Nimrichter L, Almeida IC, et al. *Histoplasma capsulatum* heat-shock 60 orchestrates the adaptation of the fungus to temperature stress. PLoS One. 2011;6(2):e14660. https://doi.org/10.1371/journal.pone.0014660.

Hartl FU, Hayer-Hartl M. Molecular chaperones in the cytosol: from nascent chain to folded protein. Science. 2002;295(5561):1852–8. https://doi.org/10.1126/science.1068408.

Haslbeck M, Braun H, Stromer T. Hsp42 is the general small heat shock protein in the cytosol of *Saccharomyces cerevisiae*. EMBO J. 2004;23(3):638–49. https://doi.org/10.15252/embj.2020105112.

Hassikou R, Hassikou K, Ouazzanni A, Badoc A, Doura A. Biologie, physiologie et pouvoir pathoge ne de quelques isolats de *Curvularia lunata* agent de la curvulariose du riz. Bull Soc Pharm Bord. 2003;142:25–44.

Helfer S. Rust fungi and global change. New Phytol. 2014;201(3):770–80. https://doi.org/10.1111/nph.12570.

Higgins SM, Lilly WW. Multiple responses to heat stress by the basidiomycete *Schizophyllum commune*. Curr Microbiol. 1993;26(3):123–7. https://doi.org/10.1007/BF01577364.

Hirt RP, Healy B, Vossbrinck CR, Canning EU, Embley TM. A mitochondrial Hsp70 orthologue in *Vairimorpha necatrix*: molecular evidence that microsporidia once contained mitochondria. Curr Biol. 1997;7(12):995–8. https://doi.org/10.1016/S0960-9822(06)00420-9.

Hjelmroos M. Relationship between airborne fungal spore presence and weather variables: *Cladosporium* and *Alternaria*. Grana. 1993;32:40–7. https://doi.org/10.1080/00173139309436418.

Ho HM, Rao CY, Hsu HH, Chiu YH, Liu CM, Chao HJ. Characteristics and determinants of ambient fungal spores in Hualien, Taiwan. Atmos Environ. 2005;39:5839–50. https://doi.org/10.1016/j.atmosenv.2005.06.034.

Hoegh-Guldberg O, Jacob D, Taylor M, Guillan Bolanos T, Bindi M, Brown S. The human imperative of stabilizing global climate change at 1.5 °C. Science. 2019;365 https://doi.org/10.1126/science.aaw6974.

Hoffmann K. Identification of the genus Absidia (Mucorales, Zygomycetes): a comprehensive taxonomic revision. In: Gherbawy Y, Voigt K, editors. Molecular identification of fungi. Berlin: Springer; 2010. p. 439–60.

Hoffmann K, Discher S, Voigt K. Revision of the genus *Absidia* (Mucorales, Zygomycetes) based on physiological, phylogenetic, and morphological characters; Thermotolerant *Absidia* spp. form a coherent group, Mycocladiaceae fam. Nov. Mycol Res. 2007;111:1169–83. https://doi.org/10.1016/j.mycres.2007.07.002.

Hollins PD, Kettewell PS, Atnson M, Stephenson DB, Corden JM, Mllington WY, Mullind J. Relationships between airborne fungal spore concentration of *Cladosporium* and summer climate in Britain. Int J Biometeorol. 2004;48:137–41. https://doi.org/10.1007/s00484-003-0188-9.

Huey RB, Kingsolver JG. Evolution of resistance to high temperature in ectotherms. Am Nat. 1993;142:S21–46. https://doi.org/10.1086/285521.

Huey RB, Stevenson RD. Integrating thermal physiology and ecology of ectotherms: a discussion of approaches. Am Zool. 1979;19(1):357–66. https://doi.org/10.1093/icb/19.1.357.

Huo Z, Li M, Wang L, Wen Q, Xiao J, Huang D, et al. Impacts of climate warming on crop diseases and pests in China. Sci Agric Sin. 2012;45(10):1926–34. https://doi.org/10.1016/j.envres.2020.110046.

Iyalla C. A review of the virulence factors of pathogenic fungi. Afr J Clin Exp Microbiol. 2017;18(1):53–8.

Jesenska Z, Pieckova E, Bernat D. Heat resistance of fungi from soil. Int J Food Microbiol. 1993;19:187–92. https://doi.org/10.1016/0168-1605(93)90076-S.

Jia L, Yuan S, Michael W, Zhigang X, Yiqing L, Yuming Y, et al. The impact of the postharvest environment on the viability and virulence of decay fungi. Crit Rev Food Sci Nutr. 2018;58(10):1681–7. https://doi.org/10.1080/10408398.2017.1279122.

Jiang H, Liu NN, Liu GL, Chi Z, Wang JM, Zhang LL. Melanin production by a yeast strain XJ5-1 of *Aureobasidium melanogenum* isolated from the Taklimakan desert and its role in the yeast survival in stress environments. Extremophiles. 2016;20:567–77. https://doi.org/10.1007/s00792-016-0843-9.

Joshi S, Bajpai A, Johri BN. Extremophilic fungi at the interface of climate change. In: Sharma VK, Shah MP, Kumar A, editors. Fungi bio-prospects in sustainable agriculture, environ-

ment and Nano-technology. London: Academic Press; 2020. p. 1–22. https://doi.org/10.1016/B978-0-12-821925-6.00001-0.
Kasprzyk I, Kaszewski BM, Weryszko-Chmielewska E, Nowak M, Sulborska A, Kaczmare J, et al. Warm and dry weather accelerates and elongates *Cladosporium* spore seasons in Poland. Aerobiologia. 2016;32(1):109–26. https://doi.org/10.1007/s10453-016-9425-7.
Kauserud H, Stige LC, Vik JO, Økland RH, Høiland K, Stenseth NC. Mushroom fruiting and climate change. Proc Natl Acad Sci U S A. 2008;105:3811–4. https://doi.org/10.1073/pnas.0709037105.
Kauserud H, Heegaard E, Semenov MA, Boddy L, Halvorsen R, Stige LC, et al. Climate change and spring-fruiting fungi. Proc Biol Sci. 2010;277:1169–77. https://doi.org/10.1098/rspb.2009.1537.
Kauserud H, Heegaard E, Büntgen U, Halvorsen R, Egli S, Senn-Irlet B.m Høiland, K.. Warming-induced shift in European mushroom fruiting phenology. PNAS. 2012;109(36):14488–93. https://doi.org/10.1073/pnas.1200789109.
Khan FI, Bisetty K, Singh S, Permaul K, Hassan MI. Chitinase from *Thermomyces lanuginosus* SSBP and its biotechnological applications. Extremophiles. 2015;19:1055–66. https://doi.org/10.1007/s00792-015-0792-8.
La Touche CJ. On a thermophile species of *Chaetomium*. Trans Brit Mycol Soc. 1950;33:1–2.
Leach MD, Budge S, Walker L, Munro C, Cowen LE, Brown AJP. Hsp90 orchestrates transcriptional regulation by Hsf1 and cell wall remodelling by MAPK signalling during thermal adaptation in a pathogenic yeast. PLoS Pathog. 2012;8(12):1003069. https://doi.org/10.1371/journal.ppat.1003069.
Li DW, Kendrick B. A year-round study on functional relationships of airborne fungi with meteorological factors. Int J Biometeorol. 1995;39:74–80. https://doi.org/10.1007/BF01212584.
Lin L, Xu J. Fungal pigments and their roles associated with human health. J Fungi. 2020;6(4):280. https://doi.org/10.3390/jof6040280.
Liu XM, Wu XL, Wei GAO, Qu JB, Qiang CHEN, Huang CY, et al. Protective roles of trehalose in *Pleurotus pulmonarius* during heat stress response. J Integr Agric. 2019;18(2):428–37.
Lockhart SR, Etienne KA, Vallabhaneni S, Farooqi J, Chowdhary A, Govender NP, et al. Simultaneous emergence of multidrug-resistant *Candida auris* on 3 continents confirmed by whole-genome sequencing and epidemiological analyses. Clin Infect Dis. 2017;64(2):134–40. https://doi.org/10.1093/cid/ciw691.
Magan N. Fungi in Extreme Environments. In: Kubicek C, Druzhinina I, editors. Environmental and microbial relationships. The Mycota. Berlin, Heidelberg: Springer; 2007. p. 85–103. https://doi.org/10.1007/978-3-540-71840-6_6.
Maheshwari R, Bharadwaj G, Bhat A, Mahalingeshwara K. Thermophilic fungi: their physiology and enzymes. Microbiol Mol Biol Rev. 2000;64(3):461–88. https://doi.org/10.1128/MMBR.64.3.461-488.2000.
Marchisio VF, Airaudi D. Temporal trends of the airborne fungi and their functional relations with the environment in a suburban site. Mycologia. 2001;93:831–40. https://doi.org/10.1080/00275514.2001.12063218.
Marín S, Companys E, Sanchis V, Ramos AJ, Magan N. Effect of water activity and temperature on competing abilities of common maize fungi. Mycol Res. 1998;102:959–64. https://doi.org/10.1017/S0953756297005613.
Martinez A, Cavello I, Garmendia G, Rufo C, Cavalitto S, Vero S. Yeasts from sub-Antarctic region: biodiversity, enzymatic activities and their potential as oleaginous microorganisms. Extremophiles. 2016;20:759–69. https://doi.org/10.1017/S0953756297005613.
Martínez-Herrera EO, Frías-De-León MG, Duarte-Escalante E, Calderón-Ezquerro MC, Jiménez-Martínez MC, Acosta-Altamirano G, et al. Fungal diversity and *aspergillus* in hospital environments. Ann Agric Environ Med. 2016;23(2):298–303. https://doi.org/10.5604/12321966.1203888.
Maya-Manzano JM, Muñoz-Triviño F, Fernández-Rodríguez S, Silva-Palacios I, Gonzalo-Garijo A, Tormo-Molina R. Airborne *Alternaria* conidia in Mediterranean rural environments in SW

of Iberian Peninsula and weather parameters that influence their seasonality in relation to climate change. Aerobiologia. 2016;32:95–108. https://doi.org/10.1007/s10453-016-9424-8.

McLean MA, Angilletta MJ, Williams KS. If you can't stand the heat, stay out of the city: thermal reaction norms of chitinolytic fungi in an urban heat island. J Therm Biol. 2005;20:384–91. https://doi.org/10.1016/j.jtherbio.2005.03.002.

Montero-Barrientos M, Hermosa R, Nicolás C, Cardoza RE, Gutiérrez S, Monte E. Overexpression of a *Trichoderma* HSP70 gene increases fungal resistance to heat and other abiotic stresses. Fungal Genet Biol. 2008;45(11):1506–13. https://doi.org/10.1016/j.fgb.2008.09.003.

Moor H, Nordén J, Penttilä R, Siitonen J, Snäll T. Long-term effects of colonization–extinction dynamics of generalist versus specialist wood-decaying fungi. J Ecol. 2020;1365 https://doi.org/10.1111/1365-2745.13526.

Moore D, Gange A, Gange E, Boddy L. In: Boddy L, Frankland J, van West P, editors. Ecology of saprotrophic basidiomycetes. Amsterdam: Elsevier; 2008. p. 79–10213.

Moreno, G, Manjón, JL, Álvarez-Jiménez J. Los hongos y el cambio climático. Kagan-Zur V., Sitrit Y, Roth-Bejerano, N, A Morte, (eds). Desert truffle. Berlín. Springer Verlag; 2013:129–135.

Morris GJ, Clarke A. Cells at low temperature. In: Grout BWW, Morris GJ, editors. The effects of low temperatures on biological systems. London: Arnold; 1987. p. 72–84.

Murgia M, Fiamma M, Barac A, Deligios M, Mazzarello V, Paglietti B. Biodiversity of fungi in hot desert sands. Microbiol Open. 2019;8(1):e00595. https://doi.org/10.1002/mbo3.595.

Naár Z, Kecskés M. Factors influencing the competitive saprophytic ability of *Trichoderma species*. Microbiol Res. 1998;153:119–29. https://doi.org/10.1016/S0944-5013(98)80029-3.

Nagao K, Ota T, Tanikawa A, Takae Y, Mori T, Udagawa S, et al. Genetic identification and detection of human pathogenic *Rhizopus* species, a major mucormycosis agent, by multiplex PCR based on internal transcribed spacer region of rRNA gene. J Dermatol Sci. 2005;39(1):23–31. https://doi.org/10.1016/j.jdermsci.2005.01.010.

Narayan A, Prakash R. Airborne thermotolerant fungi of Jabalpur, Madhya Pradesh. World J Pharm Sci. 2019;6(2):1133–45.

Newbury J, Peberdy JF. Characterization of the heat shock response in protoplasts of *aspergillus nidulans*. Mycol Res. 1996;100(11):1325–32. https://doi.org/10.1016/S0953-7562(96)80059-1.

Newsham KK, Hopkins DW, Carvalhais LC, Fretwell PT, Rushton SP, O'Donnell AG, et al. Relationship between soil fungal diversity and temperature in the maritime Antarctic. Nat Clim Chang. 2016;6(2):182–6. https://doi.org/10.1038/nclimate2806.

Nsa IY, Kareem RO, Aina OT, Akinyemi BT. Isolation of an emerging thermotolerant medically important fungus, *Lichtheimia ramosa* from soil. Afr J Microbiol Res. 2020;14(6):237–41. https://doi.org/10.5897/AJMR2020.9358.

O'Meara TR, Robbins N, Cowen LE. The Hsp90 chaperone network modulates *Candida* virulence traits. Trends Microbiol. 2017;25(10):809–19. https://doi.org/10.1016/j.tim.2017.05.003.

Paixão FRS, Fernandes ÉKK, Pedrini N. Thermotolerance of fungal conidia. In: Khan M, Ahmad W, editors. Microbes for sustainable insect Pest management. Sustainability in plant and crop protection. Cham: Springer; 2019. https://doi.org/10.1007/978-3-030-23045-6_6.

Parmesan C, Yohe G. A globally coherent fingerprint of climate change impacts across natural systems. Nature. 2003;421(6918):37–42. https://doi.org/10.1038/nature01286.

Paterson RRM, Lima N. Ecology and biotechnology of thermophilic fungi on crops under global warming. In: Fungi in extreme environments: ecological role and biotechnological significance. Berlin/Cham: Springer; 2019. p. 81–96.

Piper PW. Molecular events associated with acquisition of heat tolerance by the yeast *Saccharomyces cerevisiae*. FEMS Microbiol. 1993;11:339–56. https://doi.org/10.1111/j.1574-6976.1993.tb00005.x.

Quintero E, Rivera-Mariani F. Bolaños-Rosero B analysis of environmental factors and their effects on fungal spores in the atmosphere of a tropical urban area (San Juan, Puerto Rico). Aerobiologia. 2010;26(2):113–24. https://doi.org/10.1007/s10453-009-9148-0.

Raimalani V, Panchamia B. A look into dark side of fungal agents: researches on human diseases. In: Arya A, editor. Environment at crossroads challenges and green solutions. Scientific Publishers; 2020. p. 10–20.

Rajashekhara E, Suresh ER, Ethiraj S. Thermal death rate of ascospores of *Neosartorya fischeri* ATCC 200957 in the presence of organic acids and preservatives in fruit juices. J Food Prot. 1998;61:1358–62. https://doi.org/10.4315/0362-028x-61.10.1358.
Rajavaram RK, Sreelatha B, Sivadevuni G, Reddy SM. Incidence of thermophilic fungi from different substrates in Andhra Pradesh (India). Int J Pharma Bio Sci. 2010;1(3):34–9.
Rangel DE, Braga GU, Anderson AJ, Roberts DW. Variability in conidial thermotolerance of *Metarhizium anisopliae*, isolates from different geographic origins. J Invertebr Pathol. 2005;88(2):116–25. https://doi.org/10.1016/j.jip.2004.11.007.
Rehnstrom AL, Free SJ. The isolation and characterization of melanin-deficient mutants of *Monilinia fructicola*. Physiol Mol Plant Pathol. 1996;49:321–30. https://doi.org/10.1006/pmpp.1996.0057.
Ricke RM, Bielinsky AK. A conserved Hsp10-like domain in Mcm10 is required to stabilize the catalytic subunit of DNA polymerase-α in budding yeast. J Biol Chem. 2006;281(27):18414–25. https://doi.org/10.1074/jbc.M513551200.
Robert VA, Casadevall A. Vertebrate endothermy restricts most fungi as potential pathogens. J Infect Dis. 2009;200(10):1623–6. https://doi.org/10.1086/644642.
Rodríguez-Rajo FJ, Iglesias I, Jato V. Variation assessment of airborne *Alternaria* and *Cladosporium* spores at different bioclimatical conditions. Mycol Res. 2005;109(4):497–507. https://doi.org/10.1017/S0953756204001777.
Rojas TI, Companioni A, Batista A, Boburg S, Benítez M. Caracterización morfológica y fisiológica de cepas del Grupo Aspergillus niger. [CD-Rom]. VII Taller sobre colecciones de cultivos microbianos. La Habana, Ed. Finlay; 2010.
Roy S, Bhattacharya SG. Airborne fungal spore concentration in an industrial township: distribution and relation with meteorological parameters. Aerobiologia. 2020;36:575–87. https://doi.org/10.1007/s10453-020-09653-9.
Sabariego S, Diaz De La Guardia C, Alba F. The effect of meteorological factors on the daily variation of airborne fungal spores in Granada (southern Spain). Int J Biometeorol. 2000;44:1–5. https://doi.org/10.1007/s004840050131.
Salar RK, Aneja KR. Thermophilous fungi from temperate soils of northern India. J Agric Sci Technol. 2006;2(1):49–58.
Sánchez K, Almaguer M. Efecto de la temperatura sobre aislados de Cladosporium cladosporioides recolectados del aire de La Habana. Cuba Nova Acta Ci Compostelana. 2018;25:21–9.
Saroj A, Hemant S, Bhanu P, Naveen K. Isolation and screening of thermophilic and thermotolerant fungi for production of hemicellulases from heated environments. Mycology. 2017;8(3):125–34. https://doi.org/10.1080/21501203.2017.1337657.
Scazzocchio C. *Aspergillus*: a multifaceted genus. Encyclopedia of. Microbiology. 2009:401–21.
Ščevková J, Dušička J, Mičieta K, Somorčík J. The effects of recent changes in air temperature on trends in airborne *Alternaria, Epicoccum* and *Stemphylium* spore seasons in Bratislava (Slovakia). Aerobiologia. 2016;32(1):69–81. https://doi.org/10.1007/s10453-015-9412-4.
Schwartze VU, Jacobsen ID. Mucormycoses caused by *Lichtheimia* species. Mycoses. 2014;57:73–8. https://doi.org/10.1111/myc.12239.
Schwartze VU, Hoffmann K, Nyilasi I, Papp T, Vágvölgyi C, de Hoog S, Jacobsen ID. *Lichtheimia* species exhibit differences in virulence potential. PLoS One. 2012;7(7):e40908. https://doi.org/10.1371/journal.pone.0040908.
Seidl R, Thom D, Kautz M, Martin-Benito D, Peltoniemi M, Vacchiano G. Forest disturbances under climate change. Nat Clim Chang. 2017;7:395–402. https://doi.org/10.1038/nclimate3303.
Senanayake IC, Rathnayaka AR, Marasinghe DS, Calabon MS, Abeywickrama PD, Bhunjun CS, et al. Morphological approaches in studying fungi: collection, examination, isolation, sporulation and preservation. Mycosphere. 2020;11(1):2678–54. https://doi.org/10.5943/mycosphere/11/1/20.
Setälä H, McLean MA. Decomposition rate of organic substrates in relation to the species diversity of soil saprophytic fungi. Oecologia. 2004;139(1):98–107. https://doi.org/10.1007/s00442-003-1478-y.

Sindt C, Besancenot JP, Thibaudon M. Airborne *Cladosporium* fungal spores and climate change in France. Aerobiologia. 2016;32(1):53–68. https://doi.org/10.1007/s10453-016-9422-x.
Singh S, Sandhu DK. Thermophilous fungi in Port Blair soils. Can J Bot. 1986;64:1018–26. https://doi.org/10.1139/b86-139.
Sorger PK. Heat shock factor and the heat shock response. Cell. 1991;65(3):363–6. https://doi.org/10.1016/0092-8674(91)90452-5.
Sousa L, Camacho IC, Grinn-Gofroń A, Camacho R. Monitoring of anamorphic fungal spores in Madeira region (Portugal), 2003–2008. Aerobiologia. 2016;32(2):303–15. https://doi.org/10.1007/s10453-015-9400-8.
Stennett PJ, Beggs PJ. *Alternaria* spores in the atmosphere of Sydney, Australia, and relationships with meteorological factors. Int J Biometeorol. 2004;49(2):98–105. https://doi.org/10.1007/s00484-004-0217-3.
Suryanarayanan TS, Govindarajulu MB, Thirumalai E, Reddy MS, Money NP. Agni's fungi: heat-resistant spores from the Western Ghats, southern India. Fungal Biol. 2011;115(9):833–8. https://doi.org/10.1016/j.funbio.2011.06.011.
Takahashi T. Airborne fungal colony-forming units in outdoor and indoor environments in Yokohama, Japan. Mycopathologia. 1997;139:23–33. https://doi.org/10.1023/A:1006831111595.
Tansey MR. Isolation of thermophilic fungi from alligator nesting material. Mycologia. 1973;65:594–601. https://doi.org/10.1080/00275514.1973.12019473.
Tereschenko IE, Filonov AE. Air temperature fluctuations in Guadalajara, Mexico, from 1926 to 1994 in relation to urban growth. Int J Climatol. 2001;21:483–94. https://doi.org/10.1002/joc.602.
Tereshina VM. Thermotolerance in fungi: the role of heat shock proteins and trehalose. Microbiology. 2005;74(3):247–57. https://doi.org/10.1007/s11021-005-0059-y.
Thibaudon M, Lachasse C. *Alternaria, Cladosporium*: dispersion atmosphérique, rythmes nycthéméral et saisonnier. Rev Fr Allergoll Immunol Clin. 2006;46(3):188–96.
Tischner Z, Kredics L, Marik T, Vörös K, Kriszt B, Péter B, et al. Environmental characteristics and taxonomy of microscopic fungi isolated from washing machines. Fungal Biol. 2019;123(9):650–9. https://doi.org/10.1016/j.funbio.2019.05.010.
Tiwari S, Raman JS, Shanjar J. Role of heat-shock proteins in cellular function and in the biology of fungi. Biotechnol Res Int. 2015;2015:132635. https://doi.org/10.1155/2015/132635.
Toljander YK, Lindahl BD, Holmer L, Hogberg NOS. Environmental fluctuations facilitate species co-existence and increase decomposition in communities of wood-decay fungi. Oecologia. 2006;148:625–31. https://doi.org/10.1007/s00442-006-0406-3.
Trent JD, Gabrielsen H, Jensen B, Neuhard J, Olsen J. Acquired thermotolerance and heat shock proteins in thermophiles from the three phylogenetic domains. J Bacteriol. 1994;176:6148–52. https://doi.org/10.1128/2Fjb.176.19.6148-6152.1994.
Varela JC, Praekelt UM, Meacock PA, Planta RJ, Mager WH. The *Saccharomyces cerevisiae* HSP12 gene is activated by the high-osmolarity glycerol pathway and negatively regulated by protein kinase a. Mol Cell Biol. 1995;15(11):6232–45. https://doi.org/10.1128/MCB.15.11.6232.
Vetrovsky T, Kohout P, Kopecky M, Machac A, Man M, Bahnmann BD. A meta-analysis of global fungal distribution reveals climate-driven patterns. Nat Commun. 2019;10:1–9. https://doi.org/10.1038/s41467-019-13164-8.
Wang G, Luo Y, Wang C, Zhou Y, Mou C, Kang H, Gong YH. Hsp40 protein LeDnaJ07 enhances the Thermotolerance of *Lentinula edodes* and regulates IAA biosynthesis by interacting LetrpE. Front Microbiol. 2020;11:707. https://doi.org/10.3389/fmicb.2020.00707.
Weinstein RN, Palm ME, Johnstone K, Wynn-Williams DD. Ecological and physiological characterization of *Humicola marvinii*, a new psychrophilic fungus from fell fields in the maritime Antarctic. Mycologia. 1997;89:706–11. https://doi.org/10.2307/3761126.
Welker S, Rudolph B, Frenzel E, Hagn F, Liebisch G, Schmitz G, et al. Hsp12 is an intrinsically unstructured stress protein that folds upon membrane association and modulates membrane function. Mol Cell. 2010;39(4):507–20. https://doi.org/10.1016/j.molcel.2010.08.001.

Widden P. Seasonality of forest soil microfungi in southern Quebec. Can J Bot. 1986;64:1413–23. https://doi.org/10.1139/b86-193.

Wu YH, Chan CC, Rao C, Lee CT, Hsu HH, Chiu YH, et al. Characteristics, determinants, and spatial variations of ambient fungal levels in the subtropical Taipei metropolis. Atmos Environ. 2007;41(12):2500–9. https://doi.org/10.1016/j.atmosenv.2006.11.035.

Yang DH, Jung KW, Bang S, Lee JW, Song MH, Floyd-Averette A, et al. Rewiring of signaling networks modulating thermotolerance in the human pathogen *Cryptococcus neoformans*. Genetics. 2017;205(1):201–19. https://doi.org/10.1534/genetics.116.190595.

Zalar P, Novak M, De Hoog GS, Gunde-Cimerman N. Dishwashers–a man-made ecological niche accommodating human opportunistic fungal pathogens. Fungal Biol. 2011;115(10):997–1007. https://doi.org/10.1016/j.funbio.2011.04.007.

Zhang Y, Xu J. Molecular mechanisms of fungal adaptive evolution. In: Rampelotto P, editor. Molecular mechanisms of microbial evolution. Grand challenges in biology and biotechnology. Berlin/Cham: Springer; 2018. p. 85–95. https://doi.org/10.1007/978-3-319-69078-0_15.

Chapter 4
Impact of Climate Change on Dermatophytosis

María del Rocío Reyes-Montes, Esperanza Duarte-Escalante, María Guadalupe Frías-De-León, Angélica Pérez-Rodríguez, and Beatriz Meraz-Ríos

Contents

Abbreviation

HIV Human immunodeficiency virus

4.1 Introduction

In recent years, climate change has become a topic of great interest in the world. This term refers to the measurable changes in the climate observed over a long period. Climate change is characterized by extreme weather events, increased

M. d. R. Reyes-Montes (✉) · E. Duarte-Escalante · A. Pérez-Rodríguez · B. Meraz-Ríos
Microbiology and Parasitology Department, School of Medicine, Universidad Nacional Autónoma de México, Mexico City, Mexico
e-mail: remoa@unam.mx; dupe@unam.mx; bmeraz@unam.mx

M. G. Frías-De-León
Unit Research, Hospital Regional de Alta Especialidad de Ixtapaluca, Mexico State, Mexico

M. G. Frías-De-León et al. (eds.), *The Impact of Climate Change on Fungal Diseases*, Fungal Biology, https://doi.org/10.1007/978-3-030-89664-5_4

droughts, heatwaves, excessive rains and floods, water shortages, dust storms, tropical cyclones (typhoons or hurricanes), forest fires, and melting snow, among others (Eissa and Zaki 2011; Nava et al. 2017). Global warming also refers to the average global temperature increase in recent decades, induced by the greenhouse effect, due to the rise in CO_2 emissions produced by the growing consumption of fossil fuels worldwide, parallel to deforestation. In this context, according to future projections of climate change and global warming, the relationship between climate and health poses a growing challenge for public health. Climate change also results in epidemiologic shifts of diseases associated with alterations in ecosystems and greater population susceptibility to exposure to disease-causing agents (Ghazali et al. 2018).

Some natural phenomena resulting from climate change have expanded the distribution of animal and plant parasites and pathogens. Many of the organisms that cause diseases are strongly influenced by environmental factors such as temperature, rainfall, and humidity, which are subject to climate change. There are also other environmental change drivers, such as urbanization, loss of biodiversity, and invasive species.

Among the mycoses that may present variations in their epidemiology due to climatic change are the superficial ones. Superficial mycoses are some of the most frequent forms of infections in humans, it is estimated that they affect 20–25% of the world population, and their incidence is increasing (Das et al. 2007; Havlickova et al. 2008; Vena et al. 2012). For instance, onychomycosis represents up to 90% on toenails, while nail bed onychomycosis is responsible for up to 50% (Ghannoum et al. 2000) and corresponds to 30% of all superficial mycosis cases (Faergemann and Baran 2003). The incidence increases with age, reaching up to 60% in people over 60 years of age (Pérez et al. 2011). Likewise, it is considered one of the superficial mycoses with the greatest difficulty in diagnosing and treating (Mendoza et al. 2012).

The geographical distribution of dermatophytoses and their causal agents is influenced by several factors, such as the population type, climate, lifestyle, migration, cultural practices, and socioeconomic conditions, among others (Ameen 2010; Havlickova et al. 2008). However, it is worth mentioning that the risk of contracting certain diseases is strongly associated with the ability to manage environmental factors. If an illness is effectively controlled, the consequences of climate change could be lessened, but if control is difficult or impossible, the results could be far-reaching (Paaijmans and Thomas 2011; Parham et al. 2015).

This chapter presents evidence of the role that climate change plays in the incidence and geographic distribution of dermatophytoses.

4.2 Climate Change Vs. Pathogens

Some of the consequences of climate change include natural disasters such as earthquakes (Schneider et al. 1997), floods, tsunamis (Kawakami et al. 2012), and tornados (Neblett Fanfair et al. 2012). Their effects can disrupt on a large-scale fungal habitat, including dermatophytes, so environmental disturbance is a key factor in the spread of these organisms and their resulting potential to cause infection.

The most vulnerable regions to climate change have been shown to include temperate latitudes and are forecasted to warm disproportionately. These regions are located around the Pacific and Indian Oceans and are currently subject to significant variability in rainfall due to the El Niño/Southern Oscillation phenomenon in sub-Saharan Africa. Likewise, in expanding cities, where the urban heat island effect occurs, it is anticipated that extreme weather events could intensify. Other consequences of climate change include rising sea levels, flooding, rising water temperatures, decreasing ocean pH (due to dissolution of carbon dioxide), and invasions of nonnative species (Eissa and Zaki 2011), which contribute to the intensified severity and frequency of floods (Alderman et al. 2013). Floods have been recognized as the most frequent deadly disasters globally, as they can happen after heavy rainfall, hurricanes, storms, storm surges, or other possible causes, such as melting snow from the polar ice caps (Alderman et al. 2013; Hunter 2003).

Another impact of the global temperature increase is that microbial pathogens are expanding their habitat ranges into previously inhospitable environments (IPPC 2008). An example is emerging pathogens that are present but inactive in the environment. Even though hosts, including humans, are exposed to these microorganisms, they do not develop an immune response against them as it is not necessary (Diuk-Wasser et al. 2010; Patz and Olson 2006). However, when they are selectively activated by some physical stimulus and take advantage of changing conditions to adapt to susceptible hosts, their incidence increases rapidly in a population or a geographic range (Morse 1995). Therefore, the expansion of microorganisms is related to temperate ecosystems, which generally experience seasonal temperature changes. If these are prolonged changes, the pathogen resilience can enhance, and at the same time, the host defenses could decrease (Epstein 2001).

One of the most notable aspects of mammalian physiology is their high resistance to fungal diseases. Although fungi are relevant pathogens, there are relatively few fungal diseases associated with mammals, attributed to the combination of endothermy and adaptive immunity (Casadevall 2012). Endothermy in mammals creates a thermally restricted space for most fungal species. When combined with innate and adaptive immunity, it makes fungal diseases rare among immunologically intact humans (Robert and Casadevall 2009). However, climate change may bring new fungal pathogens that affect people, animals, and plants, as species with pathogenic potential adapt to higher temperatures (Casadevall 2018). This hypothesis postulates that many fungal species are currently not pathogenic for mammals since they cannot replicate in them; however, they have the potential to be virulent (Cohen et al. 2017), which raises the possibility of having new fungal diseases. The natural human temperature excludes most fungal species, suggesting an evolutionary connection between endothermy and protection against fungi (Bergman and Casadevall 2010). The relationship between mammals' high temperatures and their resistance to fungal diseases could have contributed to the great radiation of mammals after the cataclysm at the end of the Cretaceous period, in which high temperatures could have made them resistant to fungal pathogens (Casadevall 2005, 2012).

The rarity of fungal diseases in humans with apparent normal immunity explains why their description came afterward many bacterial diseases. Also, the intensity and pathology of a disease often depend on temperature, which occurs for two reasons. The first is that the higher the temperature, the higher the metabolism of the parasite, which increases its feeding and, often, its replication within a host, accentuating the damage (Fels and Kaltz 2006; Kirk et al. 2018). Secondly, the host experiences more environmental stress, which impairs its ability to resist the infecting pathogen.

The temperature rise could also spiral the prevalence of some skin pathologies as more people would suffer from sensitive skin and greater cutaneous xerosis due to the decrease in relative humidity. Alterations in the skin barrier function would increment the severity and prevalence of skin diseases such as dermatophytoses (Centers for Disease Control and Prevention 2005, 2006a, b; Rose and Akpinar-Elci 2015).

4.3 Influence of Climate Change on Dermatophytoses

Dermatophytoses are caused by fungi that invade and spread on the keratinized skin of mammals, including humans, often causing contagious infections (Weitzman and Summerbell 1995). Dermatophytoses are common throughout the world and are relevant in the veterinary and public health areas (Cafarchia et al. 2009, 2012; Weitzman and Summerbell 1995). The dermatophyte group comprises 52 species divided into 9 genera: *Trichophyton*, *Microsporum*, *Epidermophyton*, *Arthroderma*, *Lopophyton*, *Nannizia*, *Ctenomyces*, *Guarromyces*, and *Paraphyton* (Mercer and Stewart 2019). The fungi *T. rubrum*, *T. tonsurans*, the *T. mentagrophytes* complex, as well as *M. canis*, *M. gypseum*, and *E. floccosum* are considered the main etiological agents of dermatophytoses in humans (Das et al. 2007; Weitzman and Summerbell 1995). Based on their ecology, dermatophytes have been divided into three groups: anthropophilic, zoophilic, and geophilic. Anthropophilic dermatophytes, mainly associated with humans, can cause *tinea capitis*, *tinea corporis*, *tinea pedis*, and *tinea unguium* and rarely infect animals (Cafarchia et al. 2006; Gräser et al. 2008; Weitzman and Summerbell 1995). These fungi can cause superficial infections (dermatophytosis) in the skin of human hosts and can penetrate deeper tissues in immunocompromised hosts (Squeo et al. 1998) (Table 4.1).

The frequency of dermatophytoses tends to increase after a tsunami due to the unsanitary conditions prevailing after the disaster. Lee et al. (2006) carried out a study in which they analyzed samples obtained from patients with skin affections, 131 men and 104 women, after a tsunami caused by a 9.0 magnitude earthquake on the Richter scale that hit the South Asian and East African coastal areas. Their results showed that the most prevalent skin problems were infection-infestations (32.5%), followed by eczema (29.8%) and traumatic skin disorders (29.4%). The infestation cases corresponded to superficial fungal infections (14.3%), where tinea corporis was the most common fungal disease (5.3%).

Table 4.1 Clinical forms of dermatophytoses

Clinical form	Characteristics	Etiological agent
Tinea capitis	It is an infection that occurs mainly in children and adolescents, in which the hair and scalp are affected. According to the type of lesion, they are classified as dry or inflammatory. The dry type is divided into two varieties: Microsporic and trichophytic	Microsporic variety: *M. canis*, *M. gypseum*, and *M. audouinii*. Trichophytic variety: *T. tonsuras*, *T. mentagrophytes*, and *T. rubrum*
Tinea favosa	Favus or *tinea favosa* is characterized by being a chronic infection of the scalp mainly	The causative agents are more related to the genus *Trichophyton*, while in developing countries, they are associated with species of the genus *Microsporum*
Tinea barbae	Also called dermatophytic psychosis, it occurs almost exclusively in adolescent and adult men Two clinical variants can be diagnosed, the inflammatory and the non-inflammatory	Inflammatory variety: Zoophilic dermatophytes, such as *T. mentagrophytes*, *T. verrucosum*, *M. canis*, and *T. erinacei* Non-inflammatory variety: Anthropophilic dermatophytes such as *T. rubrum*, *T. violaceum*, and *T. tonsurans*
Tinea pedis	It is also called athlete's foot *Tinea pedis* is classified into four varieties according to their clinical characteristics: (a) intertriginous; (b) vesicular; (c) moccasin, also called chronic hyperkeratotic; and (d) acute ulceration	Intertriginosa: Anthropophilic dermatophytes like *T. rubrum* and *E. floccosum* Vesiculosa: Zoophilic dermatophyte species of the *T. mentagrophytes* complex Moccasin: Anthropophilic dermatophyte *T. rubrum* Acute ulcer: Zoophilic dermatophytes *T. interdigitale*, *T. mentagrophytes*, and *M. canis*
Tinea manuum	Uncommon infection that affects mainly the interdigital, dorsal, and palmar regions. It usually affects one hand, although sometimes both hands are affected. It has been classified as a hyperkeratotic and inflammatory variety	Hyperkeratotic variety: *T. rubrum* Inflammatory variety: Zoophilic dermatophytes *T. mentagrophytes*, *T. erinacei*, and *M. canis*, and those of anthropophilic origin like *T. rubrum* and *E. floccosum* are also involved

(continued)

Table 4.1 (continued)

Clinical form	Characteristics	Etiological agent
Tinea cruris	Also known as jock itch, it affects mainly adult men due to the occlusive function of the scrotum that produces a humid and warm environment	*T. rubrum*, *T. mentagrophytes*, and *E. floccosum*
Tinea imbricata	This ringworm is chronic and the most superficial and driest of all. It is distributed in a limited way in isolated rural communities. Some of the most affected geographical places are the islands of Oceania (Polynesia and Melanesia), where the disease is called Tokelau. With less frequency in India, central and South America; in Brazil, it is known as Chimberé	*T. concentricum*
Tinea corporis	This infection occurs on exposed hairless skin, more commonly on the neck, trunk, shoulders, and extremities and more sporadically on the legs. There are two varieties: Microsporic and trichophytic	*M. canis*, *T. rubrum*, *T. mentagrophytes*, *T. tonsurans*, and *E. floccosum*
Dermatophytic granuloma or trichophytic granuloma	Also called Majocchi's granuloma, it can occur in the dermis and subcutaneous tissue; predisposition has been seen in both immunocompetent and immunosuppressed patients, but the latter are the most affected. The areas of the body in which it is most frequently found are the scalp and the beard	*T. rubrum*, *T. tonsurans*, and *T. mentagrophytes* var. *mentagrophytes*
Tinea unguium	Also called dermatophytic onychomycosis, it occurs in both fingernails and toenails and is closely related to the presence of tinea pedis It is classified as distal-lateral subungual onychomycosis; superficial onychomycosis, which presents two modalities, white and black; proximal white subungual onychomycosis; endonyx onychomycosis; mixed pattern onychomycosis; total dystrophic onychomycosis	Distal-lateral subungual onychomycosis: *T. rubrum*, *T. mentagrophytes* var. *interdigitale*, *T. mentagrophytes* var. *nodulare* Superficial onychomycosis: *T. rubrum*, *T. mentagrophytes* var. *interdigitale* Proximal white subungual onychomycosis: *T. rubrum*, *T. mentagrophytes* var. *interdigitale* Onychomycosis endonyx: *T. soudanense* and *T. violaceum*

Likewise, Bayramgurler et al. (2002) reported that after a 7.4 magnitude earthquake on the Richter scale in the Marmara Region of Turkey, infection-infestations were the most prevalent dermatological problems, being superficial skin fungal infections the vast majority of infection-infestation cases. *Tinea pedis* was the most common superficial mycosis after the earthquake.

On the other hand, a variation in dermatophytosis infection patterns has been evidenced in different countries such as Algeria (Djeridane et al. 2006), South Africa (Vismer and Findlay 1988), Mexico (Manzano-Gayosso et al. 1994), Italy (Filipello et al. 1996), Japan (Kasai 2001), the United States (Weitzman et al. 1998), Canada (Gupta and Summerbell 1998), Brazil (Costa et al. 1999), India (Patwardhan and Dave 1999), and Australia (Coloe and Baird 1999). This heterogeneity appears to be influenced by factors such as climate (humidity, temperature), lifestyle, participation in outdoor activities, and the prevalence of underlying diseases (diabetes, human immunodeficiency virus [HIV]).

In recent years, there has been an epidemiological transformation of dermatophytes in India. *T. rubrum* remained the predominant dermatophyte; however, *T. mentagrophytes* has currently emerged throughout the country (Nenoff et al. 2019; Verma and Madhu 2017) (Fig. 4.1). This change is noticeable because the prevalence of *T. mentagrophytes* used to be 20%, and in a period of 15 years, it has increased to more than 90%. The authors suggest that the prevalence of *T. mentagrophytes* is caused by the appearance of a new genotype called the STI Type VIII genotype that has adapted to climate change. This acclimation allowed the new genotype to develop certain characteristics like a more efficient person-to-person transmission and the production of more inflammatory and eruptive lesions, with a predilection for the face (Nenoff et al. 2019; Verma and Madhu 2017). Besides, the

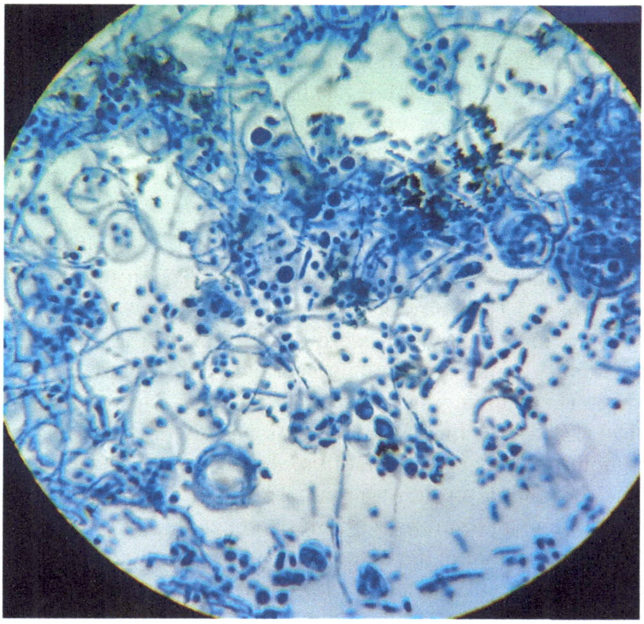

Fig. 4.1 Microconidia and macroconidia produced by *Trichophytom mentagrophytes* on potato dextrose agar in 10 days at 28 °C. Lactophenol cotton blue staining (picture taken with light microscopy at 40×)

surge of this mycosis is probably because this genotype has a clonal reproduction, which is an interesting finding since clonality is a characteristic of many fungal species that are expanding (Gladieux et al. 2015). Clonality can also be the means that favor the spread and establishment of a new population (Bazin et al. 2014; Caron et al. 2014).

On the other hand, within dermatophytoses, it is common to see an increment in the incidence of onychomycosis and tinea pedis in spring and summer worldwide due to the higher humidity and temperature conditions that favor fungal development. The latter was described by Fuentes-Rivera (2000), who conducted an onychomycosis clinical study in students from Chile, where the incidence increased in spring and summer. The study also showed an increasing persistence of onychomycosis, particularly in the autumn of 1998, explained by the typical summer climatic conditions that extended into this season due to the "El Niño" phenomenon. "El Niño" is a recurring oceanographic and climatic phenomenon that originated in the Pacific Ocean with worldwide repercussions. It occurs with an approximate 5-year periodicity and was last observed in the summer of 1998.

At the same time, in Peru, "El Niño" also caused in 1998 a marked increase in temperature and river rainfall, mainly affecting the northern part of the country. The departments of Piura and Tumbes especially suffered terrible damage in their ecosystems and human habitat. This situation resulted in a surge of dermatological diseases, including tineas (Bravo Sosu and Bravo Puccio 2001), demonstrating that temperature and humidity conditions are almost decisive in dermatophytoses spread.

Also, in Saudi Arabia, a country with a hot climate, high prevalence of onychomycosis has been reported, followed by *tinea capitis*, *tinea pedis*, *tinea cruris*, and *tinea corporis* (Abanmi et al. 2008). It has been shown that onychomycosis prevalence in hot environments is given by climatic conditions, which are a crucial requirement for fungal growth, which will then lead to acute symptomatic exacerbations. Also, human behavioral patterns are critical for the multiplication and spread of infection (Kam et al. 1997).

It is also important to highlight some in vitro studies aimed to elucidate how humidity and temperature affect fungal penetration through the stratum corneum, primarily *T. rubrum* and *T. mentagrophytes*, the main causative agents of ringworms in the stratum corneum. The studies were conducted in different humidity conditions at a 35 °C temperature. Both fungi penetrated the stratum corneum at 90 and 100% humidity. The dermatophyte invasion rate was slower at 80% humidity (Morishita et al. 2004) but increased proportionally with humidity growing (Ide et al. 1999; Ninomiya et al. 1998). Therefore, for *T. rubrum* and *T. mentagrophytes* to penetrate and proliferate in the stratum corneum, a humidity of at least 90% is required (Chittasobhon and Smith 1979; Morishita et al. 2004). On the other hand, it is essential to mention that in the case of *tinea manuum*, *tinea corporis*, and *tinea capitis*, the humidity was not the main factor contributing to the evolvement of these dermatophytoses, but rather the high-temperature environment (Ninomiya et al. 1998). Thus, two factors are necessary for the development of dermatophytoses: an environment with high temperature and humidity and sufficient time to allow the dermatophyte adhesion to the stratum corneum (Chittasobhon and Smith 1979; Sakka et al. 2015).

4.4 The Geographical Spread of Dermatophytoses as a Consequence of Climate Change

Climate change influences the formation of dust storms, which have a crucial role in pathogen dispersal around the planet and can result in a transmission route. Many microorganisms can be aerosolized, and most can be transported over short distances. Other microorganisms can associate with larger soil particles, and others can die or lose viability during transport due to dehydration, UV ray exposure, and low nutrient availability (Burrows et al. 2009; Griffin et al. 2007). However, some microorganisms can withstand adverse conditions during transport, allowing them to travel long distances and locating in other geographic areas (Burrows et al. 2009; Griffin et al. 2007). Among these microorganisms are dermatophytes since they have been isolated from dust samples. Especially members of the *Microsporum* and *Trichophyton* genera can survive because they can form spores that allow them to remain in a latent state until the non-conducive growth conditions change in the environment and become conducive (Griffin et al. 2003, 2007). Also, some of these fungi have hydrophobins, a family of highly hydrophobic secretory fungal proteins that coat the fungal walls and facilitate air dispersion favoring their survival (Heddergott et al. 2012). Thus, the dispersal of microorganisms through dust storms can be a critical factor in the distribution and geographic expansion of dermatophytes.

Finally, the increasing number of dermatophytosis in animals may be associated with relocating exotic species out of their natural environment. Such is the case for South American alpacas, exported to Europe, where they have been turned into farm and companion animals. The change in their natural habitat and climate results in them being more susceptible to other dermatophyte species, which are easily transmitted between animals and humans. It is estimated that currently, the alpaca population is 35,000 individuals in Great Britain, 20,000 in the United States, and 2000 in Poland (Twomey et al. 2014). Thus, animal importation, which must adapt to new climatic conditions and undergo lifestyle changes, results in modifying geographic distribution and dermatophyte reservoirs (Hiruma et al. 2015; Sabou et al. 2018).

4.5 Conclusion

The close relationship between dermatophytoses and climate change has been evidenced through various mechanisms such as infections with spores dispersed by dust storms; traumatic implantation of fungi in wounds through contaminated water, organic matter, and natural disasters (floods, tsunamis); the appearance of new dermatophytes; or an increased number of known dermatophyte infections due to changing characteristics of the physical environment. In the face of the emergence of new dermatophytoses caused by temperature increase and the expansion of ecological niches, it is necessary to alert physicians, as they may run the risk of inadvertently delaying diagnoses and not providing adequate pharmacotherapeutic treatment for patients.

References

Abanmi A, Bakheshwain S, El Khizzi N, Zouman AR, Hantirah S, Al Harthi F, et al. Characteristics of superficial fungal infections in the Riyadh region of Saudi Arabia. Int J Dermatol. 2008;47:229–35. https://doi.org/10.1111/j.1365-4632.2008.03563.x.

Alderman K, Turner LR, Tong S. Assessment of the health impacts of the 2011 summer floods in Brisbane. Disaster Med Public Health Prep. 2013;7:380–6. https://doi.org/10.1017/dmp.2013.42.

Ameen M. Epidemiology of superficial fungal infections. Clin Dermatol. 2010;28:197–201. https://doi.org/10.1016/j.clindermatol.2009.12.005.

Bayramgurler D, Bilen N, Namli S, Altinas L, Apaydin R. The effects of 17 august Marmara earthquake on patient admittances to our dermatology department. J Eur Acad Dermatol Venereol. 2002;16:249–52. https://doi.org/10.1046/j.1468-3083.2002.00488.x.

Bazin E, Mathe-Hubert H, Facon B, Carlier J, Ravigne V. The effect of mating system on invasiveness: some genetic load may be advantageous when invading new environ ments. Biol Invasions. 2014;16:875–86. https://doi.org/10.1007/s10530-013-0544-6.

Bergman A, Casadevall A. Mammalian endothermy optimally restricts fungi and metabolic costs. MBio. 2010;1:e00212–0. https://doi.org/10.1128/mBio.00212-10.

Bravo Sosu WS, Bravo Puccio FG. El efecto del fenomeno El Nino en las enfermedades dermatológicas. Folia Dermatol Peru. 2001;12:29–36.

Burrows SM, Elbert W, Lawrence MG, Pöschl U. Bacteria in the global atmosphere - part 1: review and synthesis of literature data for different ecosystems. Atmos Chem Phys. 2009;9:9263–80.

Cafarchia C, Romito D, Capelli G, Guillot J, Otranto D. Isolation of Microsporum canis from the hair coat of pet dogs and cats belonging to owners diagnosed with M. canis tinea corporis. Vet Dermatol. 2006;17:327–31. https://doi.org/10.1111/j.1365-3164.2006.00533.x.

Cafarchia C, Otranto D, Weigl S, Campbell BE, Parisi A, Cantacessi C, et al. Molecular characterization of selected dermatophytes and their identification by electrophoretic mutation scanning. Electrophoresis. 2009;30:3555–64. https://doi.org/10.1002/elps.200900313.

Cafarchia C, Weigl S, Figueredo LA, Otranto D. Molecular identification and phylogenesis of dermatophytes isolated from rabbit farms and rabbit farm workers. Vet Microbiol. 2012;154:395–402. https://doi.org/10.1016/j.vetmic.2011.07.021.

Caron V, Ede FJ, Sunnucks P. Unravelling the paradox of loss of genetic variation during invasion: superclones may explain the success of a clonal invader. PLoS One. 2014;9:e97744. https://doi.org/10.1371/journal.pone.0097744.

Casadevall A. Fungal virulence, vertebrate endothermy, and dinosaur extinction: is there a connection? Fungal Genet Biol. 2005;42:98–106. https://doi.org/10.1016/j.fgb.2004.11.008.

Casadevall A. Fungi and the rise of mammals. PLoS Pathog. 2012;8:e1002808. https://doi.org/10.1371/journal.ppat.1002808.

Casadevall A. Fungal diseases in the 21st century: the near and far horizons. Pathog Immun. 2018;3:183–96. https://doi.org/10.20411/pai.v3i2.249.

Centers for Disease Control and Prevention. Infectious disease and dermatologic conditions in evacuees and rescue workers after hurricane Katrina–multiple states, August-September 2005. Mortal Wkly Rep. 2005;54:961–4.

Centers for Disease Control and Prevention. Surveillance in hurricane evacuation centers-Louisiana, September-October 2005a. MMWR Morb Mortal Wkly Rep. 2006a;55:32–5.

Centers for Disease Control and Prevention. Mold prevention strategies and possible health effects in the aftermath of hurricanes and major floods. MMWR Recomm Rep. 2006b;55:1–27.

Chittasobhon N, Smith MB. The production of experimental dermatophytelesion in Guinea pig. J Invest Dermatol. 1979;73:198–201. https://doi.org/10.1111/1523-1747.ep12581683.

Cohen JM, Venesky MD, Sauer EL, Civitello DJ, McMahon TA, Roznik EA, et al. The thermal mismatch hypothesis explains host susceptibility to an emerging infectious disease. Ecol Lett. 2017;20:184–93. https://doi.org/10.1111/ele.12720.

Coloe SV, Baird RW. Dermatophyte infections in Melbourne: trends from 1961/64 to 1995/96. Pathology. 1999;31:395–7. https://doi.org/10.1080/003130299104792.

Costa TR, Costa MR, da Silva MV, Rodrigues AB, de Fernandes O, Soares AJ, et al. The etiology and epidemiology of dermatophytoses in Goiania, GO, Brazil. Rev Soc Bras Med Trop. 1999;32:367–71. https://doi.org/10.1590/s0037-86821999000400006.

Das S, Goyal R, Bhattacharya SN. Laboratory-based epidemiological study of superficial fungal infections. J Dermatol. 2007;34:245–8. https://doi.org/10.1111/j.1346-8138.2007.00262.x.

Diuk-Wasser MA, Vourc'h G, Cislo P, Gatewood Hoen A, MeltonF HSA, et al. Field and climate based model for predicting the density of host-seeking nymphal Ixodes scapularis, an important vector of tick-borne disease agents in the eastern United States. Glob Ecol Biogeogr. 2010;19:504–14. https://doi.org/10.1111/j.1466-8238.2010.00526.x.

Djeridane A, Djeridane Y, Ammar-Khodja A. Epidemiological and aetiological study on tinea pedis and onychomycosis in Algeria. Mycoses. 2006;49:190–6. https://doi.org/10.1111/j.1439-0507.2006.01230.x.

Eissa AE, Zaki MM. The impact of global climatic changes on the aquatic environment. Procedia Environ Sci. 2011;4:251–9. https://doi.org/10.1016/j.proenv.2011.03.030.

Epstein PR. Climate change and emerging infectious diseases. Microbes Infect. 2001;3:747–54. https://doi.org/10.1016/s1286-4579(01)01429-0.

Faergemann J, Baran R. Epidemiology, clinical presentation and diagnosis of onychomycosis. Br J Dermatol. 2003;149:1–4. https://doi.org/10.1046/j.1365-2133.149.s65.4.x.

Fels D, Kaltz O. Temperature-dependent transmission and latency of Holospora undulata, a micronucleus-specific parasite of the ciliate Paramecium caudatum. Proc R Soc Lond B Biol Sci. 2006;273:1031–8. https://doi.org/10.1098/rspb.2005.3404.

Filipello MV, Preve L, Tullio V. Fungi responsible for skin mycoses in Turin (Italy). Mycoses. 1996;39:141–50. https://doi.org/10.1111/j.1439-0507.1996.tb00117.x.

Fuentes-Rivera D. Epidemiología y Diagnóstico Clínico- Etiológico de Onicomicosis en un Centro Medico Universitario (Junio 97- Mayo 99). Dermatol Peruana. 2000;10:21–33.

Ghannoum MA, Hajjeh RA, Scher R, Konnikov N, Gupta AK, Summerbell R, et al. A large-scale North American study of fungal isolates from nails: the frequency of onychomycosis, fungal distribution, and antifungal susceptibility patterns. J Am Acad Dermatol. 2000;43:641–8. https://doi.org/10.1067/mjd.2000.107754.

Ghazali D, Guericolas M, Thys F, Sarasin F, Arcos Gonzalez P, Casalino E. Climate change impacts on disaster and emergency medicine focusing on mitigation disruptive effects: an international perspective. Int J Environ Res Public Health. 2018;15:1379. https://doi.org/10.3390/ijerph15071379.

Gladieux P, Feurtey A, Hood M, Snirc A, Clavel J, Dutech C, et al. The population biology of fungal invasions. Mol Ecol. 2015;24:1969–86. https://doi.org/10.1111/mec.13028.

Gräser Y, Scott J, Summerbell R. The new species concept in dermatophytes-a polyphasic approach. Mycopathologia. 2008;166:239–56. https://doi.org/10.1007/s11046-008-9099-y.

Griffin DW, Kellogg CA, Garrison VH, Lisle JT, Borden TC, Shinn EA. Atmospheric microbiology in the northern aribbean during African dust events. Aerobiologia. 2003;19:143–57. https://doi.org/10.1023/B:AERO.0000006530.32845.8d.

Griffin DW, Kubilay N, Koçak M, Gray MA, Borden TC, Shinn EA. Airborne desert dust and aeromicrobiology over the Turkish Mediterranean coastline. Atmos Environ. 2007;41:4050–62. https://doi.org/10.1016/j.atmosenv.2007.01.023.

Gupta AK, Summerbell RC. Increased incidence of Trichophyton tonsurans tinea capitis in Ontario, Canada between 1985 and 1996. Med Mycol. 1998;36:55–60.
Havlickova B, Czaika VA, Friedrich M. Epidemiological trends in skin mycoses worldwide. Mycoses. 2008;51:2–15. https://doi.org/10.1111/j.1439-0507.2008.01606.x.
Heddergott C, Bruns S, Nietzsche S, Leonhardt I, Kurzai O, Kniemeyer O, Brakhage AA. The Arthroderma benhamiae hydrophobin HypA mediates hydrophobicity and influences recognition by human immune effector cells. Eukaryot Cell. 2012;11:673–82. https://doi.org/10.1128/EC.00037-12.
Hiruma J, Kano R, Harada K, Monod M, Hiruma M, Hasegawa A, et al. Occurrence of Arthroderma benhamiae genotype in Japan. Mycopathologia. 2015;179:219–23. https://doi.org/10.1007/s11046-014-9839-0.
Hunter PR. Climate change and waterborne and vector-borne disease. J Appl Microbiol. 2003;94(Suppl):S37–46. https://doi.org/10.1046/j.1365-2672.94.s1.5.x.
Ide M, Ninomiya J, Ito Y, Teramoto T, Takiuchi I. Experimental studies on the penetration into the human stratum corneum of the dermatophyte. Nippon Ishinkin Gakkai Zasshi. 1999;40:93–7. https://doi.org/10.3314/jjmm.40.93.
IPCC 2008. 2006 IPCC Guidelines for National Greenhouse Gas Inventories – A primer, Prepared by the National Greenhouse Gas Inventories Programme, Eggleston HS, Miwa K, Srivastava N, Tanabe K (eds). Published: IGES, Japan.
Kam KM, Au WF, Cheung MM. Onychomycosis in Hong Kong. Int J Dermatol. 1997;36:757–61. https://doi.org/10.1046/j.1365-4362.1997.00048.x.
Kasai T. Epidemiological Investigation Committee for Human Mycoses in the Japanese Society for Medical Mycology. 1997 Epidemiological survey of dermatophytoses in Japan. Nippon Ishinkin Gakkal Zasshi. 2001;42:11–8. https://doi.org/10.3314/jjmm.42.11.
Kawakami Y, Tagami T, Kusakabe T, Kido N, Kawaguchi T, Omura M, et al. Disseminated aspergillosis associated with tsunami lung. Respir Care. 2012;57:1674–8. https://doi.org/10.4187/respcare.01701.
Kirk D, Jones N, Peacock S, Phillips J, Molnar PK, Krkosek M, et al. Empirical evidence that metabolic theory describes the temperature dependency of within-host parasite dynamics. PLoS Biol. 2018;16:e2004608. https://doi.org/10.1371/journal.pbio.2004608.
Lee SH, Choi CP, Eun HC, Kwon OS. Skin problems after a tsunami. J Eur Acad Dermatol Venereol. 2006;20:860–3. https://doi.org/10.1111/j.1468-3083.2006.01666.x.
Manzano-Gayosso P, Mendez-Tovar LJ, Hernandez Hernandez F, et al. Dermatophytoses in Mexico city. Mycoses. 1994;37:49–52. https://doi.org/10.1111/j.1439-0507.1994.tb00285.x.
Mendoza N, Palacios C, Cardona N, Gómez LM. Onicomicosis: afección común de difícil manejo. Rev Asoc Colomb Dermatol. 2012;20:149–58.
Mercer DK, Stewart CS. Keratin hydrolysis by dermatophytes. Med Mycol. 2019;57:13–22. https://doi.org/10.1093/mmy/myx160.
Morishita N, Ninomiya J, Sei Y, Takiuchi I. Experimental studies on the penetration of dermatophytes into the human stratum corneum and the effects of washing. Nippon Ishinkin Gakkai Zasshi. 2004;45:247–52. https://doi.org/10.3314/jjmm.45.247.
Morse SS. Factors in the emergence of infectious diseases. Emerg Infect Dis. 1995;1:7–15. https://doi.org/10.3201/eid0101.950102.
Nava A, Shimabukuro JS, Chmura AA, Luz SLB. The Impact of Global Environmental Changes on Infectious Disease Emergence with a Focus on Risks for Brazil. ILAR J. 2017;58:393–400. https://doi.org/10.1093/ilar/ilx034.
Neblett Fanfair R, Benedict K, Bos J, Bennett SD, Lo Y-C, Adebanjo T, et al. Necrotizing cutaneous mucormycosis after a tornado in Joplin, Missouri, in 2011. N Engl J Med. 2012;367:2214–25. https://doi.org/10.1056/NEJMoa1204781.
Nenoff P, Verma SB, Vasani R, Burmester A, Hipler UC, Wittig F, et al. The current Indian epidemic of superficial dermatophytosis due to Trichophyton mentagrophytes-a molecular study. Mycoses. 2019;62:336–56. https://doi.org/10.1111/myc.12878.

Ninomiya J, Ide M, Ito Y, Takiuchi I. Experimental penetration of Trichophyton mentagrophytes into human stratum corneum. Mycopathologia. 1998;141:1537. https://doi.org/10.1023/a:1006920000117.

Paaijmans KP, Thomas MB. Health: wealth versus warming. Nat Clim Chang. 2011;1:349–50.

Parham PE, Waldock J, Christophides GK, Hemming D, Agusto F, Evans KJ, et al. Climate, environmental and socio-economic change: weighing up the balance in vector-borne disease transmission. Philos Trans R Soc Lond Ser B Biol Sci. 2015;370:20130551. https://doi.org/10.1098/rstb.2013.0551.

Patwardhan N, Dave R. Dermatomycosis in and around Aurangabad. Indian J Pathol Microbiol. 1999;42:455–62.

Patz JA, Olson SH. Malaria risk and temperature: influences from global climate change and local land use practices. Proc Natl Acad Sci U S A. 2006;103:5635–6.

Pérez JE, Cárdenas C, Hoyos AM. Características clínicas, epidemiológicas y microbiológicas de la onicomicosis en un laboratorio de referencia, Manizales (Caldas), 2009. Infectio. 2011;15:168–76.

Robert VA, Casadevall A. Vertebrate endothermy restricts most fungi as potential pathogens. J Infect Dis. 2009;200:1623–6. https://doi.org/10.1086/644642.

Rose S, Akpinar-Elci M. An evaluation of the prevalence of respiratory and general symptoms among occupants of households affected by flooding in Guyana: the case of Cove & John [Internet]. Proceedings of the University of Guyana Faculty of Natural Sciences First International Conference on Sustainable Development; 2013, Aug 12–14; Georgetown, Guyana. Georgetown: Journal of Academic Research & Essays. 2015. [cited 2017 Jan 22]. http://jare.org.gy/wp-content/uploads/2015/12/Rose-S-An-evaluation-of-the-pre valence-of-respiratory-FNS.pdf.

Sabou M, Denis J, Boulanger N, Forouzanfar F, Glatz I, Lipsker D, et al. Molecular identification of Trichophyton benhamiae in Strasbourg, France: a 9-year retrospective study. Med Mycol. 2018;56:723–34. https://doi.org/10.1093/mmy/myx100.

Sakka N, Shemer A, Barziai A, Farhi R, Daniel R. Occult tinea pedís in an Israeli population and predisposing factors for the acquisition of the disease. Int J Dermatol. 2015;54:146–9. https://doi.org/10.1111/ijd.12506.

Schneider E, Hajjeh RA, Spiegel RA, Jibson RW, Harp EL, Marshall GA, et al. A coccidioidomycosis outbreak following the Northridge, calif, earthquake. JAMA. 1997;277:904–8.

Squeo RF, Beer R, Silvers D, Weitzman I, Grossman M. Invasive Trichophyton rubrum resembling blastomycosis infection in the immunocompromised host. J Am Acad Dermatol. 1998;39:379–80. https://doi.org/10.1016/s0190-9622(98)70396-2.

Twomey DF, Wu G, Nicholson R, Watson EN, Foster AP. Review of laboratory submissions from New World camelids in England and Wales (2000-2011). Vet J. 2014;200:51–9. https://doi.org/10.1016/j.tvjl.2014.01.021.

Vena GA, Chieco P, Posa F, Garofalo A, Bosco A, Cassano N. Epidemiology of dermatophytoses: retrospective analysis from 2005 to 2010 and comparison with previous data from 1975. New Microbiol. 2012;35:207–13.

Verma S, Madhu R. The great Indian epidemic of superficial Dermatophytosis: an appraisal. Indian J Dermatol. 2017;62:227–36. https://doi.org/10.4103/ijd.IJD_206_17.

Vismer HF, Findlay GH. Superficial fungal infections in Transvaal. A contemporary analysis of dermatophytosis in this region. S Afr Med J. 1988;75:587–92.

Weitzman I, Summerbell RC. The dermatophytes. Clin Microbiol Rev. 1995;8:240–59. https://doi.org/10.1128/CMR.8.2.240.

Weitzman I, Chin NX, Kunjukunju N, Della-Latta P. A survey of dermatophytes isolated from human patients in the United States from 1993 to 1995. J Am Acad Dermatol. 1998;39:255–61. https://doi.org/10.1016/s0190-9622(98)70085-4.

Chapter 5
Climate Change and Its Impact on Sporotrichosis

Concepción Toriello, Carolina Brunner-Mendoza, and Laura Parra-Jaramillo

Contents

Abbreviations

GHGs Greenhouse gases
MAPK Mitogen-activated protein-kinase
WHO World Health Organization

C. Toriello (✉) · C. Brunner-Mendoza · L. Parra-Jaramillo
Microbiology and Parasitology Department, School of Medicine, Universidad Nacional Autónoma de México, Mexico City, Mexico
e-mail: toriello@unam.mx; brunner.carolina@facmed.unam.mx

M. G. Frías-De-León et al. (eds.), *The Impact of Climate Change on Fungal Diseases*, Fungal Biology, https://doi.org/10.1007/978-3-030-89664-5_5

5.1 Introduction

Climate change is a concept that involves a shift in long-term weather patterns in the world (VijayaVenkataRaman et al. 2012). It has been suggested that greenhouse gas (GHG) emissions derived mainly from human activities are responsible for a 1 °C increase in global temperature since preindustrial times and that this increase could reach 1.5 °C between 2030 and 2052 (IPCC 2014). The greenhouse gases in the atmosphere and the rise of global temperature are responsible for other changes, such as a change in temperature and precipitation patterns; an increase in ocean temperatures, sea level, and acidity; the melting of glaciers and sea ice; and changes in the frequency, intensity, and duration of extreme weather events (EPA 2007). It has been hypothesized that an increase in global temperature could impact ecosystems and biodiversity, modifying community composition and the extension, distribution, proliferation, and extinction of several species (Zhai et al. 2018).

Human health is connected to the weather and climate since natural disasters such as droughts, floods, and cyclones directly affect nutrition and can trigger infectious diseases outbreaks (WMO 2012). Increased temperatures, precipitation extremes, extreme weather events, and a rise in sea level expose the human population to extreme heat, poor air quality, reduced food and water quality, changes in infectious agents, and population displacement, which consequently result in heat-related illnesses; cardiopulmonary illnesses; food-, water-, and vector-borne diseases; and mental health problems and stress (EPA 2007).

In recent years, research concerning the impact of climate change on infectious diseases has increased, especially research on vector-borne diseases. Despite technological advances enabling the construction of models that provide scenarios for specific parameters, climate projections are challenging and complex. The lack of data, uncertainty, and bias allow for a broad spectrum of possibilities regarding the behavior of microbial communities during climate change events (Booth 2018).

Fungal infections have been neglected infectious diseases despite being a global public health problem causing morbidity, mortality, and disability in millions of people –according to estimations, mainly affecting immunocompromised populations (Almeida et al. 2019; Casadevall 2019; Fisher et al. 2018). Rarely are potential pathogenic fungi isolated from airborne, soilborne, or water samples, and, when found, they are present in meager proportions. Its pathogenic capacity in humans, according to Köhler et al. (2015), is the result of adding four primary characteristics together: thermotolerance, the ability to penetrate or enter host tissues, the ability to use host tissues as substrate, and the evasion of the host immune response. Recent estimates suggest there are a high number of fungal infection cases related to immunocompromised populations (Rodrigues and Albuquerque 2018).

Some scenarios in which new environmental conditions may favor the presence of some etiological agents that cause mycosis have been proposed. Various physiological characteristics of human pathogenic fungi could be considered advantageous in a global warming scenario, such as thermotolerance, dimorphism, melanin production, long latency periods, stress response, and toxin production (Garcia-Solache and Casadevall 2010). In addition, the incidence of some mycoses is restricted to endemic areas. Local changes in weather patterns, land-use change, and migration can modify the distribution and composition of fungal communities by reducing particular habitats and favoring others. However, some environmental factors such as low humidity, specific nutrients, or pH could inhibit fungal growth (Deacon 2005).

5.2 *Sporothrix* Features

Some characteristics of *Sporothrix* spp. may be advantageous in global warming scenarios. This fungus is a dimorphic and saprobic ascomycete that is isolated from plants, insects, and infected mammals. According to the Index Fungorum and the Mycobank, there are approximately 100 records of *Sporothrix* species, of which only 8 species are related to clinical cases: *S. schenckii* s. str., *S. brasiliensis*, *S. globosa*, *S. luriei*, *S. mexicana*, *S. chilensis*, *S. pallida*, and *S. stenoceras* (Orofino-Costa et al. 2017).

The species related to sporotrichosis infection are transmitted predominantly by plant or soil traumatism, except for *S. brasiliensis*, linked mainly to zoonotic infections.

Many pathogenic fungi (including those belonging to the genus *Sporothrix*) have features that make them potentially threatening. Unlike bacteria and viruses, these fungi do not need a host to carry out their life cycle, survive, or spread. In addition, they can remain dormant for long periods and under extreme conditions, and some fungi can produce an infection in a broad spectrum of hosts (Casadevall 2019).

One of the most recognized pathogenic factors of *Sporothrix* spp. is dimorphism, or the ability to switch from a filamentous form to yeast in response to changes in temperature. But there are other environmental conditions such as CO_2 tension, exogenous cysteine, and estradiol that are responsible for the dimorphism reported in other thermodimorphic fungi (Boyce and Andrianopoulos 2015; Gauthier 2017). In the environment (at a temperature of 25 °C), the *Sporothrix* species grows as spore-producing filamentous fungi (Fig. 5.1), while within the host mammals (at a temperature of 35–37 °C), it acquires the yeast form and can cause an infection (Téllez et al. 2014). The mechanisms behind dimorphism in fungi are still being elucidated, but some studies have suggested the participation of heterotrimeric G proteins, calcium uptake, and the mitogen-activated protein-kinase (MAPK) signaling cascade, among others, in these mechanisms (Hou et al. 2013; Téllez et al. 2014).

Fig. 5.1 Macro and micromorphology features of *Sporothrix schenckii* strain EH-143. (**a1**) Colony on PDA (Potato Dextrose Agar), grown at 28 °C. (**a2**) Reverse of the colony. (**b**) Typical *S. schenckii* micromorphology with conidiophores arising from thin septate hyphae and conidia in clusters, 100× cotton blue

Another factor involved in the pathogenicity of this fungus is the production of melanin. This pigment protects the microorganism by reducing its susceptibility to enzymatic degradation, gamma and ultraviolet radiation, oxygen and nitrogen free radicals, heavy metal toxicity, extreme temperatures, phagocytosis, and some antifungals in the environment and within the host during infection (de S Araújo et al. 2017; Téllez et al. 2014). Other pathogenicity and virulence characteristics of *Sporothrix* are the expression of adhesins and the production of enzymes such as urease, superoxide dismutase, phospholipases, DNase, proteases, and gelatinase that degrade host tissues to obtain nutrients and protect the fungus against antimicrobial mechanisms (de S Araújo et al. 2017; Téllez et al. 2014).

Furthermore, it has been observed that this fungus can produce its siderophores in response to low environmental concentrations of iron, which could help it survive within the host (Pérez-Sánchez et al. 2010; Téllez et al. 2014). Some *Sporothrix* strains isolated from air filters exposed to hydrocarbon pollutants can assimilate volatile aromatic hydrocarbons as their sole carbon and energy source (Prenafeta-Boldu and Summerbell 2006). *S. schenkii* can survive under extreme environmental conditions, such as low temperatures, extreme osmolarity conditions, and different levels of ultraviolet radiation (Mendoza et al. 2005; Pasarell and McGinnis 1992; Téllez et al. 2014).

5.3 Epidemiology

For a long time, *S. schenkii was considered the only etiological agent of sporotrichosis.* However, phylogenetic analysis and a cladistic approach have elucidated cryptic species (*S. brasiliensis*, *S. schenckii* sensu stricto, *S. globosa,* and *S. luriei*) within *Sporothrix schenckii* (Marimon et al. 2007; Rodrigues et al. 2014a, b). Most cases of sporotrichosis are found in temperate, subtropical areas with high humidity levels, altitudes of approximately 2000 m.a.s.l., and summer rains. According to a meta-analysis, *Sporothrix spp.* have specific ecological niches within endemic areas, and they grow in soils between 6.6 and 28.84 °C, with a pH between 3.5 and 9.4, and relative humidity between 37.5 and 99.06% (Ramírez-Soto et al. 2018). They have been associated with some plants, flowers, decaying wood, and cane leaves (de Meyer et al. 2018).

An interesting phenomenon of sporotrichosis is that most cases are related to specific outbreaks. So, population genetics analyses have evidenced a particular geographical distribution for *Sporothrix* species: *S. globose* is more frequent in Asia; *S. brasiliensis,* in southeastern South America; and *S. schenckii,* in Africa, Australia, the western part of South America, and Central and North America—due to its broader range (Chakrabarti et al. 2015; Orofino-Costa et al. 2017; Zhang et al. 2015). However, these findings do not necessarily reflect the distribution of *Sporothrix* species in nature since most of them are based on studies of clinical isolates. Few studies have searched for *Sporothrix* species in soils in endemic areas (Mackinnon et al. 1969; Rodrigues et al. 2014a, b).

Recently, new clinical cases have been reported in non-endemic regions. In Inokuma et al. 2010, Inokuma et al. reported two cases in Hokkaido, Japan (a region characterized by its temperate-cold climate), where, as proposed by the authors, an increase in the incidence of sporotrichosis had been reported in the last decades, probably due to global warming.

In de Oliveira et al. 2014, de Oliveira et al. reported the first autochthonous case of human sporotrichosis by *S. globose* in an immunocompetent patient from Portugal. In the same year, the first case of human cutaneous sporotrichosis in Northern Australia was reported (Subedi et al. 2014). The reemergence of infectious diseases is linked to several factors as migration, global exchange of goods, drug resistance strains, inadequate vaccination of the population, and habitat fragmentation (NIH 2007).

5.3.1 Outbreaks

One of the first records of sporotrichosis outbreaks occurred in France in 1912, with 200 cases approximately (Zhang et al. 2015). In South Africa, in the Gauteng province, 3000 sporotrichosis cases in gold mineworkers at the Witwatersrand were reported between 1941 and 1943 (Quintal 2000). Another outbreak in 1978 in Guatemala, in an area surrounding Lake Ayarza, involved 53 patients, of which 45.3% were cases of patients who had come in contact with and handled fish (Mayorga et al. 1978).

In the United States, from 1941 to 1994, several outbreaks of sporotrichosis related to having contact with moss of the genus *Sphagnum* were reported among patients who were forest workers, botanists, and nursery workers. The largest of these outbreaks included 84 patients, and it was reported in New York in 1988 (Bravo 2012). In Australia, an outbreak of sporotrichosis associated with exposure to hay between 2000 and 2003 was also reported (Feeney et al. 2007). In China, retrospective studies have revealed large series of sporotrichosis cases (Song et al. 2013; Yu et al. 2013).

In recent years, autochthonous cases have been recorded in non-endemic areas such as the United Kingdom or certain regions in Japan (Inokuma et al. 2010; Makri et al. 2020).

5.3.2 Zoonotic Outbreaks

As already mentioned, sporotrichosis infection is caused most frequently by traumatic inoculation with contaminated soil or organic matter. However, it can also be transmitted between animals (cat-cat) or by zoonotic transmission (cat-human) from bites or scratches from infected cats (Gremião et al. 2017), other mammals (squirrels and bats) (Saravanakumar et al. 1996; Zhang et al. 2015), and arthropod bites (ants and spiders) (Miller and Keeling 2002; Moaven et al. 1999). Sporotrichosis is considered an emerging zoonosis (Bravo 2012). Its incidence has been increasing in recent decades. Southeastern Brazil is the region where the highest number of cases of feline sporotrichosis has been reported, with *S. brasiliensis* being the main causative agent. Up until 2016, 4669 cases of sporotrichosis in humans, transmitted by infected cats, had been reported in Rio de Janeiro. Outside of this country, most cases of feline sporotrichosis are caused by *S. schenkii*, and only isolated cases of transmission to humans have been reported in the United States, Mexico, Argentina, India, and Malaysia (Gremião et al. 2017).

In recent years, for the first time, cases of feline sporotrichosis have been reported in other countries. A conjunctival sporotrichosis from cat to human was reported in Thailand (Reinprayoon et al. 2020). In 2018, the first case of cutaneous sporotrichosis transmitted by a cat was reported in a 34-year-old man in Panama (Ríos et al. 2018). And in 2020, the first case of feline cutaneous sporotrichosis by *S. humicola* was reported in the United Kingdom (Makri et al. 2020).

Since a description of sporotrichosis became available, worldwide outbreaks mainly related to anthropogenic activities such as mining, agriculture, and fishing have been reported. The prevalence and incidence of sporotrichosis in endemic areas vary according to local activities. For example, in China, it is more frequent in women since they are more involved in agricultural activities, while in Mexico, it is more frequent in 0- to 15-year-old males (Toriello et al. 2021).

Recently, deep mycoses have increased due to a rise in the immunocompromised population, due to either HIV or therapeutics (immunomodulators, steroids, chemotherapy).

5.4 Clinical Presentation

Sporotrichosis is considered a neglected tropical disease (NTD) by the World Health Organization (WHO) since 2017. These diseases affect more than one billion people: primarily, populations living in tropical and subtropical climates (Malecela and Ducker 2020).

The diagnosis of this mycosis is complex since mild infections are not usually treated, epidemiological surveillance is not mandatory, and the laboratory diagnostic is inaccessible in endemic areas.

This mycosis affects the skin and lymphatic tissue and, more rarely, the lungs, meninges, and osteoarticular structures—especially in immunocompromised patients (de Lima-Barros et al. 2011; Orofino-Costa et al. 2017; Toriello et al. 2021). Clinical manifestations depend on factors such as the immunological status of the host, the inoculum amount, and the virulence of the species or strain (Arrillaga-Moncrieff et al. 2009). The most common clinical form is primary cutaneous sporotrichosis. Within this category, the lymphocutaneous, fixed cutaneous, and disseminated cutaneous sporotrichosis can be found. The lymphocutaneous form (75% of cases) most frequently affects exposed areas of the body, such as the extremities and face (de Lima-Barros et al. 2011). The fixed cutaneous form is characterized by an isolated lesion located only at the inoculation site, and it is more frequent in children (de Lima-Barros et al. 2011; Tlougan et al. 2009). The disseminated cutaneous form presents multiple lesions in noncontiguous body segments without extracutaneous manifestations. This latter form of sporotrichosis was rare and mainly associated with immunocompromised patients. However, its frequency has increased since the emergence of cases due to zoonotic transmission. Extracutaneous forms are rare. They occur more frequently in immunocompromised individuals, such as those with a history of HIV infection, alcoholism, or diabetes mellitus (de Lima-Barros et al. 2011).

5.5 Natural Disasters

It has been hypothesized that the occurrence of natural disasters such as tsunamis, floods, hurricanes, and tornadoes can displace pathogenic agents (as fungi) from their natural habitats (Benedict and Park 2014), dislocate human populations, decrease vaccination coverage, promote an inadequate infrastructure for sanitation and hygiene, and restrict access to health services (Benedict and Park 2014). In studies that review the health impact of some natural disasters, it has been observed that dermatological diseases are among those most frequently encountered during floods, along with gastrointestinal and respiratory diseases. Noninfectious and infectious dermatological diseases increased by up to 20% after the occurrence of a flood (Dayrit et al. 2018; Bandino et al. 2015).

Regarding mycoses, the infection mechanism can occur mainly through trauma or inhalation of conidia by exposition to organic matter or contaminated water. In natural disasters, these conditions are persistent, and vulnerable populations (low-income populations and older adults) have a greater chance of exposure (Benedict and Park 2014).

Due to the complexity of climate change and deep mycoses such as sporotrichosis, it is difficult to establish a direct association between them or make projections of future scenarios. However, some data suggest there is a link to possible environmental changes (climate change); for example, the emergence of sporotrichosis in non-endemic regions, the cases of sporotrichosis by low-virulence species, and the emergence of outbreaks in other mammals (feline sporotrichosis in Brazil) (Fig. 5.2).

A rise in global temperature could lead to having less restricted geographic areas of endemic mycoses, which would facilitate the interaction of environmental strains with human populations, thus leading to the emergence of a greater number of cases and new strain pathogens (de S Araújo et al. 2017).

Until now, no sporotrichosis outbreaks related to the occurrence of natural disasters have been reported. However, the evidence indicates that many fungal infections may increase with the environmental conditions caused by climate change, including some subcutaneous mycoses.

Although there is evidence that supports the link between climate change and the shifts in the epidemiology of sporotrichosis, we must consider the participation of other factors such as an increase in immunosuppressed populations, the greater availability of tools for the diagnosis, the mobilization of human and fungal

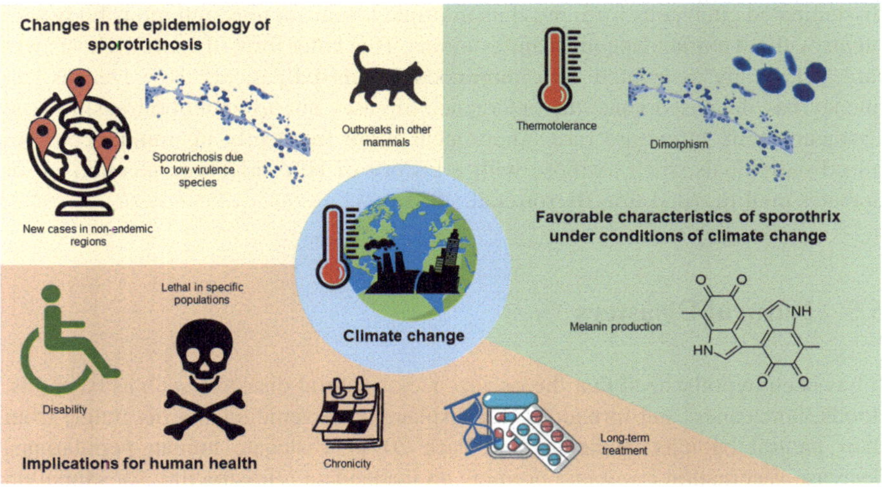

Fig. 5.2 The impact of climate change on sporotrichosis. Changes in the epidemiology of sporotrichosis, implications for human health, and the characteristics resulting from new environmental conditions

populations due to the reduction of certain habitats, and the change in land use and urbanization (de S Araújo et al. 2017). More studies are needed to determine the real impact of climate change on sporotrichosis.

5.6 Conclusion

Due to an increase in the frequency and intensity of natural disasters caused by climate change, it is necessary to remain alert for possible sporotrichosis outbreaks related to these events in the future since they represent a threat that could reach catastrophic proportions, not only for human health but also for environmental issues (phytosanitary and various animal species). Still, it has been shown that little attention has been given to mycology research.

References

Almeida F, Rodrigues ML, Coelho C. The still underestimated problem of fungal diseases worldwide. Front Microbiol. 2019;10:214. https://doi.org/10.3389/fmicb.2019.00214.

Arrillaga-Moncrieff I, Capilla J, Mayayo E, Marimon R, Mariné M, Gené J, et al. Different virulence levels of the species of Sporothrix in a murine model. Clin Microbiol Infect. 2009;15(7):651–5. https://doi.org/10.1111/j.1469-0691.2009.02824.x.

Bandino JP, Hang A, Norton SA. The infectious and noninfectious dermatological consequences of flooding: a field manual for the responding provider. Am J Clin Dermatol. 2015;16(5):399–424. https://doi.org/10.1007/s40257-015-0138-4.

Benedict K, Park BJ. Invasive fungal infections after natural disasters. Emerg Infect Dis. 2014;20(3):349. https://doi.org/10.3201/eid2003.131230.

Booth M. Climate change and the neglected tropical diseases. Adv Parasitol. 2018;100:39–126. https://doi.org/10.1016/bs.apar.2018.02.001.

Boyce KJ, Andrianopoulos A. Fungal dimorphism: the switch from hyphae to yeast is a specialized morphogenetic adaptation allowing colonization of a host. FEMS Microbiol Rev. 2015;39(6):797–811. https://doi.org/10.1093/femsre/fuv035.

Bravo TC. New observations on the epidemiology of sporotrichosis and Sporothrix schenckii complex. Sources of infection and occupational risks. Rev Mex Patol Clin Med Lab. 2012;59(2):88–100.

Casadevall A. Global catastrophic threats from the fungal kingdom. Fungal catastrophic threats. Curr Top Microbiol Immunol. 2019;424:21–32. https://doi.org/10.1007/82_2019_161.

Chakrabarti A, Bonifaz A, Gutierrez-Galhardo MC, Mochizuki T, Li S. Global epidemiology of sporotrichosis. Med Mycol J. 2015;53(1):3–14. https://doi.org/10.1093/mmy/myu062.

Dayrit JF, Bintanjoyo L, Andersen LK, Davis MD. Impact of climate change on dermatological conditions related to flooding: update from the International Society of Dermatology Climate Change Committee. Int J Dermatol. 2018;57(8):901–10. https://doi.org/10.1111/ijd.13901.

de Lima-Barros MB, Schubach AO, de Vasconcellos Carvalhaes de Oliveira R, Martins EB, Teixeira JL, Wanke B. Treatment of cutaneous sporotrichosis with itraconazole—study of 645 patients. Clin Infect Dis. 2011;5(12):e200–6. https://doi.org/10.1093/cid/cir245.

de Meyer EM, de Beer ZW, Summerbell RC, Moharram AM, de Hoog GS. Taxonomy and phylogeny of new wood- and soil-inhabiting Sporothrix species in the Ophiostoma stenoceras-Sporothrix schenckii complex. Mycologia. 2018;100:647–61. https://doi.org/10.3852/07-157R.

de Oliveira MM, Veríssimo C, Sabino R, Aranha J, Zancopé-Oliveira RM, Sampaio P, et al. First autochthone case of sporotrichosis by Sporothrix globosa in Portugal. Diagn Microbiol Infect Dis. 2014;78(4):388–90. https://doi.org/10.1016/j.diagmicrobio.2013.08.023.

de S Araújo GR, Souza WD, Frases S. The hidden pathogenic potential of environmental fungi. Future Microbiol. 2017;12(16):1533–40. https://doi.org/10.2217/fmb-2017-0124.

Deacon JW. Fungal biology. 4th ed. Oxford: Blackwell Publishing Ltd.; 2005.

EPA Environmental Protection Agency. 2007. https://www.epa.gov/climate-indicators . Accessed 20 May 2021.

Feeney KT, Arthur IH, Whittle AJ, Altman SA, Speers DJ. Outbreak of sporotrichosis. Western Australia Emerg Infect Dis. 2007;13(8):1228. https://doi.org/10.3201/eid1308.061462.

Fisher MC, Hawkins NJ, Sanglard D, Gurr SJ. Worldwide emergence of resistance to antifungal drugs challenges human health and food security. Science. 2018;360(6390):739–42. https://doi.org/10.1126/science.aap7999.

Garcia-Solache MA, Casadevall A. Global warming will bring new fungal diseases for mammals. mBio. 2010;1(1) https://doi.org/10.1128/mBio.00061-10.

Gauthier GM. Fungal dimorphism and virulence: molecular mechanisms for temperature adaptation, immune evasion, and in vivo survival. Mediat Inflamm. 2017;2017:8491383. https://doi.org/10.1155/2017/8491383.

Gremião ID, Miranda LH, Reis EG, Rodrigues AM, Pereira SA. Zoonotic epidemic of sporotrichosis: cat to human transmission. PLoS Pathog. 2017;13(1):e1006077. https://doi.org/10.1371/journal.ppat.1006077.

Hou B, Zhang Z, Zheng F, Liu X. Molecular cloning, characterization and differential expression of DRK1 in Sporothrix schenckii. Int J Mol Med. 2013;31(1):99–104. https://doi.org/10.3892/ijmm.2012.1193.

Inokuma D, Shibaki A, Shimizu H. Two cases of cutaneous sporotrichosis in continental/microthermal climate zone: global warming alert? Clin Exp Dermatol. 2010;35(6):668–9. https://doi.org/10.1111/j.1365-2230.2010.03795.x.

IPCC Intergovernmental Panel on Climate Change. Climate Change 2014 Impacts, Adaptation, and Vulnerability: Working Group II Contribution to the Fifth Assessment Report of the Intergovernmental Panel on Climate Change [Internet]. Cambridge: Cambridge University Press; 2014. [cited 2020 Oct 11]. http://ebooks.cambridge.org/ref/id/CBO9781107415379

Köhler JR, Casadevall A, Perfect J. The spectrum of Fungi that infects humans. Cold Spring Harb Perspect Med. 2015;5:a019273. https://doi.org/10.1101/cshperspect.a019273.

Mackinnon JE, Conti-Díaz IA, Gezuele E, Civila E, Silva DL. Isolation of Sporothrix schenckii from nature and considerations on its pathogenicity and ecology. Med Mycol. 1969;7(1):38–45. https://doi.org/10.1080/00362177085190071.

Makri N, Paterson GK, Gregge F, Urquhart C, Nuttall T. First case report of cutaneous sporotrichosis (Sporothrix species) in a cat in the UK. JFMS Open Rep. 2020;6(1):2055116920906001. https://doi.org/10.1177/2055116920906001.

Malecela MN, Ducker C. A road map for neglected tropical diseases 2021–2030. In: Ending the neglect to attain the sustainable development goals. World Health Organization; 2020. https://www.who.int/neglected_diseases/Ending-the-neglect-to-attain-the-SDGs%2D%2DNTD-Roadmap.pdf. Accessed 5 Sept 2020.

Marimon R, Cano J, Gené J, Sutton DA, Kawasaki M, Guarro J. Sporothrix brasiliensis, S. globosa, and S. mexicana, three new Sporothrix species of clinical interest. JCM. 2007;45(10):3198–206. https://doi.org/10.1128/JCM.00808-07.

Mayorga R, Cáceres A, Toriello C, Gutiérrez G, Alvarez O, Ramirez ME, Mariat F. Etude d'une zone d'endemie sporotrichosique au Guatemala. Med Mycol. 1978;16(3):185–98. https://doi.org/10.1080/00362177885380251.

Mendoza M, Alvarado P, de Torres ED, Lucena L, de Albornoz MC. Comportamiento fisiológico y de sensibilidad in vitro de aislamientos de Sporothrix schenckii mantenidos 18 años por dos métodos de preservación. Rev Iberoam Micol. 2005;22(3):151–6. https://doi.org/10.1016/S1130-1406(05)70029-2.

Miller SD, Keeling JH. Ant sting sporotrichosis. Cutis. 2002;69(6):439–42.

Moaven LD, Altman SA, Newnham AR. Sporotrichosis mimicking necrotising arachnidism. Med J Aust. 1999;171(11–12):685–6.

NIH National Institutes of Health (US); Biological Sciences Curriculum Study. NIH Curriculum Supplement Series [Internet]. Bethesda, MD: Understanding Emerging and Re-emerging Infectious Diseases; 2007. https://www.ncbi.nlm.nih.gov/books/NBK20370/. Accessed 25 March 2021

Orofino-Costa R, Marques MP, Rodrigues AM, Bernardes-Engermann AR. Sporotrichosis: an update on epidemiology, etiopathogenesis, laboratory and clinical therapeutics. An Bras Dermatol. 2017;92(5):606–20. https://doi.org/10.1590/abd1806-4841.2017279.

Pasarell LE, Mcginnis MR. Viability of fungal cultures maintained at-70 degrees C. J Clin Microbiol. 1992;30(4):1000–4. https://doi.org/10.1128/jcm.30.4.1000-1004.1992.

Pérez-Sánchez L, González E, Colón-Lorenzo EE, González-Velázquez W, González-Méndez R, Rodríguez-del VN. Interaction of the heterotrimeric G protein alpha subunit SSG-1 of Sporothrix schenckii with proteins related to stress response and fungal pathogenicity using a yeast two-hybrid assay. BMC Microbiol. 2010;10(1):1–7. https://doi.org/10.1186/1471-2180-10-317.

Prenafeta-Boldu FX, Summerbell R, Sybren de Hoog G. Fungi growing on aromatic hydrocarbons: biotechnology's unexpected encounter with biohazard? FEMS Microbiol Rev. 2006;30(1):109–30. https://doi.org/10.1111/j.1574-6976.2005.00007.x.

Quintal D. Sporotrichosis infection on mines of the Witwatersrand. J Cutan Med Surg. 2000;4(1):51–4. https://doi.org/10.1177/120347540000400113.

Ramírez-Soto MC, Aguilar-Ancori EG, Tirado-Sánchez A, Bonifaz A. Ecological determinants of sporotrichosis etiological agents. J Fungi (Basel). 2018;4:95. https://doi.org/10.3390/jof4030095.

Reinprayoon U, Jermjutitham M, Tirakunwichcha S, Banlunara W, Tulvatana W, Chindamporn A. Conjunctival sporotrichosis from cat to human: case report. Am J Ophthalmol Case Rep. 2020;1(20):100898. https://doi.org/10.1016/j.ajoc.2020.100898.

Ríos ME, Suarez J, Moreno J, Vallee J, Moreno JP. Zoonotic Sporotrichosis related to cat contact: first case report from Panama in Central America. Cureus. 2018;10(7):e2906. https://doi.org/10.7759/cureus.2906.

Rodrigues ML, Albuquerque PC. Searching for a change: the need for increased support for public health and research on fungal diseases. PLoS Negl Trop Dis. 2018;12(6):e0006479. https://doi.org/10.1371/journal.pntd.0006479.

Rodrigues AM, De Hoog G, Zhang Y, De Camargo ZP. Emerging sporotrichosis is driven by clonal and recombinant Sporothrix species. Emerg Microbes Infec. 2014a;3(5):e32. https://doi.org/10.1038/emi.2014.33.

Rodrigues AM, Bagagli E, de Camargo ZP, Gimenes BSM. Sporothrix schenckii sensu stricto isolated from soil in an Armadillo's burrow. Mycopathologia. 2014b;177:199–206. https://doi.org/10.1007/s11046-014-9734-8.

Saravanakumar PS, Eslami P, Zar FA. Lymphocutaneous sporotrichosis associated with a squirrel bite: case report and review. Clin Infect Dis. 1996;23(3):647–8. https://doi.org/10.1093/clinids/23.3.647.

Song Y, Li SS, Zhong SX, Liu YY, Yao L, Huo SS. Report of 457 sporotrichosis cases from Jilin province, Northeast China, a serious endemic region. J Eur Acad Dermatol Venereol. 2013;27(3):313–8. https://doi.org/10.1111/j.1468-3083.2011.04389.x.

Subedi S, Kidd SE, Baird RW, Coatsworth N, Ralph AP. Sporotrichosis from the northern territory of Australia. Am J Trop Med Hyg. 2014;91(6):1263–8. https://doi.org/10.4269/ajtmh.14-0456.

Téllez MD, Batista-Duharte A, Portuondo D, Quinello C, Bonne-Hernández R, Carlos IZ. Sporothrix schenckii complex biology: environment and fungal pathogenicity. Microbiology. 2014;160(11):2352–65. https://doi.org/10.1099/mic.0.081794-0.

Tlougan BE, Podjasek JO, Patel SP, Nguyen XH, Hansen RC. Neonatal sporotrichosis. Pediatr Dermatol. 2009;26(5):563–5. https://doi.org/10.1111/j.1525-1470.2009.00986.x.

Toriello C, Brunner-Mendoza C, Ruiz-Baca E, Duarte-Escalante E, Pérez-Mejía A, Reyes-Montes MR. Sporotrichosis in Mexico. Braz J Microbiol. 2021;52(1):49–62. https://doi.org/10.1007/s42770-020-00387-x.

VijayaVenkataRaman S, Iniyan S, Goic R. A review of climate change, mitigation and adaptation. Renew Sust Energ Rev. 2012;16(1):878–97. https://doi.org/10.1016/j.rser.2011.09.009.

World Health Organization and Meteorological Organization (WMO). Atlas of Health and Climate. Geneva: WHO Press; 2012. Accessed 25 March 2021

Yu X, Wan Z, Zhang Z, Li F, Li R, Liu X. Phenotypic and molecular identification of Sporothrix isolates of clinical origin in Northeast China. Mycopathologia. 2013;176:67–74. https://doi.org/10.1007/s11046-013-9668-6.

Zhai P, Pörtner HO, Roberts D, editors. Summary for policymakers. In: global warming of 1.5° C. An IPCC special report on the impacts of global warming of 1.5 °C above pre-industrial levels and related global greenhouse gas emission pathways. The Context of Strengthening the Global Response to the Threat of Climate Change, Sustainable Development, and Efforts to Eradicate Poverty. 2018:32.

Zhang Y, Hagen F, Stielow B, Rodrigues AM, Samerpitak K, Zhou X, et al. Phylogeography and evolutionary patterns in Sporothrix spanning more than 14 000 human and animal case reports. Persoonia. 2015;35:1. https://doi.org/10.3767/003158515X687416.

Chapter 6
Eumycetoma and Global Warming

Francisca Hernández-Hernández and Luis Javier Méndez-Tovar

Contents

Abbreviations

CONEVAL	Consejo Nacional de Evaluación de la Política de Desarrollo
HLA	Human leukocyte antigen
INEGI	Instituto Nacional de Estadística y Geografía

F. Hernández-Hernández (✉)
Departamento de Microbiología y Parasitología, Facultad de Medicina, Unidad de Micología, Universidad Nacional Autónoma de México, Ciudad de México, CDMX, Mexico

L. J. Méndez-Tovar
Servicio de Dermatología y Micología Médica, Hospital de Especialidades, Centro Médico Nacional Siglo XXI, Instituto Mexicano del Seguro Social, Ciudad de México, CDMX, Mexico
e-mail: ljmt@unam.mx

M. G. Frías-De-León et al. (eds.), *The Impact of Climate Change on Fungal Diseases*, Fungal Biology, https://doi.org/10.1007/978-3-030-89664-5_6

PCR Polymerase chain reaction
RFLP Restriction fragment length polymorphism

6.1 Introduction

Mycetoma is known as a chronic infection involving the skin and subcutaneous tissues with a tendency to invade the bones (Reis and Reis-Filho 2018). This disease is characterized by relatively indolent and deforming as well as swollen lesions. Mycetoma results from the traumatic implantation of organisms that inhabit the soil. The lesions are constituted by suppurating abscesses and draining sinuses with the presence of "grains," which are a conglomerate of the etiologic agents (Fig. 6.1). Causal agents are from exogenous origin, and they can be fungi (eumycetoma) or actinomycetes (actinomycetoma). Disease chronicity and severity depend on the timing of diagnosis. Different reports indicate that this disease affects mainly exposed body parts, particularly the feet and hands, but many cases are in vital

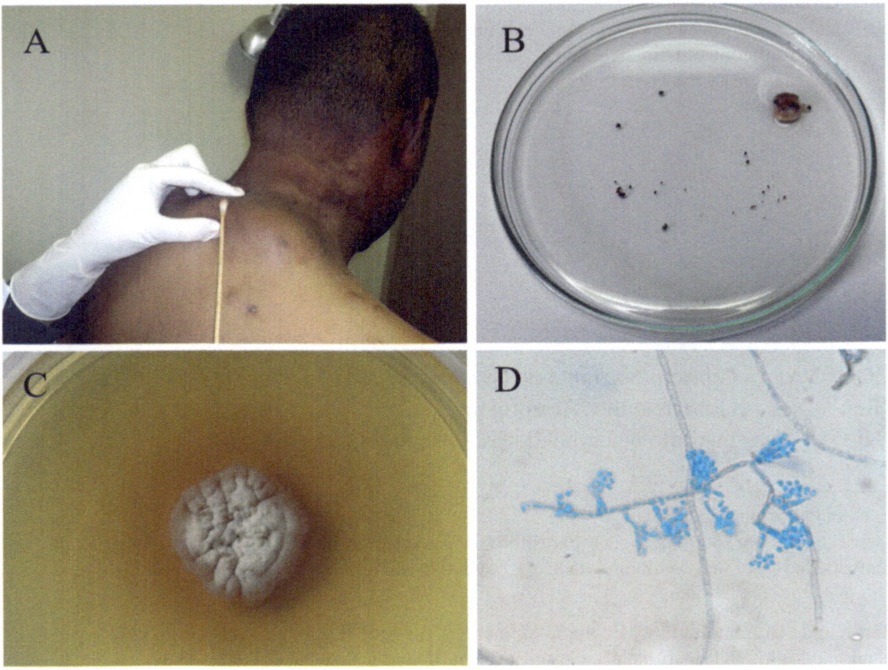

Fig. 6.1 Eumycetoma caused by *Madurella mycetomatis*. (**a**) Clinical case. (**b**) Black grains obtained from lesions. (**c**) *M. mycetomatis* grown on Sabouraud dextrose agar (2 weeks at 28 °C). (**d**) Phialides and conidia are produced in cornmeal agar (1 week at 28 °C)

zones near to central nervous system, such as the back, neck, and head. These locations increase the severity of the disease (Mahgoub et al. 1987).

Diagnosis is based on clinical data and laboratory studies for the detection of fungal structures (grains) in the infected tissue and isolation and identification of the causal agent. Phenotypic characteristics and, if possible, molecular procedures help to determine the identity of the causative agent. Treatment is different for both actinomycetoma and eumycetoma, hence the importance of establishing a timely diagnosis (Estrada-Castañón et al. 2019). The systemic antibacterial drugs are fundamental for actinomycetoma. Systemic antifungal drugs, alone or combined, are recommended for eumycetoma, and in specific cases, surgery is combined with antifungal drugs (Ameen and Arenas 2009). This chapter is focused only on eumycetoma, that is, mycetoma caused by fungi.

6.2 Geographic Distribution

Even though cases of mycetoma are reported worldwide, including countries with high economic levels, such as the United Kingdom (Hay and Mackenzie 1983) or the United States, most epidemiological studies agree that this pathology is more frequent in countries located between latitudes 15° S and 30° N. This zone is known as the "mycetoma belt" (Mariat 1963). Sudan has the greatest presence of mycetoma (Abbott 1956; Fahal et al. 2015). Other highly endemic African countries are Yibuti, Senegal (Sow et al. 2020), and Somalia (Emery and Denning 2020); India in Asia (Dubey et al. 2019); and Mexico (Bonifaz et al. 2014; Lavalle 1966; López-Martínez et al. 2013), Brazil, and Argentina in America (Emery and Denning 2020). Figure 6.2 shows the countries where eumycetoma cases are recorded.

Up to the year 2013, there were a total of 3933 mycetoma reported cases in Mexico, but only 137 were caused by fungi. Bonifaz et al. (2014) reported 482 cases, mainly distributed in states of the center of Mexico, the Gulf of Mexico, and the Pacific Ocean. The central states are temperate, while the coastal areas are tropical. Of these cases, only 38 were eumycetoma cases. Recently, we have not published the available data to know the mycetoma cases tendency over time. However, we established personal communication with collaborators responsible for mycosis diagnostic centers. Thus, the Professor and Mycologist at the Autonomous University of Sinaloa, Fernando Muñoz, informed that he attended 490 mycetoma cases during the last 20 years, from which 59 corresponded to eumycetoma (personal communication 2021). In Guerrero, México, there have been six published eumycetoma cases (five from Acapulco and one from Oaxaca) (Chávez et al. 1998). Finally, at the Civil Hospital of Acapulco in the State of Guerrero, Dr. Romero-Navarrete and Dr. Castillo Solana have diagnosed seven mycetoma cases, from which the only one was eumycetoma (personal communication 2021).

In general, the available casuistries show that eumycetomas are more numerous in African countries. For example, in Mexico, published reviews by diverse authors report a predominance of actinomycetoma, mainly caused by *N. brasiliensis*

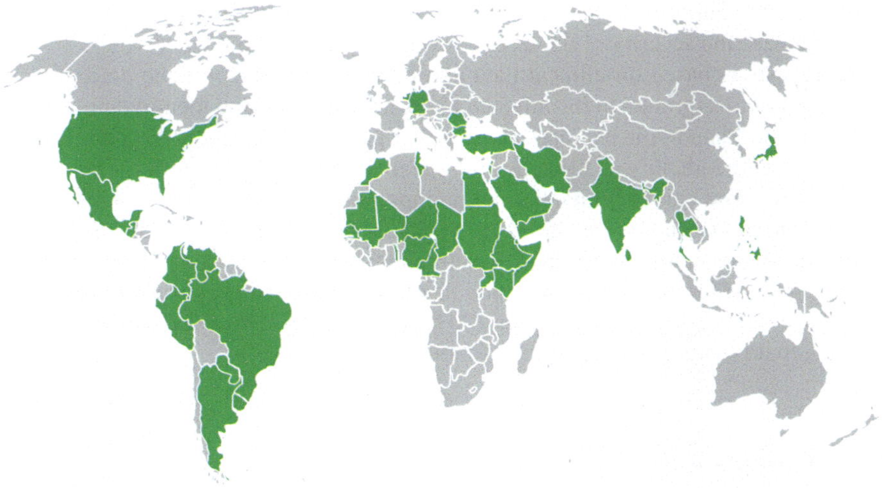

Fig. 6.2 Map showing the countries where eumycetoma cases have been published (2020)

(Bonifaz et al. 2014; Lavalle 1966; López-Martínez et al. 2013). In America, the only exception is Brazil where eumycetoma corresponds to 68% of mycetoma cases (Sampaio et al. 2017). By contrast, in Sudan, out of 6792 cases reviewed between 1991 and 2014, eumycetoma caused by *M. mycetomatis* represented 70% of patients (Ahmed et al. 2002).

6.3 Risk Factors

6.3.1 Socioeconomic Conditions

Mycetoma is part of the so-called neglected diseases that occur with most frequency in countries with low development as in many African (Kwizera et al. 2020) and Latin American countries (Mattei et al. 2013; Sampaio et al. 2015), where working conditions lead to injuries that can be contaminated with various microorganisms and where medical services are scarce and lack the equipment and personnel trained for rapid and brief diagnosis (Hay et al. 2019). According to the data on poverty measurement in Mexico gathered by the CONEVAL (Consejo Nacional de Evaluación de la Política de Desarrollo Social; www.coneval.org.mx), in 2018, the deficiencies in the Mexican population are as follows: 16.2% in health services; 57.3% in social security; 20.4% in feeding; 16.9% had educational backwardness; and 19.8% did not have access to basic services at home, particularly water. Based on INEGI (Instituto Nacional de Estadística y Geografía; inegi.org.mx) data from 2018, 5.5 million people work in the fields, concentrated in the states of Chiapas, Oaxaca, Guerrero, Puebla, and Veracruz; 56% are farmers and 44% support farmworkers (laborers and day laborers); 11% are women; 64% have an average age of 41.7 years, and 45.2% that support farmworkers are between 15 and 29 years old. Regarding the educational level, 50% do not have an academic education and only half of those who do finished primary school. In addition, most of the farmers who use rudimentary tools to work are exposed to frequent trauma. This national view explains the high incidence of mycetoma cases in Mexico.

6.3.2 Genetic Factors

Despite all the farmers who are exposed to traumatic wounds contaminated with soil or vegetable detritus, only a small proportion of them develop the disease. In Sudan, 62% of mycetoma patients have relatives with the same pathology. Regarding histocompatibility antigens, Al Dawi et al. (2013) analyzed human leukocyte antigen (HLA)-DRB1 and HLA-DQB1 allele frequencies among confirmed eumycetoma patients compared with matched controls. The HLA-DRB1 and HLA-DRB113 alleles showed significant association with eumycetoma infection ($P = 0.044$, odds ratio [OR] = 2.629).

Interestingly, the HLA-DRB102 allele had a high frequency in the control group (9.8%, $P = 0.047$), while it was absent in the eumycetoma patients, suggesting a protective role against eumycetoma. For HLA-DQB1 alleles, the HLA-DQB15 allele showed a significantly higher frequency in mycetoma patients than in healthy controls ($P = 0.029$, OR = 3.471), indicating a possible association between this allele and the development of clinical mycetoma. The authors concluded that it is necessary to increase the number of patients and controls to obtain conclusive results.

6.3.3 Immune Factors

Several studies have shown that the protective response in mycosis is cellular type (Th1). In a recent serum study, Th1- and Th2-response cytokines in two *M. mycetomatis* mycetoma groups were compared. Group 1 was treated with antifungal drugs; group 2 was treated with surgery. Those patients treated with antifungal drugs had a Th2 cytokine profile, and group 2 had a Th1 profile with a better therapeutic response (Nasr et al. 2016).

6.3.4 Hormonal Influence

Several epidemiological and clinical observations have suggested a possible relationship between mycetoma and hormonal status. Mycetoma is more frequent in men than women in a 3:1 ratio (variable depending on the endemic country). This ratio is maintained even though women from rural areas are exposed to greater injuries than women from urban areas. Some children with mycetoma develop a more severe mycetoma as they enter adolescence. This phenomenon has also been observed in pregnant women, whose severity of mycetoma decreases after delivery.

Some experimental studies have shown that human sex hormones modify the in vitro growth of some mycetoma agents. Méndez-Tovar et al. (1991) reported that progesterone reduces the in vitro growth of *Madurella mycetomatis* and *Pyrenochaeta romeroi*. The exact mechanisms of these effects are unknown, but in other pathogenic fungi, there is additional information. For example, *Trichophyton mentagrophytes* can metabolize progesterone in secondary products (Clemons et al. 1988). In *Paracoccidioides brasiliensis*, 17-β-estradiol inhibits the in vitro mycelium- or conidia-yeast transformation (Shankar et al. 2011). In *Candida albicans*, a corticosteroid-binding protein has been characterized. *Phialophora verrucosa* has a receptor with high affinity to progesterone (Hernández-Hernández et al. 1995). *Madurella grisea* (currently *T. grisea*), an important eumycetoma agent, was demonstrated to have a gene encoding for the corticosteroid-binding protein (Hernández-Hernandez et al. 1999). Therefore, more studies are necessary to show the direct effect of human sex hormones on eumycetoma agents.

6.4 Etiological Agents

Up to January 2021, based on a review of the PubMed database, there were about 21 genera of registered fungi with the ability to cause mycetoma. These include hyaline or pigmented fungi (Table 6.1). Most of these agents grow in standard laboratory conditions, and they have been reported in mycetoma highly endemic countries, indicating that the fungi find adequate climatic and environmental conditions for their growth. In recent years, new fungal species causing eumycetoma have been reported, particularly due to the greater interest of clinical mycologists and in using molecular tools to identify the etiological agents (Aguilar-Donis et al. 2011; Ahmed et al. 2014; Campos-Macías et al. 2013; Iriart et al. 2011; Mostert et al. 2005; van de Sande 2013, 2014) (Fig. 6.3). Some studies of the isolation of fungi from environmental sources are mentioned in the next paragraphs.

Table 6.1 Etiological agents of eumycetoma recorded in the literature (Medline, 1988–2020)

Genus	Species	References
Acremonium	*A. kiliense*, *A. falciforme* *A. recifei*, *A. blockii*	Dubey et al. (2019)
Aspergillus	*A. nidulans*, *A. flavus*, *A. terreus*	Dubey et al. (2019)
Alternaria	*A. alternata*	Dubey et al. (2019)
Arthrographis	*A. kalrae*	Chen-Guan Ing et al. (2015)
Candida	*C. albicans*, *C. tropicalis*	Bassiri-Jahromi (2014)
Cylindrocarpon	*C. destructans* *Cylindrocarpon* sp. *C. lichenicola*	Chazan et al. (2004), Hemashettar et al. (2000), Zoutman and Sigler (1991)
Chaetomium	*C. atrobrunneum*	Mhmoud et al. (2019)
Cladophialophora	*C. bantiana*	Gniadek et al. (2019)
Cochliobolus	*C. spicifer*	Rippon (1988)
Corynespora	*C. cassicola*	Rippon (1988)
Curvularia	*C. lunata*, *C. geniculata*	Desnos-Olivier et al. (2006), Rippon (1988)
Diaporthe	*D. phaseolorum*	Iriart et al. (2011), Mattei et al. (2013)
Exophiala	*E. jeanselmei*, *E. oligosperma*	Desnos-Olivier et al. (2006)
Emarellia	*E. grisea*, *E. paragrisea*	Borman et al. (2016)
Exserohilum (*Dreschlera*)	*E. rostrata*	Rippon (1988)
Fusarium	*F. solani*, *F. oxysporum* *F. subglutinans*, *F. verticillioides*, *F. incarnatum*	Bonifaz et al. (2017), Campos-Macías et al. (2013), Dubey et al. (2019), Rippon (1988), Yera et al. (2003)
Fonsecaea	*F. pedrosoi*	Bassiri-Jahromi (2014)
Lasiodiplodia	*L. theobromae*	Dubey (2019)
Leptosphaeria (*Falciformispora*)	*L. senegalensis*, *L. tompkinsii* (*F. senegalensis*, *F. tompkinsii*)	Desnos-Olivier et al. (2006) Ahmed et al. (2014)
Madurella	*M. mycetomatis*, *M. grisea* *M. pseudomycetomatis* *M. fahalii*, *M. tropicana*	Rippon (1988), Desnos-Olivier et al. (2006), Yan et al. (2010) de Hoog et al. (2012), van de Sande (2013)
Medicopsis (*Pyrenochaeta*)	*M. romeroi*	Desnos-Olivier et al. (2006), Dubey (2019)
Microsporum	*M. canis*	Rippon (1988), Ruiz Barbosa et al. (2020)
Neotestudina	*N. rosati*	Rippon (1988)
Nigrograna	*N. mackinnonii*	Ahmed et al. (2018)
Paecilomyces	*P. lilacinus*	Bassiri-Jahromi (2014)

(continued)

Table 6.1 (continued)

Genus	Species	References
Phaeoacremonium	*P. krajdenii*, *P. parasiticum* *P. sphrinctrophorum*	Aguilar-Donis et al. (2011), Belkin et al. (2020), Hemashettar et al. (2006)
Phanerochaete	*P. chrysosporium*	Dubey et al. (2019)
Phomopsis	*P. longicolla*	Estrada-Castañón et al. (2019)
Plenodomus	*P. avramii*	Rippon (1988)
Pleurostomophora	*P. ochracea*	Mhmoud et al. (2012)
Pseudochaetosphaeronema	*P. larense*	Rippon (1988)
Rhytidhysteron	*R. rufulum*	Emery and Denning (2020)
Sarocladium	*S. kiliense*	Dubey et al. (2019)
Scedosporium (*Pseudallescheria*)	*S. apiospermum* (*S. boydii*)	Douglas and Bigelow (1930), Cortez et al. (2008), Persaud and Holroyd (1968)
Sphaerulina	*S. rhododendricola*	Ahmed et al. (2020)
Trematosphaeria (*Madurella*)	*T. grisea* (*M. grisea*)	Ahmed et al. (2014)

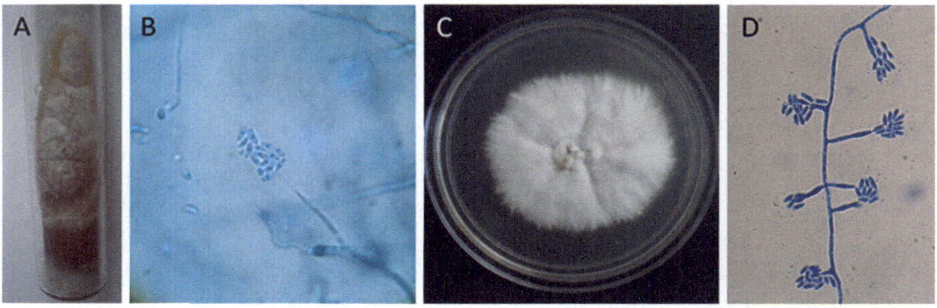

Fig. 6.3 Macro- and microscopic morphology of rare eumycetoma agents grown on Sabouraud dextrose agar. (**a**) and (**b**), *Phaeoacremonium parasiticum*. (**c**) and (**d**), *Fusarium subglutinans*

6.5 Ecological Conditions

In geographical regions where mycetoma is present, the weather and ecological conditions are highly variable. Mexico has all kinds of weather (ecotone). In the review published by López-Martínez et al. (2013), among the 32 states of the Mexican Republic, there are registered cases of mycetoma in 24 states. In 2019, Peña-Corona-Villanueva (2019) found 36 cases of mycetoma treated at a specialty hospital (Instituto Mexicano del Seguro Social) between 1993 and 2016. Regarding the mapping of mycetoma in Mexico, the states of Sonora, Chihuahua, and Baja California Sur were added but not previously reported. In this work, one case of *Madurella mycetomatis* mycetoma was found; the agent is only reported in states of Mexico with more wet weather.

In African countries with a high frequency of mycetoma, climate variability is also evident. Uganda is crossed by a mountain range 1000 to 1500 m above sea level, which has temperatures ranging from 20 to 25 °C, and has an 88.8% prevalence of eumycetomas. Kwizera et al. (2020) analyzed on the frequency of the pathology in each decade, showing that in the period from 1970 to 1979, there were more cases of mycetoma.

Sudan has the oldest and most numerous cases of this condition in the world (Lynch 1964). In Sudan, mycetoma is a public health problem, and therefore, it has specialized hospitals for the care of this pathology. The country has a strip of humid heat on the shores of the Red Sea, with average temperatures of 26–28 °C. In this strip, the rains occur between October and January, and the rest of the country is desert. Toward the center of the country, the climate is desertic, and near the border with Egypt, it shares the Sahara Desert, which has the highest temperatures and sunniest land in the world. In 2015, Fahal et al. published a casuistry of 6792 cases; the epidemiological characteristics of age, gender, and occupation were like other reports. Regarding the type of mycetoma, 4754 cases (70%) were eumycetoma, with a clear predominance of *M. mycetomatis* as an etiological agent.

Senegal is another endemic area of mycetoma. There are two very marked types of climates during the year: dry, from the end of May to October, and abundant rains, from the end of October to mid-May. Humidity is much more abundant in the south, where due to the African monsoon, the territory receives up to 1500 mm of water per year. There are desert areas in the northern part of the country with dunes and sandstorms and completely arid soils. In this country, Sow et al. (2020) published a series of 193 mycetoma cases, from which 47.2% corresponded to eumycetoma, 36.8% actinomycetoma, and 16.1% not determined.

Countries close to the African continent, such as Israel, also have reports of mycetoma cases. Regional conditions in this country vary considerably, with humid summers and mild winters on the coast, dry summers and moderately cold winters in mountainous areas, very hot summers and pleasant winters in the Jordan Valley, and semidesert conditions throughout the year in the Negev. Bitan et al. (2017) published a series of mycetoma cases diagnosed between 1942 and 2015; 21 patients were registered, of which 7 were eumycetomas, 13 were actinomycetoma, and the agent was not determined in 1 case.

A few dozen cases of mycetoma have been reported in Brazil. In Rio de Janeiro, Sampaio et al. (2015) reported 21 cases diagnosed between 1991 and 2014: 61.9% were eumycetoma, and 38.1% were actinomycetoma. Among the fungi identified, *Scedosporium apiospermum* and *M. mycetomatis* predominated, and *Nocardia* spp. were the most frequent actinomycetes. In Brazil, the climate is of humid tropical type; temperatures range between 23 and 27 °C with high humidity caused by rain volumes of between 1000 and 1500 mm per year and occasional monsoon-type storms.

6.6 Relationship Between Mycetoma and Environmental Etiological Agents

Throughout the history of mycetoma, various studies have been carried out to identify the relationship between the disease and the patient's exposure to the causative agents in their work environment. In the next section, we will refer to some studies about isolation and identification of environmental fungi that cause mycetoma in highly endemic areas to show a direct relationship between the disease and the presence of the fungus in the patient's workplace or habitat.

In 1968, Segretain and Mariat recovered *Leptosphaeria senegalensis* (currently *Falciformispora senegalensis*) from thorns of *Acacia* species in West and Central sub-Saharan Africa.

In 2002, given that Sudan is mycetoma highly endemic area, Ahmed et al. conducted a study to isolate *M. mycetomatis* from environmental sources. They processed a total of 74 soil and thorn samples by culture, followed by PCR (polymerase chain reaction) and RFLP (restriction fragment length polymorphism). Further authors isolated DNA directly from soil and thorns, followed by PCR-RFLP. Isolates obtained by culture did not correspond to *M. mycetomatis*. By molecular methods, only 17 (23%) out of 74 soil samples, and 5% out of 22 thorn samples, were positive for *M. mycetomatis*.

Phaeoacremonium spp., infrequent fungi cause of mycetoma, have been found as environmental reservoirs in infected woody plants. To our knowledge, the species associated with eumycetoma are *P. parasiticum*, *P. sphinctrophorum*, and *P. krajdenii* (Hemashettar et al. 2006).

Sybren de Hoog (2013) performed a phylogenetic study of several clinical and environmental *Chaetomium* and *Madurella* isolates to predict the natural habitat of *Madurella mycetomatis*, the main fungus causing eumycetoma in several countries. The authors found that *M. mycetomatis* belongs to the family *Chaetomiaceae*, often found in dung and manure enriched soil.

Several *M. mycetomatis* isolation attempts from environmental sources have been performed without success. In the high eumycetoma endemic area in Sudan, despite several attempts, *M. mycetomatis* could not be cultivated from the environment. The difficulty in recovering *M. mycetomatis* from the soil might indicate that pure soil is not the natural habitat for this fungus. Therefore, association with cattle dung seems to be an alternative option. The Gezira region in Sudan is highly endemic for eumycetoma by *M. mycetomatis*. Most of its inhabitants live on cattle and camel husbandry and agriculture. Cows are kept in pens surrounded by walls made of mud or thorny bushes. The floors of the pens are paved with dry faces, thorns, and trash. The family house is usually in direct contact with the pen. Inhabitants of the villages are mostly barefoot among the thorny bushes. If cow dung is an essential factor in inoculation by *M. mycetomatis*, preventive measures may involve the use of appropriate footwear in addition to restructuring the villages with stricter separation of animal husbandry and human settlement to reduce the frequency of contact with mycetoma etiological agents (Parthasarathi et al. 2007).

In Mexico, Espinosa-Texis et al. (2017) aimed to isolate and identify fungi and actinomycetes causing subcutaneous infections, including eumycetoma, from soil and plants in the State of Puebla. From 281 fungal isolates, the authors found *M. mycetomatis* in 7.6% and 4.7% from soil and plants samples, respectively. Soil characteristics associated with mycetoma agents were regezol (a very weakly developed mineral soil in unconsolidated materials) and vertisol (the most productive soils in Mexico and the world, because of their high natural fertility, which is the product of cation exchange capacity and high moisture retention) type soils. Plants related to mycetoma agents were *Jacaranda mimosifolia*, *Prosopis* sp., *Pithecellobium* sp., *Saccharum officinarum*, *Festiuca tolusensis*, *Baccharis conferta*, *Simsia amplexicaulis*, *Zea mays*, *Agave* sp., *Rosa centifolia*, *Schinus mole*, and *Opuntia* sp.

6.7 Repercussions of Climate Change on the Frequency and Geographic Distribution of Eumycetoma

Mycetoma is part of the neglected diseases group, i.e., parasitic, viral, or fungal diseases that have been present for hundreds or thousands of years in regions of greatest poverty. In Mexico, as in many other countries, poverty has increased significantly due to various factors. This has caused fewer economic resources for healthcare centers, including medical personnel, personnel trained to perform laboratory and cabinet diagnosis, and medical and surgical treatment of this syndrome.

Mycetoma is undoubtedly closely linked to poverty and its consequences: hygienic deficiency, prone to infection, malnutrition, poor immune response, lack of protective footwear, exposure to injuries, and finally, the development of mycetoma. Poverty also implies a lack of availability of tools and machinery to carry out farm work.

Concerning global warming and according to the analysis of the main international data compiled by the World Meteorological Organization, 2019 was the second warmest year on record after 2016.

In 2019, the average global temperature was 1.1 °C higher than the average in the period 1850–1900, corresponding to preindustrial conditions. In this way, 2016 was the warmest year due to a very fort "El Niño" event, inducing an unusual increase in temperature and the climatic change for a long time. According to Petteri Taalas (Secretary-General of Meteorology), with our actual carbon dioxide emission, the temperature will increase by 3–5 °C by the end of the century. The most worrying news is that the trend will continue to increase due to the emission of greenhouse gases into the atmosphere, endangering the life of the planet. These combined events have significantly impacted both the health and well-being of humans and the environment.

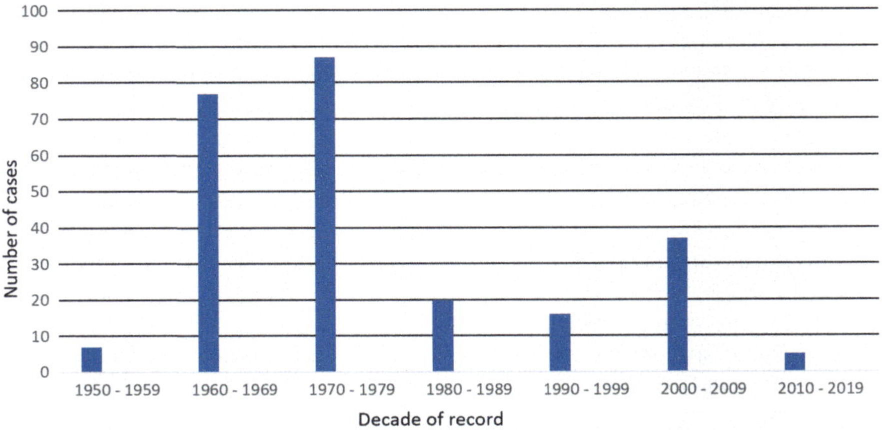

Fig. 6.4 Cases of mycetoma registered in Uganda in the period 1950–2019. Data from Kwizera R et al. (2020)

This warming has undoubtedly caused great changes and catastrophes on the planet, such as the increase in hurricanes, melting of glaciers, and an increase in desertification. However, this temperature variability has not yet shown an impact on the frequency of the eumycetoma because their agents could be isolated in geographical regions with high variability of temperatures and humidity. It is true that epidemiological reports indicate an increase in some fungal infections, but regarding eumycetoma, this augmentation could be attributed to factors inherent to the host, a greater interest in their identification, and greater availability of diagnostic tools, especially in developed countries. In recent years, molecular tools have contributed to the rapid and accurate identification of eumycetoma agents, including new agents. It is now possible to detect fungal DNA or RNA directly from the infected tissues of patients, with a low number of grains, allowing the diagnosis to be established even in the absence of cultures.

A review of the cases of mycetoma in countries with a high and low frequency of mycetoma shows that cases can occur in all climates and temperature conditions. We consider that the global warming reported in the last five decades has not had a direct impact on the number of mycetoma cases in the world; at least, there is no evidence in the reports on mycetoma frequency.

In Uganda, there are decades of data on mycetoma, with its incidence decreasing in recent years (Fig. 6.4). However, evolutionary changes are slow but progressive, and it is likely that in the coming years, we will see different behavior in the incidence of fungal infections, including mycetoma, associated with global warming.

6.8 Conclusion

There is no strong evidence that global warming has directly increased the number of reported cases of eumycetoma. However, the increase of natural disasters associated with global warming, which increases poverty in many countries of the world, is and will increase the number of reported cases of eumycetoma because of the higher incidence of this fungal infection.

References

Abbott P. Mycetoma in the Sudan. Trans R Soc Trop Med Hyg. 1956;50(1):11–24. https://doi.org/10.1016/0035-9203(56)90004-9.

Aguilar-Donis A, Torres-Guerrero E, Arenas-Guzmán R, Hernández-Hernández F, López-García L, Criales-Vera S, et al. Mycetoma caused by Phaeoacremonium parasiticum –a case confirmed by B-tubulin sequence analysis. Mycoses. 2011;54(5):e615–e8. https://doi.org/10.1111/j.1439-0507.2010.01929.x.

Ahmed SA, Adelmann D, Fahal AH, Verbrugh H, van Belkum A, de Hoog S. Environmental occurrence of Madurella mycetomatis, the major agent of human eumycetoma in Sudan. J Clin Microbiol. 2002;40(3):1031–6. https://doi.org/10.1128/JCM.40.3.1031-1036.2002.

Ahmed SA, van de Sande WW, Stevens DA, Fahal AH, van Diepeningen AD, Menken SB, et al. Revision of agents of black-grain eumycetoma in the order Pleosporales. Persoonia. 2014;33:141–54. https://doi.org/10.3767/003158514X684744.

Al Dawi AF, Mustafa MI, Fahal AH, El Hassan AM, Bakhiet S, Magoub ES, et al. The association of HLA-DRB1 and HLA-DQB1 alleles and the occurrence of eumycetoma in Sudanese patients. Khartoum Med J. 2013;6(3):923–9.

Ameen M, Arenas R. Developments in the management of mycetomas. Clin Exp Dermatol. 2009;34(1):1–7. https://doi.org/10.1111/j.1365-2230.2008.03028.x.

Bitan O, Wiener-Well Y, Segal R, Schwartz E. Mycetoma (Madura foot) in Israel: recent cases and a systematic review of the literature. Am J Trop Med Hyg. 2017;96(6):1355–61. https://doi.org/10.4269/ajtmh.16-0710.

Bonifaz A, Tirado-Sánchez A, Calderón L, Saúl A, Araiza J, Hernández M, et al. Mycetoma: experience of 482 cases in a single center in Mexico. PLoS Negl Trop Dis. 2014;8(8):e3102. https://doi.org/10.1371/journal.pntd.0003102.

Campos-Macías P, Arenas-Guzmán R, Hernández-Hernández F. Fusarium subglutinans: a new eumycetoma agent. Med Mycol Case Rep. 2013;2:128–31. https://doi.org/10.1016/j.mmcr.2013.06.004.

Chávez G, Arenas R, Pérez-Polito A, Torres B, Estrada R. Micetomas eumicéticos por Madurella mycetomatis. Informe de seis casos. Rev Iberoam Micol. 1998;15(2):90–3.

Clemons KV, Schär G, Stover EP, Feldman D, Stevens DA. Dermatophyte-hormone relationships: characterization of progesterone-binding specificity and growth inhibition in the genera Trichophyton and Microsporum. J Clin Microbiol. 1988;26(10):2110–5.

Dubey N, Capoor MR, Hasan AS, Gupta A, Ramesh V, Sharma S, et al. Epidemiological profile and spectrum of neglected tropical disease eumycetoma from Delhi, North India. Epidemiol Infect. 2019;147:e294. https://doi.org/10.1017/S0950268819001766.

Emery D, Denning DW. The global distribution of actinomycetoma and eumycetoma. PLoS Negl Trop Dis. 2020;14(9):e0008397. https://doi.org/10.1371/journal.pntd.0008397.

Espinosa-Texis AP, Castillo-Hernández D, Diaz-Hernández M, Hernández-Hernández F. Presencia de agentes potenciales causantes de infecciones subcutáneas humanas en suelo y plantas del estado de Puebla, México. Terra Latinoam. 2017;35(2):113–22.

Estrada-Castañón R, Estrada-Chávez G, Chávez-López MG. Diagnosis and management of fungal neglected tropical diseases in community settings—mycetoma and sporotrichosis. Trop Med Infect Dis. 2019;4(2):81. https://doi.org/10.3390/tropicalmed4020081.

Fahal AH, Mahgoub ES, Ahmed M, Hassan EL, Abdel-Rahman ME. Mycetoma in the Sudan: an update from the Mycetoma research Centre, University of Khartoum, Sudan. PLoS Negl Trop Dis. 2015;9(3):e0003679. https://doi.org/10.1371/journal.pntd.0003679.

Hay RJ, Mackenzie DW. Mycetoma (madura foot) in the United Kingdom – a survey of forty-four cases. Clin Exp Dermatol. 1983;8(5):553–62. https://doi.org/10.1111/j.1365-2230.1983.tb01823.x.

Hay R, Denning DW, Bonifaz A, Queiroz-Telles F, Beer K, Bustamante B, et al. The diagnosis of fungal neglected tropical diseases (fungal NTDs) and the role of investigation and laboratory tests: an expert consensus report. Trop Med Infect Dis. 2019;4(4):122. https://doi.org/10.3390/tropicalmed4040122.

Hemashettar BM, Siddaramappa B, Munjunathaswamy BS, Pangi AS, Pattan J, Andrade AT, et al. Phaeoacremonium krajdenii, a cause of white grain eumycetoma. J Clin Microbiol. 2006;44(12):4619–22. https://doi.org/10.1128/JCM.01019-06.

Hernández-Hernández F, de Bièvre C, Camacho-Arroyo I, Cerbón MA, Dupont B, López-Martínez R. Sex hormone effects on Phialophora verrucosa in vitro and characterization of progesterone receptors. J Med Vet Mycol. 1995;33(4):235–9.

Hernández-Hernandez F, López-Martínez R, Camacho-Arroyo I, Mendoza-Rodríguez CA, Cerbón MA. Detection and expression of corticosteroid binding protein gene in human pathogenic fungi. Mycopathologia. 1999;143(3):127–30. https://doi.org/10.1023/a:1006997029459.

Iriart X, Binois R, Fior A, Blanchet D, Berry A, Cassaing S, et al. Eumycetoma caused by Diaporthe phaseolorum (Phomopsis phaseoli): a case report and a mini-review of Diaporthe/Phomopsis spp invasive infections in humans. Clin Microbiol Infect. 2011;17(10):1492–4. https://doi.org/10.1111/j.1469-0691.2011.03568.x.

Kwizera R, Bongomin F, Meya DB, Denning DW, Fahal AH, Lukande R. Mycetoma in Uganda: a neglected tropical disease. PLoS Negl Trop Dis. 2020;14(4):e0008240. https://doi.org/10.1371/journal.pntd.0008240.

Lavalle P. Nuevos datos sobre la etiología del micetoma en México y sobre su patogenia. Gac Med Mex. 1966;96(6):545–74.

López-Martínez R, Méndez-Tovar LJ, Bonifaz A, Arenas R, Mayorga J, Welsh O, et al. Actualización de la epidemiología del micetoma en México. Revisión de 3,933 casos. Gac Med Mex. 2013;149(5):586–92.

Lynch JB. Mycetoma in the Sudan. Ann R Coll Surg Engl. 1964;35(6):319–40.

Mahgoub ES, Mariat F, Segretain G, Berrueta S. Memorias. II Simposio Internacional del Micetoma. Taxco, Guerrero: México; 1987.

Mariat F. Sur la distribution géographique et la repartition des agents de mycétomes. Bull Soc Path Exot. 1963;56:33–55.

Mattei AS, Severo CB, Guazzelli LS, Oliveira FM, Gené J, Guarro J, et al. Cutaneous infection by Diaporthe phaseolorum in Brazil. Med Mycol Case Rep. 2013;2:85–7. https://doi.org/10.1016/j.mmcr.2013.03.001.

Méndez-Tovar LJ, de Bièvre C, López-Martínez R. Effets des hormones sexuelles humaines sur le développement in vitro des agents d'eumicétomes. J Mycol Med. 1991;118:141–3.

Mostert L, Groenewald JZ, Summerbell RC, Robert V, Sutton DA, Padhye AA, et al. Species of Phaeoacremonium associated with infections in humans and environmental reservoirs in infected woody plants. J Clin Microbiol. 2005;43(4):1752–67. https://doi.org/10.1128/JCM.43.4.1752-1767.2005.

Nasr A, Abushouk A, Hamza A, Siddig E, Fahal AH. Th-1, Th-2 cytokines profile among Madurella mycetomatis eumycetoma patients. PLoS Negl Trop Dis. 2016;10(7):e0004862. https://doi.org/10.1371/journal.pntd.0004862.

Parthasarathi K, Ranganathan LS, Anandi V, Zeyer J. Diversity of microflora in the gut and casts of tropical composting earthworms reared on different substrates. J Environ Biol. 2007;28(1):87–97.

Peña-Corona-Villanueva A. Estudio epidemiológico del micetoma en el Servicio de Dermatología y Micología Médica del Hospital de Especialidades CMN SXXI, IMSS. Tesis de Especialidad en Dermatología. CDMX: UNAM; 2019.

Reis CMS, Reis-Filho EGM. Mycetomas: an epidemiological, etiological, clinical, laboratory and therapeutic review. An Bras Dermatol. 2018;93(1):8–18. https://doi.org/10.1590/abd1806-4841.20187075.

Sampaio FM, Galhardo MC, De Farias-Cardoso R, de Oliveira-Coelho JM, Lyra RM, do Valle AC, et al. Eumycetoma on the foot caused by Madurella mycetomatis: amputation after significant worsening during pregnancy. Acta Derm Venereol. 2015;95(3):374–5. https://doi.org/10.2340/00015555-1963.

Sampaio FM, Wanke B, Freitas DF, Coelho JM, Galhardo MC, Lyra MR, et al. Review of 21 cases of mycetoma from 1991 to 2014 in Rio de Janeiro, Brazil. PLoS Negl Trop Dis. 2017;11(2):e0005301. https://doi.org/10.1371/journal.pntd.0005301.

Segretain G, Mariat F. Recherches sur la présence d'agents de mycétomes dans le sol et sur les épineux du Sénégal et de la Mauritanie. Bull Soc Pathol Exot Filiales. 1968;61(2):194–202.

Shankar J, Restrepo A, Clemons KV, Stevens DA. Hormones and the resistance of women to paracoccidioidomycosis. Clin Microbiol Rev. 2011;24(2):296–313. https://doi.org/10.1128/CMR.00062-10.

Sow D, Ndiaye M, Sarr L, Kanté MD, Ly F, Dioussé P, et al. Mycetoma epidemiology, diagnosis management, and outcome in three hospital centres in Senegal from 2008 to 2018. PLoS One. 2020;15(4):e0231871. https://doi.org/10.1371/journal.pone.0231871.

van de Sande WJ. Global burden of human mycetoma: a systematic review and meta-analysis. PLoS Negl Trop Dis. 2013;7(11):e2550. https://doi.org/10.1371/journal.pntd.0002550.

van de Sande WJ, Fahal AH, Goodfellow M, Mahgoub ES, Welsh O, Zijlstra EE. Merits and pitfalls of currently used diagnostic tools in mycetoma. PLoS Negl Trop Dis. 2014;8(7):e2918. https://doi.org/10.1371/journal.pntd.0002918.

Chapter 7
Climate Change Impact on Chromoblastomycosis

Alexandro Bonifaz, Arturo Robles-Tenorio, and Andrés Tirado-Sánchez

Contents

Abbreviations

5-FC	5-Flucytosine
CBM	Chromoblastomycosis
HLA-A29	Human leukocyte antigen A29
IL-10	Interleukin-10
KOH	Potassium hydroxide
PCR	Polymerase chain reaction

A. Bonifaz (✉)
Dermatology Department, Hospital General de México "Dr. Eduardo Liceaga", CDMX, Mexico City, Mexico

Mycology Laboratory, Hospital General de México "Dr. Eduardo Liceaga", CDMX, Mexico City, Mexico

A. Robles-Tenorio
Instituto Dermatológico de Jalisco "Dr. José Barba Rubio", Zapopan, Mexico

A. Tirado-Sánchez
Dermatology Department, Hospital General de México "Dr. Eduardo Liceaga", CDMX, Mexico City, Mexico

Clinical Medicine Department, Instituto Mexicano del Seguro Social, CDMX, Mexico City, Mexico

M. G. Frías-De-León et al. (eds.), *The Impact of Climate Change on Fungal Diseases*, Fungal Biology, https://doi.org/10.1007/978-3-030-89664-5_7

7.1 Introduction

Chromoblastomycosis (CBM) is a typical implantation mycosis caused by melanized fungi. CBM mainly occurs in tropical and subtropical regions. Worldwide, the leading etiological agents are *Fonsecaea pedrosoi* and *Cladophialophora carrionii*. The disease usually presents as chronic, nodular, or verrucous lesions affecting the skin and subcutaneous tissues of the lower extremities (Bonifaz et al. 2019; Queiroz-Telles et al. 2009, 2017).

This chapter aims to reflect on the possible effects of climate change in CBM, considering the impact of geographic mobility, deforestation, urbanization, and global warming. The taxonomy and reproduction of the main etiological agents will be discussed. The etiopathogenesis, epidemiology, clinical features, diagnosis, and therapy will also be revised.

7.2 Etiopathogenesis

CBM is caused by a spectrum of dimorphic, melanized fungi. The presence of septate, muriform cells in the host tissue represents the hallmark of infection. Most of the causative agents belong to the *Ascomycota* phylum, *Chaetothyriales* order, *Herpotrichiellaceae* family and do not show a teleomorphic stage (Brandt and Warnock 2003; Bonifaz et al. 2019; Brito and Bittencourt 2018; De Hoog et al. 2000; Queiroz-Telles 2017).

F. pedrosoi is the most widely reported species worldwide, followed by *C. carrionii*. The first predominates in tropical, warm, humid climates, while the second is more often found in warmer, semiarid regions (Bonifaz et al. 2019; de Hoog et al. 2007; González et al. 2013; Queiroz-Telles et al. 2017; Torres-Guerrero et al. 2012). Other less frequently reported agents include *F. monophora*, *F. nubica*, *Phialophora verrucosa*, *P. americana*, *P. macrospora*, *C. yegresii*, *Rhinocladiella aquaspersa*, *R. richardsiae*, *Exophiala dermatitidis*, and *E. spinifera* (Table 7.1) (Ahmed et al. 2021; Badalin et al. 2010; Brito and Bittencourt 2018; Gomes et al. 2016; Marques et al. 2004; Najafzadeh et al. 2010; Queiroz-Telles et al. 2017; Son et al. 2010; Tomson et al. 2006).

CBM develops after traumatic inoculation of pathogenic hyphae and conidia in the skin. Injuries with wood, grain stalks, grass, thorns, palms, bamboo, coconut shells, and cacti are common precipitants. Lesions have resulted from insect stings, cow stomps, rooster pecking, caterpillar contact, and leech bite, as well as from contact with agricultural tools like hoes, axes, knives, and mills (Queiroz-Telles et al. 2017).

Individuals harboring the human leukocyte antigen A29 (HLA-A29) have shown increased genetic susceptibility. Pyridinoline produced by melanized fungi is responsible for collagen cross-linking, causing important fibrosis and decreased antifungal penetration. Other virulence factors include melanin and proteolytic enzymes, which facilitate immune system evasion and fungal invasion, respectively. Interestingly, it has been demonstrated that progesterone and testosterone can

Table 7.1 Main etiological agents of chromoblastomycosis

More frequent	Less frequent	Exceptional
Fonsecaea pedrosoi	*Cladophialophora yegresii*	*Aureobasidium pullulans*
Cladophialophora carrionii	*Exophiala dermatitidis*	*Chaetomium funicola*
	Exophiala jeanselmei	*Catenulostroma chromoblastomycosum*
	Exophiala spinifera	*Cyphellophora ludoviensis*
	Fonsecaea monophora	*Rhinocladiella tropicalis*
	Fonsecaea nubica	*Rhinocladiella similis*
	Fonsecaea pugnacius	*Fonsecaea pugnacious*
	Phialophora Americana	*Cladophialophora samoensis*
	Phialophora chinensis	*Exophiala psychrophila*
	Phialophora macrospora	*Cladosporium lageronii*
	Phialophora verrucosa	*Phoma insulana*
	Rhinocladiella aquaspersa	
	Rhinocladiella richardsiae	

activate *P. verrucosa* cytosolic receptors, hindering fungal growth in vitro. However, the in vivo implications of hormonal influences on CBM remain uncertain (Ameen 2010; Bonifaz et al. 2019; López-Martínez 2007; Queiroz-Telles et al. 2011; Torres-Guerrero et al. 2012; Tsuneto et al. 1989).

7.3 Epidemiology

Melanized fungi are saprobiotic organisms found in soil, organic matter, plants, and wood. Most agents thrive in areas of high humidity, with a temperature ranging from 25 to 30 °C, average precipitation of 1500 mm per year, mainly between latitudes 30 °N and 30 °S. Nonetheless, species like *C. carrionii* and *C. yegresii* are well-adapted to semidesert conditions (Atoche et al. 2016; Bonifaz et al. 2019; González et al. 2013; de Hoog et al. 2000, 2004; Queiroz-Telles et al. 2009; Queiroz-Telles 2017).

Some CBM agents have been directly isolated from natural sources. *F. pedrosoi* and *C. carrionii* have been found in decomposing vegetation, word bark, wasp hives, and various palms. Other species have also been recovered from cacti (*C. yegresii*), plant thorns (*F. minima* and *F. erecta*), and wood pulp (*P. verrucose*) (Marques et al. 2006; Queiroz-Telles et al. 2017; Salgado et al. 2004).

Most cases occur in tropical and subtropical climates. Madagascar and Brazil are the countries with the highest prevalence worldwide. Mexico is the most affected country in North America. Costa Rica, Honduras, Dominican Republic, and Cuba also show considerably high prevalence in Central America and the Caribbean. Puerto Rico, Guatemala, Colombia, and Venezuela are affected to a lesser extent. Other countries such as India, Sri Lanka, Australia, Congo, Taiwan, and China have also reported considerable case series (Fig. 7.1) (Agarwal et al. 2017; Attapattu 1997; Bonifaz et al. 2004; Correira et al. 2010; Esterre et al. 1997; Queiroz-Telles

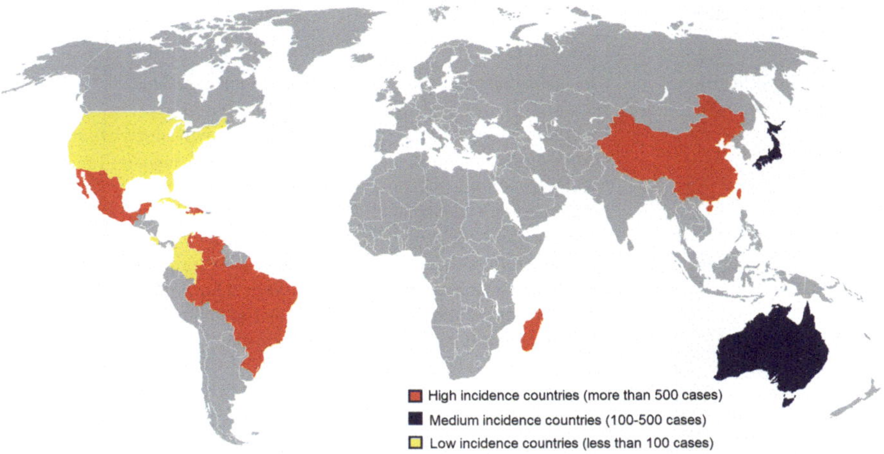

Fig. 7.1 Endemic zones of chromoblastomycosis

et al. 2017; Sharma et al. 2010; Torres-Guerrero et al. 2012; Weedon et al. 2013; Yang et al. 2013; Yang et al. 2018).

CBM is more frequent in males around the second and fourth life decade. Occupation is still a well-known risk factor in farmers, lumberjacks, and gardeners (Bonifaz et al. 2001; Brito et al. 2018; López-Martínez et al. 2007; Minotto et al. 2001; Pires et al. 2012; Queiroz-Telles et al. 2017). There is only one case series documenting CBM by *C. carrionii* in children from 2 to 19 years old in Venezuela (Pérez-Blanco et al. 2006). Race affinity has not been determined.

Since CBM is not a notifiable disease, incidence and prevalence may be underestimated. In Madagascar, 1343 cases were documented in a 40-year retrospective study, with an estimated prevalence of 1/200,000 inhabitants (Esterre et al. 1996). The figures in Venezuela are around 16 cases per 1000 inhabitants (Richard-Yegres et al. 1992; Yegres and Yegres 2005). The Brazilian Amazon is another important region where possibly many cases go unnoticed due to insufficient healthcare access (Avelar-Pires et al. 2013; Gomez et al. 2016; Londero and Ramos 1976; Minotto et al. 2001; Queiroz-Telles et al. 2009; Santos et al. 2020). In the northwest of Mexico, more than 250 cases have been reported, constituting the most affected region in North America (Bonifaz et al. 2001, 2019; Rojas-García et al. 2019; Romero-Navarrete et al. 2014).

7.4 Climate Change Effects of Chromoblastomycosis

Since fungal diseases are often regional, there is no doubt that climate change will have an impact on the expansion, delimitation, and possibly change of the main etiological agents of CBM. Therefore, it is crucial to study fungal habitats and vectors to predict these inevitable changes. For example, *Coccidioides immitis* and *Cryptococcus neoformans* have already expanded from their endemic zones due to changes in the hydrologic cycle (Lundgren 2018). In his work "Clues to the Presence of Pathogenic Fungi in Certain Environments," Restrepo et al. (2000) state that fungi have a broad presence in nature, with a tremendous capacity for adaptation and survival in many different substrates. A profound understanding of the ever-changing ecology of such habitats is only the first step to deduct the possible impact of climate change on pathogenic and opportunistic fungi. Notably, large-scale farming and the use of agro-industrial fungicides may facilitate the emergence of highly resistant "superspecies," which could lead to catastrophic public health crises (Casadevall 2019; Hernández and Martínez 2018).

More specifically, CBM-causing organisms comprise several saprophytic fungi that only grow under certain conditions. Venezuela can illustrate how different ecosystems tailor organismal prevalence. Tropical regions favor the abundance of *F. pedrosoi* cases, while *C. carrionii* reports are limited to the semiarid, northeastern region. Remarkably, *C. carrionii* muriform cells have been isolated from cacti tissues, demonstrating that it can survive in the hostility of the desert (de Hoog et al. 2000; Richard-Yegres et al. 1992). From these observations, it may be assumed that desertification could lead to an upsurge of *C. carrionii* cases, while *F. pedrosoi* would have to adapt to unusual conditions or inevitably show a downward trend.

Madagascar is another example of selective fungal populations in different niches. *F. pedrosoi* predominates in the tropical climate of the north and northeastern regions, with an average precipitation of 1500 mm per year, mean temperatures above 15 °C, and a wide diversity of palms, orchids, bamboo, and fruit trees. On the contrary, *C. carrionii* is far more abundant in the south, with precipitations ranging from 300 to 600 mm per year, mean temperatures above 20 °C, and a desert flora composed of cacti, thorny plants, and eucalyptus trees.

In India, *F. pedrosoi* is the leading causative organism of CBM, followed by *C. carrionii*. Although the latter is not found in specific, natural regions of the country (as in Venezuela or Madagascar), most cases arise from desertification areas. This confirms that *C. carrionii* is well-adapted to a dry environment (Agarwal et al. 2017; Sharma et al. 2010; Verma et al. 2018). Accordingly, urbanization and mass migration may lead to the development of diseases in non-endemic areas, as it often happens by the displacement of Latin Americans to the United States or Africans to Europe (Hay 1979; Lisi 1983).

Modern molecular biology and genomics are now a fundamental approach to understand changes in species diversity and its relationship with evolutionary biology. It has been recently shown that previous *F. pedrosoi* cases were in fact *F. nubica*, the current principal agent of the northern region of Madagascar (Esterre et al. 1996, 1997; Esterre and Queiroz-Telles 2006; Najafzadeh et al. 2011; Rasamoelina et al. 2017, 2020). In an elegant sequencing study, Deng et al. (2015) discovered that haplotype diversity was higher in African *C. carrionii* and *F. pedrosoi* strains compared with South American populations. Furthermore, with increased geographic location, *Fonsecaea* showed greater species divergence and fewer haplotypes than *C. carrionii*, suggesting a long evolutionary history. Whether these changes occur as a result of environmental or host adaptation will be hard to determine. Nonetheless, these experiments shed light on fungal species evolution and survival, which can be analyzed for future environmental and epidemiological studies.

7.5 Clinical Manifestations

CBM is usually pruritic and slightly painful. Disease manifestations are classified as nodular, verrucous, tumor, cicatricial, plaque, and mixed forms (Queiroz-Telles et al. 2017; Queiróz et al. 2018). Satellite lesions are a common finding. The infection typically begins as a nodule which evolves into large exophytic, verrucous plaques that may ulcerate. It can also present as erythematous, hyperkeratotic, scaly plaques, similar to psoriasis. Later stages of infection display achromic areas, prominent scarring, and lymph stasis.

The foot and the leg are the most commonly involved areas (50–80%), but it can also affect the upper extremities, trunk, face, and ears, in decreasing order. The infection is often asymmetric and confined to the subcutaneous tissue. However, immunocompromised patients can show extraordinary, disseminated forms that affect the central nervous system (Bonifaz et al. 2019; Santos et al. 2020). Bacterial aggregation is a common complication. Bone and visceral involvement occurs rarely. Chronicity may result in the development of squamous cell carcinoma (Figs. 7.2 and 7.3) (Ameen 2010; Bonifaz et al. 2001, 2019; López-Martínez et al. 2007; Queiroz-Telles et al. 2017).

The main differential diagnoses include tuberculosis verrucosa cutis, psoriasis, Bowen's disease, sporotrichosis, leishmaniasis, paracoccidioidomycosis, phaeohyphomycosis, coccidioidomycosis, blastomycosis, tertiary syphilis, atypical mycobacterial infections, sarcoidosis, and squamous cell carcinoma (Bonifaz et al. 2019).

7.6 Diagnosis

A smear with 20% potassium hydroxide (KOH) is a fast and convenient method to diagnose CBM. Muriform cells can be observed with optical microscopy resembling "coffee grains" (Fig. 7.4). These are brown-colored, septate, double-membrane structures, measuring from 4 to 10 μm. Scales from hyperkeratotic lesions can particularly exhibit thick, septate, darkly pigmented filaments originating from muriform cells, a phenomenon not observed in phaeohyphomycosis.

Cultures are performed in Sabouraud agar with or without antibiotics at 25–28 °C. Most species grow over 20–40 days and develop black or olive, radiated colonies (Fig. 7.5) (Bonifaz et al. 2019; Torres-Guerrero et al. 2012). The assessment of its reproductive form is helpful as a first approximation to identify the causative organism. For example, *F. pedrosoi* shows blastoconidia arranged in hormodendrum (3–5 μm chained conidia) or phialides (3–4 μm elliptical conidia) (Fig. 7.6), identical to *F. monophora* and *F. nubica* morphology (Bonifaz et al. 2019; Yaguchi et al. 2007). As a distinguishing feature, *C. carrionii* shows 4–8 μm conidia grouped in long chains of 8–10 units that arise from a conidiogenous cell (Fig. 7.7) (Bonifaz et al. 2019; Badali et al. 2008).

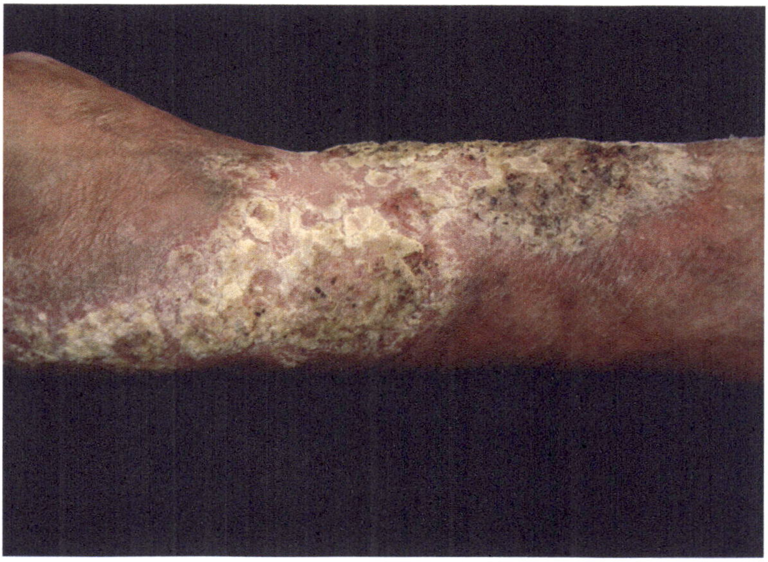

Fig. 7.2 Verrucous chromoblastomycosis

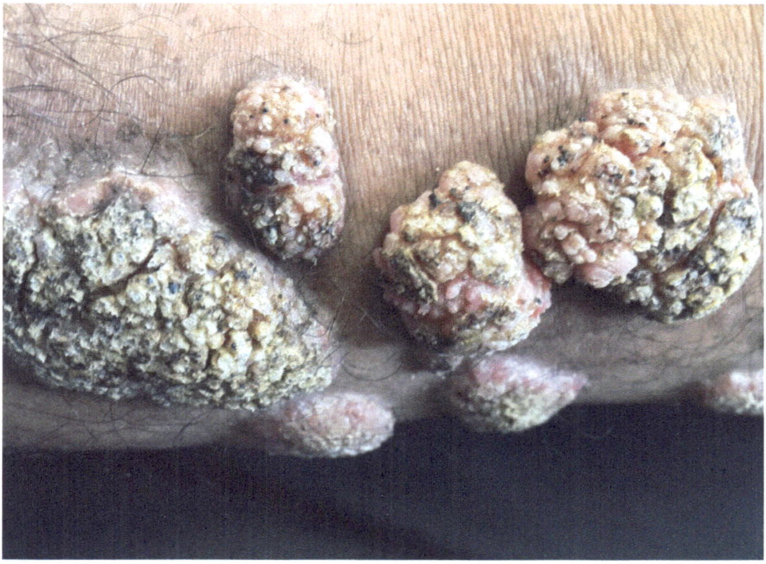

Fig. 7.3 Close-up view of verrucous chromoblastomycosis

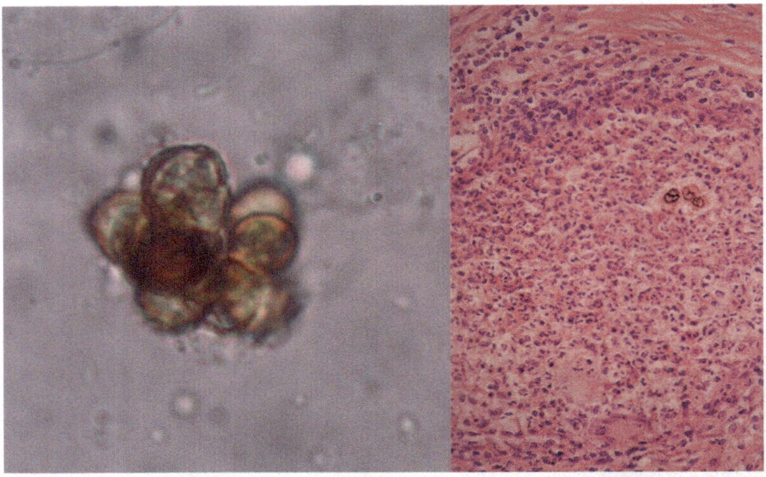

Fig. 7.4 Muriform cells observed on 20% KOH smear and hematoxylin and eosin histopathology (40×)

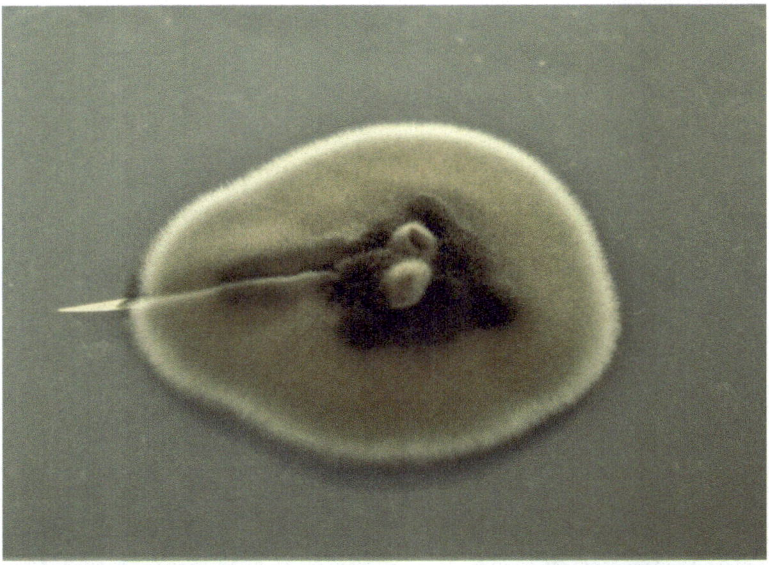

Fig. 7.5 Culture in Sabouraud agar without antibiotics at 28 °C

Less common agents include *P. verrucosa*, *Rhinocladiella*, and *Acrotheca*. *P. verrucosa* is a complex comprised of *P. americana*, *P. europea*, *P. chiiesnis*, and *P. macrospora*, among other species. It exhibits characteristic flask-shaped, 6–8 μm phialides and round-to-elliptical, 1–3 μm conidia. *Rhinocladiella* and *Acrotheca*

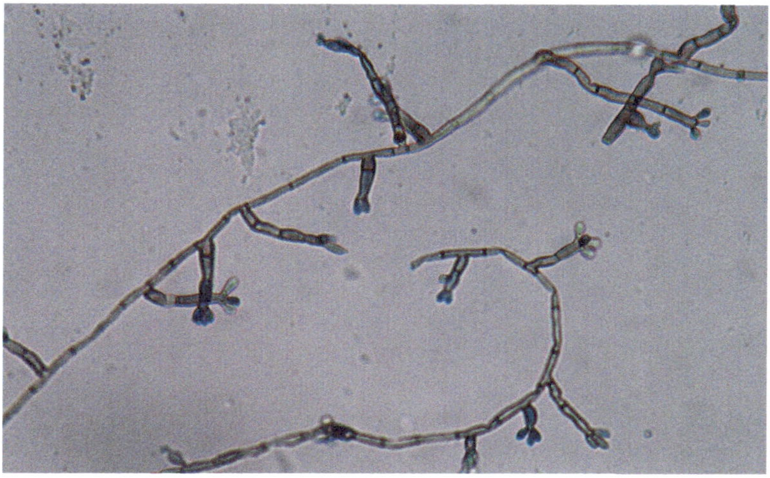

Fig. 7.6 *Fonsecaea pedrosoi*. Conidia arranged in phialides

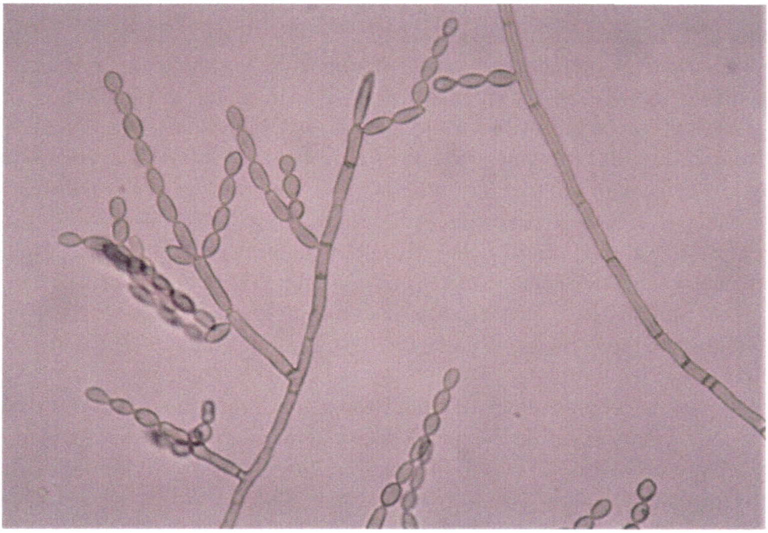

Fig. 7.7 *Cladophialophora carrionii*. Conidia arranged in hormodendrum

display 3–5 μm conidia that emerge from the ends of hyphae or conidiophores (Ahmed et al. 2021; Bonifaz et al. 2019).

Histopathology shows hyperkeratosis, parakeratosis, and acanthosis in the epidermis. A suppurative granuloma is a frequent finding in the superficial and mid-dermis (80%), composed of Langhans cells, epithelioid cells, and lymphocytes. Muriform cells can be observed within the infiltrate, in microabscesses, or inside

Langhans cells without special staining due to their pigment (Avelar-Pires et al. 2013; Bonifaz et al. 2019; López-Martínez 2007).

Serological tests are not reliable since they cross-react with a variety of pathogenic (*Sporothrix schenckii* and *Blastomyces dermatitidis*) and contaminant fungi (*Cladosporium,* and *Alternaria*).

Stringent species identification can only be achieved with polymerase chain reaction (PCR) or genomic techniques (Bonifaz et al. 2019; Najafzadeh et al. 2018; 2018; Queiroz-Telles et al. 2017).

7.7 Treatment and Prophylaxis

Treatment for CBM must be personalized based on individual factors, extension, and depth of the infection. In general, itraconazole is the most widely used antifungal therapy. It is reportedly more effective against *C. carrionii* than against *F. pedrosoi*. Recommended doses range from 200 to 400 mg/day for a period of 8–12 months, achieving cure rates from 15 to 80%. To claim a cure, control biopsies must be performed at 3- to 4-month intervals for 2 years (Bonifaz et al. 2004; Queiroz-Telles and Santos 2013; Queiroz-Telles et al. 2017).

Other alternatives include terbinafine, fluconazole, voriconazole, posaconazole, 5- flucytosine (5-FC), and amphotericin B. The dose for terbinafine is 250 to 500 mg/day and can be prescribed for longer periods than azoles. Fluconazole doses range from 200 to 400 mg/day. The use of newer triazoles has only been reported recently. 5-FC used at 100 to 150 mg/kg is one of the most promising drugs, yet it is only available in a few countries. Administration of intravenous, intra-arterial, and intralesional amphotericin B has also shown an adequate response. However, the frequency of adverse effects is often a limitation. There are reports of a combination of antifungals (e.g., itraconazole plus terbinafine) or systemic therapy with photodynamic therapy (Bassas-Vila et al. 2014; Bonifaz et al. 1997; Queiroz-Telles et al. 2017).

Cryosurgery is a convenient treatment for small, localized infections or when systemic therapy not possible. Nonetheless, systemic antifungal therapy and cryosurgery are preferably used in combination to reduce the risk of lymphatic dissemination by cryosurgery alone (Bonifaz et al. 2004, 2005).

As an adjunct to itraconazole, the use of (1 → 3)-β-polyglucoside has reportedly shown a partial response, possibly due to increasing the Th2 immune response and interleukin-10 (IL-10) (Queriroz-Telles et al. 2017).

The usage of appropriate footwear is the most cost-effective prophylactic intervention. Avoiding trauma to other exposed body areas in high-risk occupations may also be achieved by wearing protective equipment (Bonifaz et al. 2019; Vázquez-González et al. 2013). Table 7.2 summarizes the clinical features, diagnosis, and treatment of CBM.

Table 7.2 Clinical features, diagnosis, and treatment of chromoblastomycosis

Disease forms	Diagnosis	Main treatments
• Nodular • Verrucous • Tumor • Plaques • Lymphangitic • Cicatricial	• KOH smear: Muriform cells • Culture: Black, slow-growing colonies • Biopsy: Suppurative granuloma with muriform cells • Molecular and genomic techniques: Species identification	• Itraconazole: 200–400 mg/day • Terbinafine: 250–500 mg/day • Fluconazole: 200–400 mg/day • 5-Fluorocitosine: 100–150 mg/kg/day *Can be used in combination with:* • Simple excision • Cryosurgery • Photodynamic therapy

7.8 Conclusion

CBM is a worldwide hyperendemic disease that especially compromises middle-aged males who work in outdoor activities. *Fonsecaea pedrosoi* is the main etiological agent in tropical climates. Verrucous and infiltrative plaques are the most common dermatological patterns. The disease shows a chronic behavior and has low mortality but high morbidity. Since endemic mycoses occur in very specific environments, climate change will undoubtedly have an impact on the expansion, delimitation, and possibly change of the main etiological agents, including those related with CBM.

References

Agarwal R, Singh G, Ghosh A, Verma KK, Pandey M, Xess I. Chromoblastomycosis in India: review of 169 cases. PLoS Negl Trop Dis. 2017;11(8):e0005534. https://doi.org/10.1371/journal.pntd.0005534.

Ahmed SA, Bonifaz A, González GM, Moreno LF, Menezes da Silva N, Vicente VA, et al. Chromoblastomycosis caused by Phialophora-proven cases from Mexico. J Fungi (Basel). 2021;7(2):95. https://doi.org/10.3390/jof7020095.

Ameen M. Managing chromoblastomycosis. Trop Dr. 2010;40:65–7. https://doi.org/10.1258/td.2009.090264.

Atoche-Diéguez CE, Torres-Guerrero E, López-González R, Arenas R. Panorama epidemiológico de la cromoblastomicosis en el Centro Dermatológico de Yucatán. Dermatol Rev Mex. 2016;60(6):459–66.

Attapattu MC. Chromoblastomycosis: a clinical and mycological study of 71 cases from Sri Lanka. Mycopathologia. 1997;137:145–51. https://doi.org/10.1023/a:1006819530825.

Avelar-Pires C, Simoes-Quaresma JA, Moraes-de Macedo GM, Brasil-Xavier M, Cardoso-de BA. Revisiting the clinical and histopathological aspects of patients with chromoblastomycosis from the Brazilian Amazon region. Arch Med Res. 2013;44:302–6. https://doi.org/10.1016/j.arcmed.2013.04.008.

Badali H, Gueidan C, Najafzadeh MJ, Bonifaz A, Gerrits van den Ende AHG, De Hoog GS. Biodiversity of the genus Cladophialophora. Stud Mycol. 2008;61:175–91. https://doi.org/10.3114/sim.2008.61.18.

Badali H, Bonifaz A, Barrón-Tapia T, Vázquez-González D, Estrada-Aguilar L, Cavalcante Oliveira NM, et al. Rhinocladiella aquaspersa, proven agent of verrucous skin infection and a novel type of chromoblastomycosis. Med Mycol. 2010;48:696–703. https://doi.org/10.3109/13693780903471073.

Bassas-Vila J, Fuente MJ, Guinovart R, Ferrándiz C. Chromoblastomycosis: response to combination therapy with cryotherapy and terbinafine. Actas Dermosifiliogr. 2014;105:196–8. https://doi.org/10.1016/j.ad.2013.02.008.

Bonifaz A, Martínez-Soto E, Carrasco-Gerard E, Peniche J. Treatment of chromoblastomycosis with itraconazole, cryosurgery and combination of both. Int J Dermatol. 1997;36:542–7. https://doi.org/10.1046/j.1365-4362.1997.00085.x.

Bonifaz A, Carrasco E, Saúl A. Chromoblastomycosis: clinical and mycological experience of 51 cases. Mycoses. 2001;44:1–7. https://doi.org/10.1046/j.1439-0507.2001.00613.x.

Bonifaz A, Paredes-Solís V, Saúl A. Treating chromoblastomycosis with systemic antifungals. Expert Opin Pharmacother. 2004;5:247–54. https://doi.org/10.1517/14656566.5.2.247.

Bonifaz A, Saúl A, Paredes-Solís V, Araiza J, Fierro-Arias L. Treatment of chromoblastomycosis with terbinafine: experience with four cases. J Dermatolog Treat. 2005;16:47–51. https://doi.org/10.1080/09546630410024538.

Bonifaz A, Saldaña M, Mena L. Chromoblastomycosis. Cap 26. En Fundamentos de las micosis humanas. 1ª. Ed. Medellín, Colombia. Fondo Editorial Corporación para las Investigaciones Biológicas. 2019.pp: 235–41.

Brandt ME, Warnock DW. Epidemiology, clinical manifestations, and therapy of infections caused by dematiaceous fungi. J Chemother. 2003;(Suppl 2):36–47. https://doi.org/10.1179/joc.2003.15.Supplement-2.36.

Brito AC, Bittencourt MJS. Chromoblastomycosis: an etiological, epidemiological, clinical, diagnostic, and treatment update. An Bras Dermatol. 2018;93(4):495–506. https://doi.org/10.1590/abd1806-4841.20187321.

Casadevall A. Global catastrophic threats from the fungal kingdom: fungal catastrophic threats. Curr Top Microbiol Immunol. 2019;424:21–32. https://doi.org/10.1007/82_2019_161.

Correia RT, Valente NY, Criado PR, Martins JE. Chromoblastomycosis: study of 27 cases and review of medical literature. An Bras Dermatol. 2010;85:448–54. https://doi.org/10.1590/s0365-05962010000400005.

De Hoog GS, Queiroz-Telles F, Haase G, Fernandez-Zeppenfeldt G, Attili Angelis D, Gerrits Van Den Ende AH, et al. Black fungi: clinical and pathogenic approaches. Med Mycol. 2000;38(suppl. 1):243–5.

De Hoog GS, Attili-Angelis D, Vicente VA, van den Ende AH, Queiroz-Telles F. Molecular ecology and pathogenic potential of Fonsecaea species. Med Mycol. 2004;42:405–16. https://doi.org/10.1080/13693780410001661464.

De Hoog GS, Nishikaku AS, Fernández-Zeppenfeldt GD, Padín-González C, Burger E, Badali H, et al. Molecular analysis and pathogenicity of the Cladophialophora carrionii complex, with the description of a novel species. Stud Mycol. 2007;58:219–34. https://doi.org/10.3114/sim.2007.58.08.

Deng S, Tsui CKM, Gerrits van den Ende AHG, Yang L, Najafzadeh MJ, Badali H, et al. Global spread of human Chromoblastomycosis is driven by recombinant Cladophialophora carrionii and predominantly clonal Fonsecaea species. PLoS Negl Trop Dis. 2015;9(10):e0004004. https://doi.org/10.1371/journal.pntd.0004004.

Esterre P, Queiroz-Telles F. Management of chromoblastomycosis: novel perspectives. Curr Opin Infect Dis. 2006;19:148–52. https://doi.org/10.1097/01.qco.0000216625.28692.67.

Esterre P, Andriantsimahavandy A, Ramarcel ER, Pecarrere JL. Forty years of chromoblastomycosis in Madagascar: a review. Am J Trop Med Hyg. 1996;55:45–7. https://doi.org/10.4269/ajtmh.1996.55.45.

Esterre P, Andriantsimahavandy A, Raharisolo C. Natural history of chromoblastomycosis in Madagascar and the Indian Ocean. Bull Soc Pathol Exot. 1997;90(5):312–7.

Gomes RR, Vicente VA, Azevedo CM, Salgado CG, da Silva MB, Queiroz-Telles F, Marques SG, et al. Molecular epidemiology of agents of human Chromoblastomycosis in Brazil with the description of two novel species. PLoS Negl Trop Dis. 2016;10(11):e0005102. https://doi.org/10.1371/journal.pntd.0005102.

González GM, Rojas OC, Bocanegra-García V, González JG, Garza-González E. Molecular diversity of Cladophialophora carrionii in patients with chromoblastomycosis in Venezuela. Med Mycol. 2013;51:170–7. https://doi.org/10.3109/13693786.2012.695457.

Hay RJ. Mycoses imported from the West Indies. A report of three cases. Postgrad Med J. 1979;55(647):603–4. https://doi.org/10.1136/pgmj.55.647.603.

Hernandez H, Martinez LR. Relationship of environmental disturbances and the infectious potential of fungi. Microbiology (Reading). 2018;164(3):233–41. https://doi.org/10.1099/mic.0.000620.

Lisi P, Caraffini S. Imported skin diseases-a case of subcutaneous chromomycosis caused by Phialophora gougerotii. Int J Dermatol. 1983;22(3):180–1. https://doi.org/10.1111/j.1365-4362.1983.tb03360.x.

Londero AT, Ramos CD. Chromomycosis. A clinical and mycological study of 35 cases observed in the hinterland of Rio Grande do sul, Brazil. Am J Trop Med Hyg. 1976;25:132–5. https://doi.org/10.4269/ajtmh.1976.25.132.

López-Martínez R, Méndez Tovar LJ. Chromoblastomycosis. Clin Dermatol. 2007;25:188–94. https://doi.org/10.1016/j.clindermatol.2006.05.007.

Lundgren AD. Climate change and skin disease. Cutis. 2018;101(4):E12–4.

Marques SG, Pedrozo Silva CM, Resende MA, Andreata LS, Costa JM. Chromoblastomycosis caused by Rhinocladiella aquaspersa. Med Mycol. 2004;42:261–5. https://doi.org/10.1080/13693780310001597700.

Marques SG, Silva Cde M, Saldanha PC, Rezende MA, Vicente VA, Queiroz-Telles F, et al. Isolation of Fonsecaea pedrosoi from the shell of the babassu coconut (Orbignya phalerata Martius) in the Amazon region of Maranhao. Brazil Nippon Ishinkin Gakkai Zasshi. 2006;47:305–11. https://doi.org/10.3314/jjmm.47.305.

Minotto R, Bernardi CD, Mallmann LF, Edelweiss MI, Scroferneker ML. Chromoblastomycosis: a review of 100 cases in the state of Rio Grande do Sul, Brazil. J Am Acad Dermatol. 2001;44:585–9. https://doi.org/10.1067/mjd.2001.112220.

Najafzadeh MJ, Sun J, Vicente V, Xi L, Van den Ende AH, De Hoog GS. Fonsecaea nubica spp. nov, a new agent of human chromoblastomycosis revealed using molecular data. Med Mycol. 2010;48:800–6. https://doi.org/10.3109/13693780903503081.

Najafzadeh MJ, Sun J, Vicente VA, Klaassen CH, Bonifaz A, Gerrits van den Ende AH, et al. Molecular epidemiology of Fonsecaea species. Emerg Infect Dis. 2011;17:464–9. https://doi.org/10.3201/eid1703.100555.

Najafzadeh MJ, Gerrits van den Ende AHG, Vicente VA, Dolatabadi S, Sun J, de Hoog GS. Identification of chromoblastomycosis agents by PCR based reverse line blot (PCR-RLB) hybridization assay. Microb Pathog. 2018;125:43–7. https://doi.org/10.1016/j.micpath.2018.09.002.

Pérez-Blanco M, Hernández Valles R, García-Humbria L, Yegres F. Chromoblastomycosis in children and adolescents in the endemic area of the falcon state. Venezuela Med Mycol. 2006;44:467–71. https://doi.org/10.1080/13693780500543238.

Pires CA, Xavier MB, Quaresma JA, Macedo GM, Sousa BR, Brito AC. Clinical, epidemiological and mycological report on 65 patients from the eastern Amazon region with chromoblastomycosis. An Bras Dermatol. 2012;87:555–60. https://doi.org/10.1590/s0365-05962012000400006.

Queiróz AJR, Pereira Domingos F, Antônio JR. Chromoblastomycosis: clinical experience and review of literature. Int J Dermatol. 2018;57:1351–5. https://doi.org/10.1111/ijd.14185.

Queiroz-Telles F, Santos DW. Challenges in the therapy of chromoblastomycosis. Mycopathologia. 2013;175:477–88. https://doi.org/10.1007/s11046-013-9648-x.

Queiroz-Telles F, Esterre P, Pérez-Blanco M, Vitale RG, Salgado CG, Bonifaz A. Chromoblastomycosis: an overview of clinical manifestations, diagnosis and treatment. Med Mycol. 2009;47:3–15. https://doi.org/10.1080/13693780802538001.

Queiroz-Telles F, Nucci M, Colombo AL, Tobón A, Restrepo A. Mycoses of implantation in Latin America: an overview of epidemiology, clinical manifestations, diagnosis and treatment. Med Mycol. 2011;49:225–36. https://doi.org/10.3109/13693786.2010.539631.

Queiroz-Telles F, de Hoog S, Santos DW, Salgado CG, Vicente VA, Bonifaz A, et al. Chromoblastomycosis. Clin Microbiol Rev. 2017;30:233–76. https://doi.org/10.1128/CMR.00032-16.

Rasamoelina T, Raharolahy O, Rakotozandrindrainy N, Ranaivo I, Andrianarison M, Rakotonirina B, et al. Chromoblastomycosis and sporotrichosis, two endemic but neglected fungal infections in Madagascar. J Mycol Med. 2017;27(3):312–24. https://doi.org/10.1016/j.mycmed.2017.08.003.

Rasamoelina T, Maubon D, Andrianarison M, Ranaivo I, Sendrasoa F, Rakotozandrindrainy N, et al. Endemic Chromoblastomycosis caused predominantly by Fonsecaea nubica, Madagascar. Emerg Infect Dis. 2020;26(6):1201–11. https://doi.org/10.3201/eid2606.191498.

Restrepo A, Baumgardner DJ, Bagagli E, Cooper CR Jr, McGinnis MR, Lázera MS, et al. Clues to the presence of pathogenic fungi in certain environments. Med Mycol. 2000;38(Suppl 1):67–77.

Richard-Yegres N, Yegres F, Zeppenfeldt G. Cromomicosis: endemia rural, laboral y familiar en Venezuela. Rev Iberoam Micol. 1992;9:38–41.

Rojas-García OC, García-Martínez JM, Carrión-Álvarez D. Chromoblastomicosis en Mexico. Una enfermedad olvidada. Salud Publica Mex. 2019;61:3–4.

Romero-Navarrete M, Arenas R, Muñoz-Estrada VF, Atoche-Diéguez A, Mayorga J, Bonifaz A, et al. Cromoblastomicosis en México. Revisión de 603 casos en siete décadas. Dermatol CMQ. 2014;12:87–93.

Salgado CG, da Silva JP, Diniz JA, da Silva MB, da Costa PF, Teixeira C, et al. Isolation of Fonsecaea pedrosoi from thorns of Mimosa pudica, a probable natural source of chromoblastomycosis. Rev Inst Med Trop São Paulo. 2004;46:33–6. https://doi.org/10.1590/s0036-46652004000100006.

Santos DWCL, Vicente VA, Weiss VA, de Hoog GS, Gomes RR, Batista EMM, et al. Chromoblastomycosis in an endemic area of Brazil: a clinical-epidemiological analysis and a worldwide haplotype network. J Fungi (Basel). 2020;6(4):204. https://doi.org/10.3390/jof6040204.

Sharma A, Hazarika NK, Gupta D. Chromoblastomycosis in subtropical regions of India. Mycopathologia. 2010;169:381–6. https://doi.org/10.1007/s11046-009-9270-0.

Son YM, Kang HK, Na SY, Lee HY, Baek JO, Lee JR, et al. Chromoblastomycosis caused by Phialophora richardsiae. Ann Dermatol. 2010;22:362–6. https://doi.org/10.5021/ad.2010.22.3.362.

Tomson N, Abdullah A, Maheshwari MB. Chromomycosis caused by Exophiala spinifera. Clin Exp Dermatol. 2006;31:239–44. https://doi.org/10.1111/j.1365-2230.2005.02006.x.

Torres-Guerrero E, Isa-Isa R, Isa M, Arenas R. Chromoblastomycosis. Clin Dermatol. 2012;30:403–8. https://doi.org/10.1016/j.clindermatol.2011.09.011.

Tsuneto LT, Arce-Gomez B, Petzl-Erler ML, Queiroz-Telles F. HLA-A29 and genetic susceptibility to chromoblastomycosis. J Med Vet Mycol. 1989;27(3):181–5. https://doi.org/10.1080/02681218980000241.

Verma S, Thakur BK, Raphael V, Thappa DM. Epidemiology of subcutaneous mycoses in Northeast India: a retrospective study. Indian J Dermatol. 2018;63(6):496–501. https://doi.org/10.4103/ijd.IJD_16_18.

Weedon D, Van Deurse M, Allison S, Rosendahl C. Chromoblastomycosis in Australia: an historical perspective. Pathology. 2013;45(5):489–91. https://doi.org/10.1097/PAT.0b013e32836326a1.

Yaguchi T, Tanaka R, Nishimura K, Udagawa S. Molecular phylogenetics of strains morphologically identified as Fonsecaea pedrosoi from clinical specimens. Mycoses. 2007;50:255–60. https://doi.org/10.1111/j.1439-0507.2007.01383.x.

Yang YP, Li W, Huang WM, Zhou Y, Fan YM. Chromoblastomycosis caused by Fonsecaea: clinicopathology, susceptibility and molecular identification of seven consecutive cases in southern China. Clin Microbiol Infect. 2013;19:1023–8. https://doi.org/10.1111/1469-0691.12067.
Yang CS, Chen CB, Lee YY, Yang CH, Chang YC, Chung WH, et al. Chromoblastomycosis in Taiwan: a report of 30 cases and a review of the literature. Med Mycol. 2018;56(4):395–405. https://doi.org/10.1093/mmy/myx075.
Yegres RN, Yegres JF. La endemia de cromomicosis en Venezuela. Una estrategia para su control Vitae. 2005;7:24. http://caibco.uvc.ve

Chapter 8
Effect of Climate Change on the Incidence and Geographical Distribution of Coccidioidomycosis

Esperanza Duarte-Escalante, María del Rocío Reyes-Montes, María Guadalupe Frías-De-León, and Beatriz Meraz-Ríos

Contents

Abbreviations

CDC	Centers for Disease Control
IPCC	Intergovernmental Panel on Climate Change
NDVI	Normalized difference vegetation index
UND	Undetermined

8.1 Introduction

Coccidioidomycosis, also known as San Joaquín Valley fever, is a mycosis caused by the species *Coccidioides immitis* and *C. posadasii* (Fig. 8.1) (Fisher et al. 2002) and considered an endemic disease of the American continent (Laniado-Laborín

E. Duarte-Escalante (✉) · M. d. R. Reyes-Montes · B. Meraz-Ríos
Microbiology and Parasitology Department, School of Medicine, Universidad Nacional Autónoma de México, Mexico City, Mexico
e-mail: dupe@unam.mx; remoa@unam.mx; bmeraz@unam.mx

M. G. Frías-De-León
Unit Research, Hospital Regional de Alta Especialidad de Ixtapaluca, Mexico State, Mexico

M. G. Frías-De-León et al. (eds.), *The Impact of Climate Change on Fungal Diseases*, Fungal Biology, https://doi.org/10.1007/978-3-030-89664-5_8

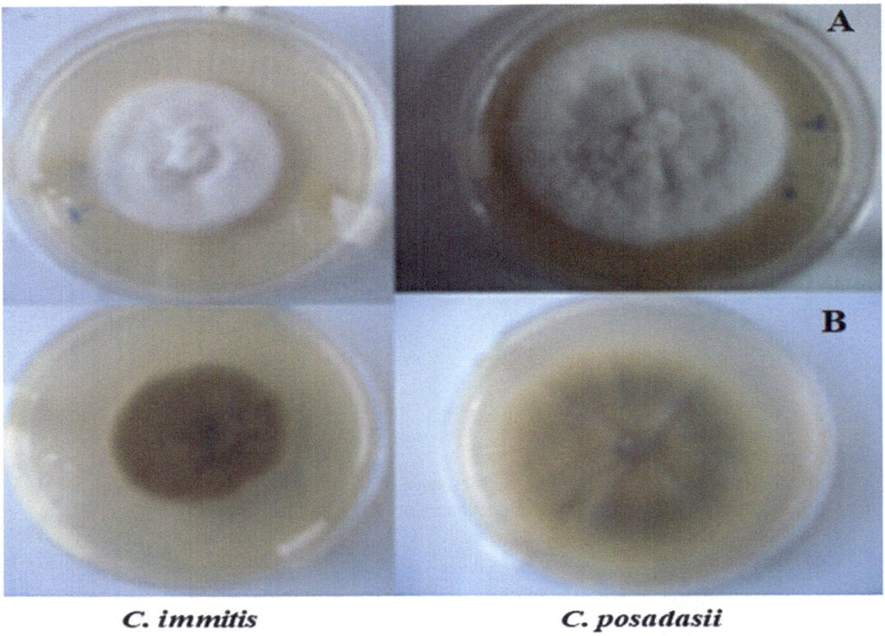

Fig. 8.1 Colonial morphology of *Coccidioides immitis* and *Coccidioides posadasii*. The isolates were seeded in Agar-Mycobiotic and were incubated at 28 °C, for 10 days. (**a**) The obverse shows hairy colonies, with smooth edges and a powdery appearance. (**b**) The reverse side shows dark brown pigment (Duarte-Escalante E)

et al. 2019; Ashraf et al. 2020). The infection is acquired by inhaling arthroconidia, the asexual reproductive structures, which infect the host and are mainly dispersed by air currents. Depending on the individual's immune status and other factors, including the number of arthroconidia inhaled, the infection can progress to asymptomatic, mild, moderate, severe, or disseminated coccidioidomycosis (Saubolle et al. 2007). In addition to the cases reported in humans, many other mammals, both domestic and wild, are susceptible to this infection (Alvarado et al. 2018; Reyes-Montes et al. 2016; Laniado-Laborín et al. 2019).

Epidemiological information on coccidioidomycosis has shown that the number of cases has increased throughout the continent in the last two decades. In the year 2000, it was estimated that, in the United States, it affected more than 100,000 people. In 2017, 14,364 cases were reported to the Centers for Disease Control (CDC), most from Arizona and California. However, these numbers could be underestimated because many cases are not correctly diagnosed (Gorris et al. 2019; https://www.cdc.gov/fungal/diseases/).

In other countries of the American continent, the actual incidence and prevalence rates of this mycosis are unknown due to the underreporting of cases, among other reasons, since it is not a notifiable disease (Ashraf et al. 2020; Laniado-Laborín et al. 2019).

Some authors mention two fundamental aspects in the incidence and prevalence of coccidioidomycosis: The first aspect is the geographical distribution of the disease associated with desert environments, and the second one is the correlation with climatic factors such as temperature, precipitation, and wind (Alvarado et al. 2018; Baptista-Rosas et al. 2010; Brito Silva et al. 2019; Comrie 2005; Gorris et al. 2018; Kolivras et al. 2001; Kolivras and Comrie 2003; Kollath et al. 2019; Laniado-Laborín et al. 2019; Maddy and Coccozza 1964; Shriber et al. 2017;Talamantes et al. 2007; Tong et al. 2017; Weaver and Kolivras 2018).

Climatic factors are also of great importance in the distribution of other biological species and their interaction with the environment. Therefore, the imminent global climate change, associated with the temperature increase in certain regions of the planet, prolonged droughts, and floods, among others, are events that generate an ecological response from living beings. However, the real impact on living organisms and biodiversity is still unknown (Palmer et al. 2017). In recent years, studies based on mathematical models have emerged to predict the response of biodiversity to climate change and design strategies to reduce its effect. Nevertheless, these studies include several underlying assumptions. Also, the resulting predictions differ according to the model used, so it is essential to consider its limitations, and they should be interpreted with caution (Bellard et al. 2018).

It is crucial to consider the close relationship between climatic factors and the incidence of coccidioidomycosis in light of the increase in the number of cases of this disease in recent years and the imminent climate change. For this reason, some factors that can be modified with climate change and impact the incidence of coccidioidomycosis and the geographical distribution of *Coccidioides* spp. are addressed in this chapter.

8.2 Habitat of *Coccidioides* spp.

Coccidioidomycosis is considered an endemic mycosis of the American continent, with areas of high endemicity in the Southwestern United States, Northwestern Mexico, and some areas of Central and South America (Ashraf et al. 2020; Baptista-Rosas and Riquelme 2007; Laniado-Laborín et al. 2019; Sondermeyer et al. 2016) (Table 8.1). The endemic regions of this mycosis are commonly described as areas with a semiarid climate that facilitates the dispersal of *Coccidioides* spores in the air (Drutz and Catanzaro 1978; Pappagianis 1988; Smith et al. 1946). Other characteristics associated with its habitat have also been described, like high soil salinity (Egeberg et al. 1964; Elconin et al. 1964, 1967; Lacy and Swatek 1974) and sandy soils that are rich in organic matter and salts (Drutz and Catanzaro 1978; Einstein and Johnson 1993), which could be requirements for the fungus development. Fisher et al. (2007) evidenced the ability of this fungus to grow in almost any type of desert soil, including those with high pH levels, and to tolerate extreme air temperatures ranging from −40.0 to 48.8 °C and soil temperatures in the range of −6.5 to 60.5 °C. Also, under laboratory conditions, *C. immitis* has shown the ability to

Table 8.1 Endemic areas of coccidioidomycosis in the American continent

Country	State/city
USA	California Washington Arizona Texas Utah
Mexico	**Northern region** Baja California Baja California Sur Sonora Chihuahua Coahuila Nuevo León Tamaulipas **Pacific coast** Extends from northern region to Michoacán **Central region** Extends from the northeastern border of Coahuila to Michoacán
Guatemala	Guatemala Motagua Valley
Honduras	Motagua Valley
Nicaragua	A coccidioidomycosis case diagnosed in a dog that traveled to Norway from Nicaragua raised the possibility of the disease in humans
Colombia	UND
Venezuela	Luria Lara Falcón
Bolivia	UND
Brazil	Piaui Maranhao Ceara Bahía
Paraguay	Chaco region
Argentina	Catamarca province La Rioja province San Luis Santiago del Estero province Cordoba province

UND undetermined
Ashraf et al. (2020) and Laniado-Laborín et al. (2019)

survive in seawater and saturated saltwater for up to 6 weeks (Dzawachiszwili et al. 1964). Such capability can be used to adapt and expand to new habitats. Other studies carried out in South American countries also describe the *Coccidioides* habitat as arid and semiarid zones, with alkaline soils that have little xerophytic vegetation, high average temperatures, and low average annual rainfall (Alvarado et al. 2018; Baptista-Rosas et al. 2015; Brito Silva et al. 2019; de Araújo et al. 2018; Fisher et al. 2007; Laniado-Laborín et al. 2019).

Although there is broad agreement in recognizing that desert regions, where up to 50 °C can be reached, are the habitat of *Coccidioides* spp., it has been mentioned that the optimal temperature range for its development is restricted. It has been revealed that arthroconidia can survive up to 6 months at temperatures between 15 and 37 °C and can only survive for 1 week at temperatures of 50 °C. Thus, the existence of other microclimates that promote fungal growth has been proposed, such as rodent burrows (Brown et al. 2017). At the same time, the influence of other climatic elements such as rainfall has also been considered as they can modify the soil and the topographic characteristics, thus creating new favorable micro-niches for the development of *Coccidioides* spp. (Baptista-Rosas et al. 2010; Gorris et al. 2018; Laniado-Laborín 2006; Laniado-Laborín et al. 2019). Therefore, it can be suggested that *Coccidioides* spp. will easily adapt to other environmental conditions caused by the impending climate change.

8.3 Climatic Factors and Their Role in the Incidence of Coccidioidomycosis

For many years, various studies have shown significant evidence about climatic factors and their close relationship with coccidioidomycosis incidence. Several authors have documented that the most relevant climatic factors associated with the incidence of this disease are temperature, precipitation, humidity, and dust storms (Brito Silva et al. 2019; Comrie 2005; Gorris et al. 2018; Kolivras et al. 2001; Kolivras and Comrie 2003; Kollath et al. 2019; Maddy and Coccozza 1964; Shriber et al. 2017; Talamantes et al. 2007; Tong et al. 2017; Weaver and Kolivras 2018).

To assess the relationship between disease and climatic conditions, Kolivras and Comrie (2003) used data from coccidioidomycosis cases in Pima County, Arizona, USA, to develop multivariate monthly models of relationships between disease frequency, climatic conditions, and variability. Their results revealed that temperature and precipitation in different seasons were important predictors of incidence. Also, to define the disease seasonality, Comrie (2005) developed a method to estimate the date of exposure to *Coccidioides* spp. Based on incidence data of coccidioidomycosis in Pima County, Arizona, USA, their results revealed a marked disease seasonality, showing that the peaks of fungus exposure occurred between June–July and October–November, which corresponded to the driest and dustiest months of the year. Talamantes et al. (2007) examined the effect of precipitation, surface temperature, and wind speed in Kern County, Arizona, USA, through simple linear correlation analysis and a generalized autoregressive moving average model, and they suggested that fluctuations in incidence were likely due to human activities (such as construction) rather than climatic fluctuations.

On the other hand, a more recent study carried out by Tong et al. (2017) mentioned that computer models predict that, as the Earth warms, the Southwestern

United States becomes drier, which means an increase both in dust storms and in the number of coccidioidomycosis cases according to US CDC records.

Another study that related climatic conditions with the incidence of coccidioidomycosis is that of Weaver and Kolivras (2018). They used data from 2000 to 2015 and regression analysis to model and evaluate the seasonal relationships between the incidence of coccidioidomycosis and climatic variables, including concurrent and delayed precipitation, temperature, and wind speed, among others. Their results showed a statistically significant relationship between incidence and climatic conditions in Kern County, California, USA, which were consistent with previous studies. The findings also suggested that previous precipitation is an important predictor of the disease.

On the other hand, Shriber et al. (2017) quantified the vulnerability to coccidioidomycosis, particularly in the states of Arizona and California, USA, and suggested that vulnerability is associated with the incidence of the disease so that regions with greater climatic variability are the most vulnerable.

Gorris et al. (2018), using a regional database of the disease, conducted a larger study. They evaluated environmental controls on a spatial extent, seasonal dynamics, and long-term trends in coccidioidomycosis incidence. They evidenced that air temperature, precipitation, soil moisture, surface dust concentrations, normalized difference vegetation index (NDVI), and cropland area provided information on spatial regulation and temporal patterns of this mycosis incidence. This analysis revealed that the incidence was higher in regions with warmer surface air temperatures and drier soils. Furthermore, the seasonal pattern of coccidioidomycosis incidence differed in the Southwestern United States. It appeared to be related to precipitation, including the influence of the North American Pacific and monsoon storm systems. They also identified that coccidioidomycosis in California and South Central Arizona had a peak incidence after dry spells, supporting earlier findings, showing that a wet season followed by a dry season increases incidence.

On the other hand, Kollath et al. (2019) hypothesized that *Coccidioides* respond to soil moisture, and when moisture is abundant, the fungus grows as mycelium. However, as it undergoes water stress during the dry season, the arthroconidia (small spores comparable in size to powdered aerosols) are dispersed, which is when humans are most at risk of inhaling infectious propagules. The authors showed that, in Arizona, USA, there is a positive correlation between a higher incidence and high levels of precipitation in the winter and spring months and an increase in coccidioidomycosis cases in the spring months after heavy winter rains.

Even though most of these studies have been performed in the United States, there is also evidence of a correlation between climatic factors and coccidioidomycosis in other countries. Maddy and Coccozza (1964) analyzed climatic data obtained from meteorological stations in various parts of Mexico. Data included the average temperature of the hottest month, the average temperature of the coldest month, and the average temperature of the rainy months to predict *C. immitis* propagation's probable potential in the environment. Their analysis showed that the areas with the greatest potential for fungal spread were Sonora, Chihuahua, Coahuila, Nuevo León, Baja California, and Baja California Sur, coinciding with the areas considered endemic of the disease. Likewise, Brito Silva et al. (2019) used time series to evaluate the statistical correlations between the climate and hospitalizations for mycoses in several Brazilian cities, and they found three cases of coccidioidomycosis associated with a hot period in the State of Ceará, thus supporting the evidence on the close relationship between climatic factors and coccidioidomycosis. Also, in Brazil, de Araújo et al. (2018) reported 30 cases of coccidioidomycosis associated with a drought in the State of Pernambuco, a region where the disease had never been recorded previously.

Perhaps one of the climatic factors associated with coccidioidomycosis that has received more attention is the wind, since it constitutes a dispersal mechanism, not only for *Coccidioides* spp. (Tong et al. 2017) but also for other microorganisms, even among widely dispersed habitats (Prospero et al. 2005).

Coccidioides spp. is the best-known human pathogen associated with dust storms (Freedman et al. 2018; Gorris et al. 2018; Querol et al. 2019). It has been reported in recent years that significant population growth has occurred in regions of Southern Arizona, USA, an area considered endemic for coccidioidomycosis. The latter has led to changes, such as reduction of plant cover, forest fires, diversions of water currents, changes in land use, and droughts, which promote dust emissions and favor the movement of *Coccidioides* spp. spores (Tong et al. 2017). There is also evidence that the aerosolization and dispersal of *Coccidioides* spp. arthroconidia can be favored by the alteration of the soil caused by natural disasters, such as earthquakes, or by anthropogenic activities, like construction (Benedict and Park 2014; Gorris et al. 2018; Griffin 2007).

On the other hand, Gorris et al. (2018) mentioned that, according to US CDC data, there was an increase in coccidioidomycosis incidence between 1998 and 2011 in the same regions where the incidence of dust storms had risen. These storms could transport the fungi through the air from the desert areas of North America to the neighboring areas, including the southern regions of the United States and Mexico and even reaching the state of Washington, which is considered a non-endemic area.

Thus, the role of wind in the incidence of coccidioidomycosis appears to be very important. It has been shown that high-energy wind conditions in arid regions can result in the mobilization of substantial amounts of soils in the atmosphere, and dust storms can generate significant dispersion, either at a continental or transoceanic level (Benedict and Park 2014; Gorris et al. 2018; Griffin 2007; Querol et al. 2019). Therefore, although coccidioidomycosis is recognized as endemic to the American continent, there is evidence of this disease in very distant geographical regions, such as Asia and Africa. A retrospective analysis of coccidioidomycosis in China performed by Tetro (2019) revealed 38 cases with no history of travel to endemic areas and suggested that the fungus may already be spreading globally. On the other hand, Jambalang et al. (2010) reported the first case of coccidioidomycosis in chickens in Nigeria, a very distant place from the considered endemic areas. In Mexico, a case of coccidioidomycosis was reported in a patient native of the State of Campeche, a region deemed non-endemic to the disease, and who did not register any visits to endemic areas (Duarte Escalante et al. 2013).

Likewise, Petersen et al. (2004) documented a coccidioidomycosis outbreak among workers in an archaeological site in the State of Utah, in the Northern United States, far from the recognized endemic areas. Also, three cases of confirmed coccidioidomycosis were documented during 2010–2011 among residents of the South Central part of the State of Washington, USA (Litvintseva et al. 2015; Marsden-Haug et al. 2013, 2014), showing that the disease's geographic distribution is expanding toward the north of the country, far from the formally recognized endemic areas.

So, perhaps, long-distance dust storms may be the fungus' dispersal mechanism to reach these regions. It has been shown that, for instance, mineral dust generated in Asian areas is generally transported long distances in the lower troposphere. However, it can rise to the upper troposphere, about 8–10 km above the Earth's surface, and be transported worldwide. This phenomenon can also be the transport route for billions of microorganisms in the air, with different destinations along their way (Behzad et al. 2018; Uno et al. 2009).

Nevertheless, the fungal dispersion could also involve mechanisms other than the wind, explaining the coccidioidomycosis cases in remote regions from the known endemic areas. Fisher et al. (2013) stated that humans could mediate the intercontinental dispersal of pathogenic fungi by exchanging commercial products around the world, which could be the beginning of a new dispersal mechanism; however, there is still no information on this mechanism regarding *Coccidioides* spp.

8.4 Climate Change and Its Effects on the Expansion of *Coccidioides* spp. Geographical Distribution

At the beginning of the 1990s, climate change was established as a formal field of study to understand the magnitude and risks to human health within the framework of a constantly changing climate, mainly caused by anthropogenic activities. In 1988, the World Meteorological Organization and the United Nations Environment Program created the Intergovernmental Panel on Climate Change (IPCC). This Panel's primary function is to provide scientific assessments of climate change and its possible consequences for human and natural systems and propose options for their mitigation (Ebi and Hess 2017).

In the last century, biophysical factors and anthropogenic activity have caused changes in the planet's climatic variables. The average temperature of the Earth has increased by almost 0.6 °C due to anthropogenic activities, and it is estimated that by the end of the twenty-first century, the temperature of the Earth's surface could increase between 2.6 and 4.8 °C. In addition, the mean sea level could rise between 45 and 82 cm. Likewise, if atmospheric pollution continues due to greenhouse gas emissions, such as carbon dioxide or methane, the planet will continue to warm with serious implications. Changes in the rainfall regime will occur coupled with extreme events like droughts, rainfall, and more intense hurricanes. These changes in the climate will affect all living beings, including humans, and they could also affect the distribution of numerous pathologies, mainly emerging and reemerging infectious diseases, since the climate could impact the size of the pathogen population, vectors, and hosts; the duration of disease transmission; and the persistence of epidemic outbreaks (Cox et al. 2012; Palmer et al. 2017). On the other hand, forecasting the response of biodiversity to climate change has become a highly relevant issue. Predictive models play an essential role in alerting scientists and those involved in decision-making about possible future risks and support the development of strategies to reduce the impacts of climate change on biodiversity (Bellard et al. 2012).

In recent years, a consistent drought trend has been projected in the Southwestern United States as a result of various global climate models and regional climate simulations (Cayan et al. 2013; Tong et al. 2017). Thus, considering the imminent climate change, some authors have mentioned that the global temperature increase will allow *Coccidioides* spp. to expand its geographic range northward into areas that would not have been considered suitable for the survival of the fungus previously (Brown et al. 2017; Engelthaler et al. 2016; Gorris et al. 2018, 2019; Van Leukena et al. 2016).

Engelthaler et al. (2016) mentioned that the geographic range of coccidioidomycosis is expanding, according to recent epidemiological and population genetic data.

Also, Achakulwisut et al. (2019) suggested a growing increase in aridity in the Southwestern United States due to climate change, based on the analysis of the impact of drought conditions on dust activity in this region in recent years. Meanwhile, Cayan et al. (2013) described the possible climatic scenarios predicted for the twenty-first century in the Southwestern United States. They included an increase in the Earth's surface temperature, variations in temperatures and rainfall during short periods, less rain in the southern part of the southwestern region and increased rainfall in the northern part, as well as a reduction in the snow cover of the southwestern mountains between the months of February and May during the period from 2001 to 2100. Therefore, as the land warms, the Southwestern United States becomes drier and more prone to dust storms, causing an increase in the coccidioidomycosis incidence (Tong et al. 2017).

A recent study by Gorris et al. (2019) also supported that climate projections for the Western United States indicate that temperatures will increase, and precipitation patterns will change, altering coccidioidomycosis dynamics. Besides, they estimated the areas that could potentially become endemic of coccidioidomycosis during the twenty-first century, using a climate niche model derived from data on the disease incidence. By the year 2100, in a high global warming scenario, his model predicts that the climate-limited area of endemicity will be more than double. The number of affected states will also grow, and the number of coccidioidomycosis cases will increase by 50%. The disease will also spread northward into the dry western states, including Idaho, Wyoming, Montana, Nebraska, South Dakota, and North Dakota, revealing how climate change may influence the geographic expansion distribution of the coccidioidomycosis in the United States.

As previously discussed, coccidioidomycosis is a disease that shows an increased incidence associated with climate change, mainly in the Southwestern United States and in countries such as Mexico, Brazil, and Argentina, which present similar behavior. Interestingly enough, Matlock et al. (2019) revealed that public health agencies in the United States do not yet have efficient communication and health promotion strategies about this disease directed at vulnerable and affected communities to reduce this mycosis prevalence. Given the cross-border nature of the climate influence on the risk of contracting coccidioidomycosis, a concerted effort and a more effective health communication strategy are necessary than the current practices, which is also imperative for the rest of the countries with endemic areas for this mycosis. Berry et al. (2018) declared that health authorities need information on climate conditions and vulnerable populations, among other aspects, to reduce exposures, recommend necessary protection measures, and build climate-resilient health systems, which can constitute a considerable challenge in each country.

On the other hand, it is essential to explore other aspects that may be crucial for *Coccidioides* spp. and for the disease it causes. Bellard et al. (2012) pose the interesting question of whether the species will be able to adapt quickly enough to climate change to maintain themselves or not, especially considering the case of *Coccidioides* spp. Although there is still no answer, given the evidence of the fungus geographical expansion outside its habitat in endemic areas, everything seems to indicate that this species could have a high probability of adapting to new environmental conditions and remaining in the environment.

8.5 Conclusion

The impact of climate change on the incidence and geographical expansion of coccidioidomycosis is evident. Data show a growing incidence and an increase in its geographical extension, mainly in the Southwestern United States. However, it is suggested that countries such as Mexico, Brazil, and Argentina could display similar behavior. Therefore, it is essential to call on all health authorities in countries with endemic areas of this mycosis to build climate-resilient health systems.

References

Achakulwisut P, Anenberg SC, Neumann JE, Penn SL, Weiss N, Crimmins A, et al. Effects of increasing aridity on ambient dust and public health in the U.S. southwest under climate change. GeoHealth. 2019;3(5):127–44. https://doi.org/10.1029/2019GH000187.

Alvarado P, Teixeira MM, Andrews L, Fernandez A, Santander G, Doyle A, et al. Detection of Coccidioides posadasii from xerophytic environments in Venezuela reveals risk of naturally acquired coccidioidomycosis infections. Emerg Microbes Infect. 2018;7:46. https://doi.org/10.1038/s41426-018-0049-6.

Ashraf N, Kubat RC, Poplin V, Adenis AA, Denning DW, Wright L, et al. Re-drawing the maps for endemic mycoses. Mycopathologia. 2020;185:843–65. https://doi.org/10.1007/s11046-020-00431-2.

Baptista-Rosas R, Riquelme M. Epidemiología de la Coccidioidomicosis en México. Rev Iberoam Micol. 2007;24:100–5.

Baptista-Rosas RC, Arellano E, Hinojosa A, Riquelme M. Bioclimatología de la Coccidioidomicosis en Baja California. México Invest Geog. 2010;71:21–30.

Baptista-Rosas RC, Catalán-Dibene J, Romero-Olivares AL, Hinojosa A, Cavazos T, Riquelme M. Molecular detection of Coccidioides spp. from environmental samples in Baja California: linking Valley Fever to soil and climate conditions. Fungal Ecol. 2015;5(2):177–90. https://doi.org/10.1016/j.funeco.2011.08.004.

Behzad H, Mineta K, Gojobori T. Global Ramifications of dust and sandstorm microbiota. Genome Biol Evol. 2018;10(8):1970–87. https://doi.org/10.1093/gbe/evy13.

Bellard C, Bertelsmeier C, Leadley P, Thuiller W, Courchamp F. Impacts of climate change on the future of biodiversity. Ecol Lett. 2012;15(4):365–77. https://doi.org/10.1111/j.1461-0248.2011.01736.x.

Bellard C, Jeschke JM, Leroy B, Mace GM. Insights from modeling studies on how climate change affects invasive alien species geography. Ecol Evol. 2018;8:5688–700. https://doi.org/10.1002/ece3.409.

Benedict K, Park BJ. Invasive fungal infections after natural disasters. Emerg Infect Dis. 2014;20(3):349–55. https://doi.org/10.3201/eid2003.131230.

Berry P, Enright PM, Shumake-Guillemot J, Villalobos Prats E, Campbell-Lendrum D. Assessing health vulnerabilities and adaptation to climate change: a review of International Progress. Int J Environ Res Public Health. 2018;15(12):2626. https://doi.org/10.3390/ijerph15122626.

Brito Silva F, Nascimento Santos JR, Chagas da Silva L, Costa Gomes W, MendesVillis PC, dos SantosGomes E, et al. Climate drivers of hospitalizations for mycoses in Brazil. Sci Rep. 2019;9:6902. https://doi.org/10.1038/s41598-019-43353-w.

Brown HE, Mu W, Khan M, Tsang C, Liu J, Tong D. Spatial scale in environmental risk mapping: A Valley fever case study. J Public Health Res. 2017;6:886.

Cayan D, Tyree M, Kunkel KE, Castro C, Gershunov A, Barsugli J, et al. Future climate: projected average. In: Garfin G, Jardine A, Merideth R, Black M, LeRoy S, editors. Assessment of Climate Change in the Southwest United States: A Report Prepared for the National Climate Assessment, A report by the Southwest Climate Alliance. Washington, DC: Island Press; 2013. p. 101–25.

Comrie AC. Climate factors influencing coccidioidomycosis seasonality and outbreaks. Environ Health Perspect. 2005;113(6):688–92. https://doi.org/10.1289/ehp.7786.

Cox R, Revie CW, Sanchez J. The use of expert opinion to assess the risk of emergence or re-emergence of infectious diseases in Canada associated with climate change. PLoS One. 2012;7(7):e41590. https://doi.org/10.1371/journal.pone.0041590.

de Araújo PSR, de Souza Junior VR, Padilha CE, de Oliveira MI, Arraes LC, Vieira R, et al. Coccidioidomycosis: first cases reported in Pernambuco, Brazil. Rev Inst Med Trop São Paulo. 2018;60:e75. https://doi.org/10.1590/S1678-9946201860075.

Drutz DJ, Catanzaro A. Coccidioidomycosis. Parts I Am Rev Respir Dis. 1978;117(3):559–85.

Duarte Escalante E, Zúñiga G, Frías De León MG, Canteros C, Castañón Olivares LR, Reyes Montes MR. AFLP analysis reveals high genetic diversity but low population structure in Coccidioides posadasii isolates from Mexico and Argentina. BMC Infect Dis. 2013;13:411. https://doi.org/10.1186/1471-2334-13-411.

Dzawachiszwili N, Landau JW, Newcomer VD, Plunkett OA. The effect of sea water and sodium chloride on the growth of fungi pathogenic to man. J Invest Dermatol. 1964;43:103–9. https://doi.org/10.1038/jid.1964.

Ebi KL, Hess JJ. The past and future in understanding the health risks of and responses to climate variability and change. Int J Biometeorol. 2017;61(Suppl 1):S71–80. https://doi.org/10.1007/s00484-017-1406-1.

Egeberg RO, Elconin AE, Egeberg MC. Effect of salinity and temperature on Coccidioides immitis and three antagonistic soil saprophytes. J Bacteriol. 1964;88(2):473–6.

Einstein HE, Johnson RH. Coccidioidomycosis: New aspects of epidemiology and therapy. Clin Infect Dis. 1993;16(3):349–54. https://doi.org/10.1093/clind/16.3.349.

Elconin AF, Egeberg RO, Egeberg MC. Significance of soil salinity on the ecology of Coccidioides immitis. J Bacteriol. 1964;87(3):500–3.

Elconin AF, Egeberg MC, Bald JG, Matkin AO, Egeberg RO. A fungicide effective against Coccidioides immitis in the soil. In: Ajello L, editor. Coccidioidomycosis. Tucson: University of Arizona Press; 1967. p. 319–21.

Engelthaler DM, Roe CC, Hepp CM, Teixeira M, Driebe EM, Schupp JM, et al. Local population structure and patterns of western hemisphere dispersal for Coccidioides spp., the fungal cause of Valley Fever. MBio. 2016;7(2):e00550–16. https://doi.org/10.1128/mBio.00550-16.

Fisher MC, Koenig GL, Taylor JW. Molecular and phenotypic description of Coccidioides posadasii sp. nov., previously recognized as the non-California population of Coccidioides immitis. Mycologia. 2002;94:73–84.

Fisher FS, Bultman MW, Johnson SM, Pappagianis D, Zaborsky E. Coccidioides niches and habitat parameters in the southwestern United States: a matter of scale. Ann N Y Acad Sci. 2007;1111:47–72. https://doi.org/10.1196/annals.1406.031.

Fisher MC, Henk DA, Briggs CJ, Brownstein JS, Madoff LC, McCraw SL, et al. Emerging fungal threats to animal, plant and ecosystem health. Nature. 2013;484(7393):186–94. https://doi.org/10.1038/nature10947.

Freedman M, Jackson BR, McCotte O, Benedict K. Coccidioidomycosis outbreaks, United States and worldwide, 1940–2015. Emerg Infect Dis. 2018;24(3):417–24. https://doi.org/10.3201/eid2403.170623.

Gorris ME, Cat LA, Zender CS, Treseder KK, Randerson JT. Coccidioidomycosis dynamics in relation to climate in the southwestern United States. GeoHealth. 2018;2:6–24. https://doi.org/10.1002/2017GH000095.

Gorris ME, Treseder KK, Zender CS, Randerson JT. Expansion of coccidioidomycosis endemic regions in the United States in response to climate change. GeoHealth. 2019;3(10):308–27. https://doi.org/10.1029/2019GH000209.

Griffin DW. Atmospheric movement of microorganisms in clouds of desert dust and implications for human health. Clin Microbiol Rev. 2007;20(3):459–77. https://doi.org/10.1128/CMR.00039-06.

Jambalang AR, Ogo IN, Ibu JO, Ghislaine M, Bertu W, Jwander L, et al. Coccidioidomycosis in chicken pullets in Jos, Plateau State, Nigeria: a case report. Nig Vet J. 2010;31(3):249–51.

Kolivras KN, Comrie AC. Modeling valley fever (coccidioidomycosis) incidence on the basis of climate conditions. Int J Biometeorol. 2003;47:87–101. https://doi.org/10.1007/s00484-002-0155-x.

Kolivras KN, Johnson PS, Comrie AC, Yool SR. Environmental variability and coccidioidomycosis (valley fever). Aerobiologia. 2001;17:31–42.

Kollath DR, Miller KJ, Barker BM. The mysterious desert dwellers: Coccidioides immitis and Coccidioides posadasii, causative fungal agents of coccidioidomycosis. Virulence. 2019;10(1):222–33. https://doi.org/10.1080/21505594.2019.1589363.

Lacy GH, Swatek FE. Soil ecology of Coccidioides immitis at amerindian middens in California. Appl Microbiol. 1974;27(2):379–88.

Laniado-Laborín R. Coccidioidomicosis. Más que una enfermedad regional. Rev Inst Nal Enf Resp Mex. 2006;19(4):301–8.

Laniado-Laborín R, Arathoon EG, Canteros C, Muniz-Salazar R, Rendon A. Coccidioidomycosis in Latin America. Med Mycol. 2019;57:S46–55. https://doi.org/10.1093/mmy/myy037.

Litvintseva AP, Marsden-Haug N, Hurst S, Hill H, Gade L, Driebe EM, et al. Valley fever: finding new places for an old disease: coccidioides immitis found in Washington state soil associated with recent humanlnfection. Clin Infect Dis. 2015;60(1):e1–3. https://doi.org/10.1093/cid/ciu681.

Maddy K, Coccozza J. The probable geographic distribution of Coccidioides immitis in Mexico. Bol Oficina Sanit Panam. 1964;57:44–54.

Marsden-Haug N, Goldoft M, Ralston C, Limaye AP, Chua J, Hill H, et al. Coccidioidomycosis acquired in Washington State. Clin Infect Dis. 2013;56(6):847–50. https://doi.org/10.1093/cid/cis1028.

Marsden-Haug N, Hill H, Litvintseva AP, Engelthaler DM, Driebe EM, Roe CC, et al. Coccidioides immitis identified in soil outside of its known range—Washington, 2013. MMWR Morb Mortal Wkly Rep. 2014;63(20):450.

Matlock M, Hopfer S, Ogunseitan OA. Communicating risk for a climate-sensitive disease: A case study of Valley Fever in Central California. Int J Environ Res Public Health. 2019;16:3254. https://doi.org/10.3390/ijerph16183254.

Palmer G, Platts PJ, Brereton T, Chapman JW, Dytham C, Fox R, et al. Climate change, climatic variation and extreme biological responses. Phil Trans R Soc B. 2017;372:20160144. https://doi.org/10.1098/rstb.2016.0144.

Pappagianis D. Epidemiology of coccidioidomycosis. Curr Top Med Mycol. 1988;2:199–238. https://doi.org/10.1007/978-1-4612-3730-3_6.

Petersen LR, Marshall SL, Barton C, Hajjeh RA, Lindsley MD, Warnock DW, et al. Coccidioidomycosis among workers at an archeological site, northeastern Utah. Emerg Infect Dis. 2004;10(4):637–42. https://doi.org/10.3201/eid1004.030446.

Prospero JM, Blades E, Mathison G, Naidu R. Interhemispheric transport of viable fungi and bacteria from Africa to the Caribbean with soil dust. Aerobiologia. 2005;21:1–19. https://doi.org/10.1007/s10453-004-5872-7.

Querol X, Tobías A, Pérez N, Karanasiou A, Amato F, Stafoggia M, et al. Monitoring the impact of desert dust outbreaks for air quality for health studies. Environ Int. 2019;2019(130):104867.

Reyes-Montes MR, Pérez-Huitrón A, Ocaña J, Frías De León MG, Martínez-Herrera EO, Arenas R, et al. The habitat of Coccidioides spp. and the role of animals as reservoirs and disseminators in nature. BMC Infect Dis. 2016;16(1):550. https://doi.org/10.1186/s12879-016-1902-7.

Saubolle MA, McKellar PP, Sussland D. Epidemiologic, clinical, and diagnostic sspects of coccidioidomycosis. J Clin Microbiol. 2007;45(1):26–30. https://doi.org/10.1128/JCM.02230-06.

Shriber J, Conlon KC, Benedict K, McCotter OZ, Bell JE. Assessment of vulnerability to coccidioidomycosis in Arizona and California. Int J Environ Res Public Health. 2017;14:680. https://doi.org/10.3390/ijerph14070680.

Smith CE, Beard RR, Rosenberger HG, Whiting EG. Effect of season and dust control on coccidioidomycosis. JAMA. 1946;132(14):833–8. https://doi.org/10.1001/jama.1946.02870490011003.

Sondermeyer LG, Lee L, Gilliss D, Vugia JD. Coccidioidomycosis-associated deaths in California, 2000–2013. Public Health Rep. 2016;131:531–53.

Talamantes J, Behseta S, Zender CS. Fluctuations in climate and incidence of coccidioidomycosis in Kern County, California. A review. Ann N Y Acad Sci. 2007;1111:73–82. https://doi.org/10.1196/annals.1406.028.

Tetro JA. From hidden outbreaks to epidemic emergencies: the threat associated with neglecting emerging pathogens. Microbes Infect. 2019;21:4–9. https://doi.org/10.1016/j.micinf.2018.06.004.

Tong DQ, Wang JXL, Gill TE, Lei H, Wang B. Intensified dust storm activity and Valley fever infection in the southwestern United States. Geophys Res Lett. 2017;44:4304–12. https://doi.org/10.1002/2017GL073524.

Uno I, Eguchi K, Yumimoto K, Takemura T, Shimizu A, Uematsu M, et al. Asian dust transported one full circuit around the globe. Nat Geosci. 2009;2(8):557–60.

Van Leukena JPG, Swart AN, Havelaar AH, Van Pul A, Van der Hoek W, Heederik D. Atmospheric dispersion modelling of bioaerosols that are pathogenic to humans and livestock–A review to inform risk assessment studies. Microb Risk Anal. 2016;1:19–39. https://doi.org/10.1016/j.mran.2015.07.002.

Weaver EA, Kolivras KN. Investigating the relationship between climate and Valley Fever (Coccidioidomycosis). EcoHealth. 2018;15:840–52. https://doi.org/10.1007/s10393-018-1375-9.

Chapter 9
Impact of Climate Change on Opportunistic Molds Infections

Gloria M. González

Contents

Abbreviations

CSF Cerebrospinal fluid
RVVC Recurrent vulvovaginal candidiasis

G. M. González (✉)
School of Medicine, Autonomous University of Nuevo Leon,
Monterrey, Nuevo Leon, Mexico

M. G. Frías-De-León et al. (eds.), *The Impact of Climate Change on Fungal Diseases*, Fungal Biology, https://doi.org/10.1007/978-3-030-89664-5_9

9.1 Introduction

9.1.1 Fungal Diversity

Fungi are a group of highly diversified eukaryotic organisms that provide essential functions to maintain our planet. There are an enormous number of potential sites for the establishment of fungi in any location in the world. About two decades ago, in a highly referenced manuscript, the magnitude of fungal diversity was calculated as 1.5 million species in the world living in a wide variety of habitats (Hawksworth 2001). Only 70,000 fungal species were adequately described, and approximately 1 group of 300 species was recognized as pathogenic to humans at that time (Taylor et al. 2001). In 2017, the update of the global estimate of fungal diversity was published, using extrapolations of the fungus: plant ratio and information generated from molecular sequencing in a variety of environmental samples. The current estimated range is 2.2–3.8 million fungal species, with 120,000 known species. Only 3–8% have been named (Hawksworth and Lucking 2017). Currently, studies like this one face the problem of what is the conceptual recognition of a fungal species. This fact has a direct impact on the number of formally recognized species. In the pre-molecular era, knowledge of fungal species was based entirely on culture characteristics, microscopic morphology, and biochemical evidence. For the molecular age, a reevaluation of the species concept is required (Richards 2010). Despite various efforts, there is currently no single criterion that can be widely applied. There are outstanding efforts that are directed at the use of mechanistic systems for the recognition of species in phylogenetic trees, but the concept of a consolidated species has many other aspects that must be considered. The concept of a consolidated species that is being increasingly handled by mycologists adopts a multifaceted approach that considers morphological, ecological, and phylogenetic concepts. This fact may lead, on the one hand, to the recognition of more species or, on the other hand, to the fusion of previously recognized separate species (Quaedvlieg et al. 2014). An important consideration to mention is that the speciation process in general is an active process. A considerable proportion of species that are being studied today are not included in the updates of fungal diversity. Some researchers follow an incomplete approach, which makes it difficult to properly resolve the species and its diversity (Hawksworth and Lucking 2017). Many geographic areas and habitats have not been studied, and there are ecologically cryptic fungi in the environment that usually only develop hyphae and cryptic species hidden under well-established names. One problem today is that no site has a complete inventory of the fungi present and many habitats remain little explored. For now, only 34,878 species have sequences in GenBank representing only 29% of the 120,000 known species. Obtaining sequences for the remaining 71% of species is essential to determine the new species that are recovered from the environment, specimens, or diverse isolates (Hawksworth and Lucking 2017).

9.2 Human Fungal Infections

9.2.1 The Global Burden of Fungal Diseases

Fungal diseases have been underestimated for many years despite the great influence they have on plants, animals, and humans. Morbidity due to some fungi is significant; for example, superficial infections are very common in humans and are estimated to affect around 25% of the world's population. Similarly, it is estimated that 50–75% of women of reproductive age will suffer at least one episode of *Candida vulvovaginitis* (Table 9.1) (Bongomin et al. 2017).

Invasive fungal diseases have a lower incidence than superficial and mucosal infections. However, they represent a major global problem because they are associated with high mortality. Knowledge about the global incidence of invasive fungal diseases is very uneven due to several factors, including the scarcity of national surveillance systems and the fact that it is not mandatory to report diseases caused by these microorganisms. Also, it is very common to observe a lack of clinical suspicion, except in specialized units. Even today, there are few places where cultures and other tests are carried out for the diagnosis of mycoses, not to mention that some fungal diseases have been recognized until very recently (Denning 2016).

A rough estimate of the resources available to various countries establishes that if the medical units had the availability of diagnostic tests and antifungal drugs, a high percentage of patients (~80%) with some invasive mycosis would survive the disease. However, early recognition and proper management of severe diseases remain a challenge, especially because of limited resources since many conventional diagnostic tests take time to establish an accurate diagnosis, and antifungal treatment is often expensive and/or toxic. These resources are not available in all countries of the world (Global Action Fund for Fungal Infections 2017).

Table 9.2 includes opportunistic fungal diseases (candidiasis, aspergillosis, cryptococcosis, mucormycosis, and pneumocystosis) which have a high mortality (46–75%, 30–95%, 20–70%, 30–90%, and 20–80%, respectively).

Many fungi species are responsible for these invasive diseases, which kill 1.4 million people each year. This amount is similar to that caused by tuberculosis

Table 9.1 The global partial burden of some fungal diseases

Fungal disease	Global burden
Superficial diseases	~1,000,000,000
Keratitis	~1,000,000
Vulvovaginal candidiasis and RVVC	~134,000,000
Allergic manifestations	~23,306,675
Severe chronic diseases[a]	~3,076,000

RVVC recurrent vulvovaginal candidiasis

[a] Chronic pulmonary aspergillosis, mycetoma, chromoblastomycosis, coccidioidomycosis, paracoccidioidomycosis, blastomycosis, histoplasmosis

Table 9.2 The global partial burden of some invasive fungal diseases

Fungal disease	Annual incidence
Invasive candidiasis	~750,000
Invasive aspergillosis	>300,000
P. jirovecii pneumonia (AIDS and non-AIDS)	~500,000
Cryptococcosis (AIDS)	~223,000
Mucormycosis	>10,000

Bongomin et al. (2017). Global and multinational prevalence of fungal disease estimate precision. J Fungi 3:57. doi:https://doi.org/10.3390/jof3040057

(World Health Organization, Tuberculosis https://www.who.int/news-room/fact-sheets/detail/tuberculosis).

In the last decades, opportunistic fungal infections have acquired outstanding importance in the clinical setting. *Candida* spp. and *Cryptococcus* spp. are the most frequently reported yeasts, and *Aspergillus* spp., *Scedosporium* spp., *Fusarium* spp., and Zygomycetes are the most frequently described filamentous fungi. Invasive infections have increased making them a current public health problem. This has brought a prevailing demand to expand the knowledge of medical mycology in healthcare providers (Enoch et al. 2006).

Some of the reasons that have been proposed to explain this increase in invasive fungal diseases are the wide use of immunosuppressive and antineoplastic drugs, medical devices, prolonged treatment with broad-spectrum antibiotics, more aggressive surgeries, and prolonged stays in intensive care units. In a word, the increase in immunocompromised individuals places them at a high risk of acquiring a fungal infection (Vivek et al. 2019). What is notable in the aforementioned is that invasive fungal infections are associated with deterioration of the function of the immune system. In fact, fungi that cause diseases in humans became notorious only recently. The increase in oral thrush due to *Candida* is associated with the introduction of antibiotics and the increase in cryptococcosis and aspergillosis that occurs with HIV (Vincent et al. 2009).

On the other side of the spectrum, there are thousands of fungi that produce disease in insects, amphibians, reptiles, and plants. There are few fungal infections in mammals, which is attributed to the combination of endothermy and adaptive immunity. There are an estimated 270,000 and 50,000 fungal species that cause fungal disease in plants and insects, respectively (Hawksworth and Rossman 1997).

Despite the wide fungal diversity mentioned above, very few fungi cause disease in humans. The list of pathogenic fungi reviewed in most educational programs for training health personnel includes dermatophytes, *Malassezia* spp., *Trichosporon* spp., *Madurella* spp., *Trematosphaeria* spp., *Sporothrix* spp., *Fonsecaea* spp., *Rhinocladiella* spp., *Phialophora* spp., *Coccidioides* spp., *Histoplasma* spp., *Paracoccidioides* spp., *Blastomyces* spp., *Candida* spp., *Cryptococcus* spp., *Pneumocystis jirovecii*, *Aspergillus* spp., Phaeohyphomycosis (*Cladophialophora* spp., *Exophiala* spp.,

Exophialophora spp.), Hyalohyphomycosis (*Fusarium* spp., *Scedosporium* spp., *Acremonium* spp., *Paecilomyces* spp., *Scopulariopsis* spp., *Penicillium* spp.), and Zygomycetes (*Mucor* spp., *Rhizopus* spp., *Lichtheimia* spp.).

There are a large number of environmental sources of fungi that infect humans; for example, house pets are a potential source of fungal infections, as well as the soil on ornamental plants. This also includes all those conditions that have to do with the spread of fungi from one habitat to another and are important causes of fungal diseases, such as demolitions, excavations, constructions, dust accumulations, water leaks, etc. Most fungal infections are acquired from the environment by inhalation or skin trauma. In general, the fungi specified above rarely cause disease in immunocompetent hosts. So it seems that an innate and adaptive defense system in mammals is crucial for resistance to fungal infections. When we are infected by a pathogenic fungus, it can produce what is called virulence factors to survive in the complex environment that is the human body. The most studied virulence factors are a capsule, lytic enzymes, such as phospholipases, keratinases, proteases, and hemolysins, adhesion capacity, thermotolerance, the morphological change that occurs from the commensal phase to the pathogen phase, dimorphism, chemical changes of the cell wall, etc. Humans have defense mechanisms that prevent an infection from turning into disease: phagocytosis, complement, inflammatory response, and the rapid activation of an adaptive response that involves cell-mediated immunity and humoral immunity. In warm-blooded animals, such as mammals and birds, endothermy prevents the establishment of many fungi that are unable to grow at high temperatures. Endothermy is considered a very efficient evolutionary strategy that excludes most fungal species from a mammalian host. Temperature increases in response to infections like fever limit the survival of a pathogen (Hillenius and Ruben 2004). Endothermy is related to metabolic and thermodynamic benefits, but these benefits come at a high cost since endothermic vertebrates require around ten times more oxygen to carry out metabolism than ectothermic vertebrates (Ruben 1995). If the environment has a large number of fungal species and only a few are pathogenic to humans and other animals, it is suggested with some certainty that the existence of a complex defense mechanism and a high and stable temperature limit the establishment and growth of pathogenic fungi, thus providing relative resistance to systemic infections (Biegańska 2014).

9.3 Global Warming and Its Health Influence

9.3.1 Global Warming and the Impact on Fungi

Climate change is one of humanity's most pressing problems. Today's complex world is characterized among other things by globalization; mass travel; communications; migration networks; enormous economic differences; economic, political, and social crises; a lack of work and education; and demographic changes. Climate

Table 9.3 Rights threatened by climate change

Human rights and social justice	Civil and political rights	Economic, social, and cultural rights	Elimination of discrimination
Security	Right to life	Self-determination	Women
Adequate living (health, well-being)	Culture	Political status	
	Religion	Economic, social, and cultural development	
	Language	Physical and mental health	
		Education	

change will amplify many of these factors (McMichael 2015). The consequences on the environment and health will mainly affect poor people from all the countries in the world. This fact leads to profound effects on human rights; social justice; political, economic, social, and cultural rights; and the elimination of discrimination, especially in women living in rural areas (Table 9.3).

The climate of our planet has changed throughout its history, with periods of cold global temperature followed by periods of warming (Hoffman et al. 1998). The shift from a mainly agricultural economy to an industrial economy has had a major influence on the population, which experienced rapid growth, especially in urban areas. The Industrial Revolution was a turning point in history, modifying and influencing all aspects of everyday life in one way or another. This transition began toward the end of the eighteenth century in the textile industry, as well as in the extraction and use of coal.

The generation of greenhouse gases, such as carbon dioxide, methane, and nitrous oxide, will increase the global temperature by 2–5 °C in the coming years. This will take the Earth to the warmest period in the last 40 million years (Intergovernmental Panel on climate change 2007). Fungi can grow in a temperature range from 12 to 30 °C. A wide tolerance to temperature can be observed among the different species, with some growing at −10 and others up to 65 °C (Marquez et al. 2007). However, at temperatures above 30 °C, there is a sudden decrease in both growth and viability in fungal species (Robert and Casadevall 2009).

The rise in temperature in the world will lead to more drastic and frequent meteorological conditions; increased frequency and/or duration of heat waves; heavy precipitation events; intensity and/or duration of droughts; greater melting of ice; intense tropical cyclone activity; sea-level changes; alteration of distribution patterns of diseases carried by water, food, and vectors; and many other environmental changes (Wheeler and Watts 2018). Those conditions entail soil degradation, loss of productivity of agricultural land, loss of biodiversity, degradation of ecosystems, reduced freshwater resources, acidification of oceans, increased pollution, and the disruption and depletion of the ozone layer (Rockstrom et al. 2009). Humans depend on natural environments with clean air, safe and nutritious food and water, and a stable place to live (Watts et al. 2015); with these rapid and drastic environmental changes, there are a series of consequences for the health and well-being of humans. One such consequence may be the gradual change in fungal infections, which can become a threat to humans. The effects of ambient temperature and humidity on fungal sporulation and dispersal could favor an increase of emerging thermotolerant fungi that will be able to survive warmer temperatures. These changes would have great implications on human health because there are few antifungal drugs and it would be necessary to implement new diagnostic tests (Bromuro et al. 2010).

Very little has been studied about global warming and its impact on fungi, but it is believed that it could have an important effect on these microorganisms. A warmer ambient temperature will surely change the distribution of susceptible and heat-tolerant species favoring those that are more heat-tolerant and creating the conditions for environmental fungi to spread and come into close contact with humans; a good example is *P. brasiliensis*, *C. gattii*, and *A. fumigatus*.

A. fumigatus stands out as the main opportunistic pathogenic filamentous fungus in immunocompromised patients, especially in transplanted individuals. It is a saprophyte that has qualities that make it highly competitive in the environment, since it can adapt to a wide range of temperatures and pH, as well as various nutrients. However, thermotolerance is referred to as a potential virulence factor for this microorganism. *A. fumigatus* is capable of growing rapidly at 37 °C and tolerates temperatures of 60 °C. This fact contrasts with the majority of pathogenic fungi for humans that have optimal temperatures for their growth in the range of 25 and 35 °C (Barrozo et al. 2009). Thermotolerance in *A. fumigatus* could play a very important role in allowing it to survive in a climate-changed world.

On the other hand, there will be greater selective pressure, and the prevalence of heat-tolerant species could increase. For every 1 °C increase in body temperature to the range of 30–42 °C, approximately 6% of fungal species are excluded as potential pathogens (Fig. 9.1). Global warming will be responsible for reducing the thermal gradient between ambient temperature and the temperature of mammals. The current gradient is 22 °C, and for every degree that the average global temperature increases, the gradient is reduced by about 5% (Robert and Casadevall 2009).

It seems that the virulence of the fungi parallels the emergence of thermotolerance. In addition, fungal species that are more closely related to known pathogens, but are not thermotolerant, will be the ideal candidates for the emergence of new pathogens. At this point, there is an enormous risk of the emergence of new fungal pathogens (Robert and Casadevall 2009).

9.4 Fungal Infections and Natural Disasters

9.4.1 Natural Disasters

Natural disasters also play an important role in the increase of some infectious diseases in general, and fungal diseases do not go unnoticed. The World Health Organization defines a natural disaster as a disruption of a society that results in widespread human, material, or environmental loss that exceeds the ability of the

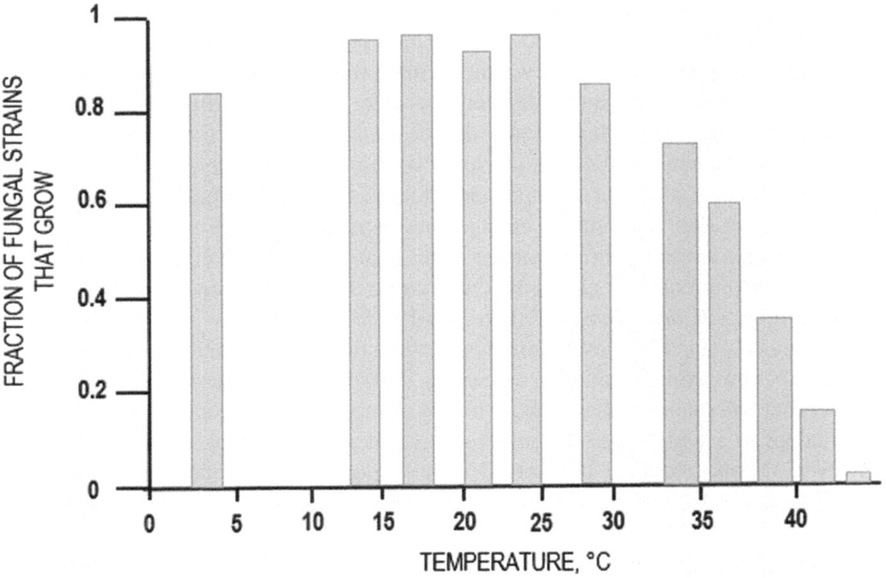

Fig. 9.1 Frequency histogram of thermal growth tolerance for 4802 fungal strains (bars)

affected society to cope through the use of local resources (WHO definitions: emergencies, cited 2021 Jun 12).

Economic, social, and cultural adjustments create a unique set of circumstances in each type of natural disaster, just as the immediate causes of illness or death vary in each type of event (hit with objects, lacerations, crushing, suffocation, and drowning) (Noji 2000).

Postimpact characteristics and recovery phases from a natural disaster, such as displacement of human populations, reduction in vaccine coverage for preventable diseases, deficiencies in sanitation, and limited access to hospitals, can lead to an increase in infectious diseases that were previously controlled or occurred infrequently in the affected area. There is a strong association between fungal infections and natural disasters. This is attributed to the fact that in the impact phase of a given disaster, the fungal conidia that are in the environment can be inhaled, aspirated, or inoculated into the skin of humans. On the other hand, after a natural disaster, conidia can spread to other regions. This would lead to an increase in their environmental concentration in a given region or the entrance of conidia into habitats where they are not normally present. Both situations lead to a probability of making contact with injured people and causing fungal disease. The direct role that climate change has in the mechanisms that pathogenic fungi have regarding environmental growth, dispersal, and subsequent distribution is not fully understood. However, indirectly, climate change can lead to an increase in natural disasters which is associated with an increase in fungal infections (Benedict and Park 2014).

Natural disasters can be classified into three groups: geophysical (earthquakes, volcanic eruptions, and tsunamis), hydrometeorological (floods, hurricanes, and tornados), and geomorphological (landslides and avalanches) (Floret et al. 2006). The most common routes of fungal infection in natural disasters are inhalation, semi-drowning, and subcutaneous tissue trauma.

Inhalation

This is a very common route to acquire a fungal infection. Fungi can cause respiratory infections ranging from asymptomatic to severe, life-threatening manifestations in humans. This wide spectrum of clinical manifestations depends on the host and the characteristics of the pathogen. *Coccidioides* spp. is a fungus that grows in semiarid soil and is endemic in the Southwestern United States and Northern Mexico. This fungus is one of the best examples that associate the impact of natural disasters related to the removal of dust and the subsequent appearance of fungal diseases. A classic report of a coccidioidomycosis outbreak occurred after a dust storm in the San Joaquin Valley in California in 1977. The storm started in Bakersfield, a city well recognized as endemic for this fungus. The dispersion of dust with conidia caused the establishment of fungi in other cities such as Sacramento, considered non-endemic. After this dust storm, more than 100 cases of coccidioidomycosis were reported (Flynn et al. 1979). Another very illustrative report on the effect of natural disaster conditions and the appearance of fungal diseases was the earthquake in Northridge, California, in 1994. As a consequence, *Coccidioides* conidia were dispersed by the dust produced, and later outbreaks of coccidioidomycosis occurred in cities that were not considered endemic (Schneider et al. 1997).

Semi-drowning

Semi-drowning and drowning are very common conditions during a natural disaster related to flooding from hurricanes and tsunamis. However, tsunamis have a greater impact on the human population. Most immediate deaths usually result from drowning. However, other people remain alive with varying degrees of damage to their health. It is important to remember that these floods do not consist only of water, but they also contain a large amount of debris mixed with the water, traveling at a variable speed according to each event (Keim 2013) In general, three stages that impact human health during floods are caused by tsunamis: (a) In the first stage that occurs in the first minutes of the disaster, people are hit by a large number of diverse materials, such as construction material, metal, or glass sheets, or entangled in materials such as cables, wires, etc. Victims usually succumb to immediate damage to their bodies. (b) In the second stage, which occurs in the following minutes to hours after the disaster, complications also considered immediate appear as a result of wounds: bleeding, lung collapse, etc. and (c) in the third stage, which occurs days after the disaster, late complications occur that are mainly associated with infectious diseases (wound infections and aspiration pneumonia) (Noji 2000).

Aspiration of water contaminated with various debris can lead to sinus and lung infections. The aspiration pneumonia that occurs after this natural disaster is often referred to as a tsunami lung. The pathophysiology of the tsunami lung involves inflammation induced by chemicals, mechanical processes, and microorganisms. With regard to the latter, this entity can be caused by bacteria, fungi, or both (Table 9.4). Most often, a polymicrobial lung infection develops.

Allworth (2005) reported necrotic pneumonia in a group of survivors of the Asian tsunami in 2004. Approximately 1 month after the immersion event, patients developed fever, chronic nonproductive cough, and radiological evidence of bilateral necrotizing pneumonia with cavitation. A notable characteristic of these patients was the subacute presentation of their disease, the persistence of symptoms despite broad-spectrum antibiotic therapy, and the development of clinical and radiological manifestations of necrosis with pleural involvement.

Nakamura et al. (2011) reported the case of a 59-year-old woman who had been washed away by the tsunami that hit the Sanriku Coast in northeast Japan in 2011. She breathed in saltwater contaminated with soil and oil spilled into the sea by a ship that had overturned. She swam and reached land again, was in a gym as a shelter, and then was transferred to a hospital where she developed respiratory failure. On day 69 of hospitalization, *Scedosporium apiospermum* was identified from bronchoalveolar lavage.

The genus *Scedosporium* appears to be the most common opportunistic fungus associated with situations that lead to semi-drowning. However, the information is limited to case reports only. In these reports, the tropism of the microorganism to reach the central nervous system has also been documented, a situation that can occur even in immunocompetent individuals. The way in which *Scedosporium apiospermum* reaches the central nervous system is believed to be due to local dissemination from sites near the brain. On the one hand, there are paranasal sinuses or cribriform plate and, on the other hand, hematogenous dissemination from the lungs (Nakamura et al. 2011). This genus consists of cosmopolitan saprophytic fungi in

Table 9.4 Etiological agents frequently reported in the tsunami lung

Bacteria	Fungi
Aeromonas spp.	*Candida albicans*
Pseudomonas spp.	*Aspergillus fumigatus*
Acinetobacter baumannii	*Scedosporium apiospermum*
Staphylococcus aureus	
Stenotrophomonas maltophilia	
Burkholderia pseudomallei	
Burkholderia cepacia	
Legionella pneumophila	
Escherichia coli	

the soil of regions with temperate and tropical climates. Its isolation is frequent in soils with high human activity such as agricultural soils, urban gardens, playgrounds, sports, and domestic gardens. This is probably due to organic contamination (Elizondo-Zertuche et al. 2017).

Another case is that of a 68-year-old woman who experienced a semi-drowning episode during the tsunami that followed the 2011 Japan earthquake. She was diagnosed with a tsunami lung approximately 7 hours after the disaster. Before the incident, she had no significant history. On admission to the hospital, she was conscious; however, she gradually developed dyspnea and mental deterioration and was transferred to another hospital where she was intubated. Blood cultures were negative for bacteria and fungi. The 1,3 B-D-glucan test, which was also performed, was positive. *A. fumigatus* was identified in sputum cultures (Yutaka et al. 2012). This report is a warning of aspergillosis as an important complication of the tsunami lung and how even previously healthy individuals can contract a severe *Aspergillus* spp. infection. Therefore, it is important to consider the administration of an effective antifungal as empirical treatment in the early phase of hospitalization in survivors of a natural disaster, even in immunocompetent patients.

Subcutaneous Tissue Trauma

Wounds contaminated with water, dirt, or other debris is a very common situation after a natural disaster. Although the majority of soft tissue infections after a natural disaster are bacterial, fungal infections can also occur. Fungal infections are poorly recognized, especially early in the disease because they are clinically similar to bacterial infections.

Mucormycosis, caused by fungi belonging to the *Mucorales* order, is the most recognized fungi of post-disaster soft tissue infections. The natural disasters most implicated with this type of infection are volcano eruptions, tornados, and tsunamis. In some of these reports, the clinical challenge associated with soft tissue mucormycosis is highlighted since they are initially indistinguishable from other types of wound infections, but mucormycosis requires more aggressive antifungal treatment and surgical debridement. Post-disaster cases of mucormycosis have also been reported, especially in people working in the reconstruction stage, with cleaning brush and lifting fallen trees near their houses. Other fungi that have been reported in soft tissue fungal infections include *Fusarium* spp. and *Cladophialophora bantiana*.

Petrini et al. (2006) reported two Swedish individuals with severe soft tissue damage during the 2004 Thai tsunami. They subsequently received skin grafts after multiple surgeries. Four to 6 weeks after the event, they showed signs of purulent infection in non-traumatized skin and outside the margins of the resection. Skin biopsies revealed granulomatous reactions and pigmented hyphae in one of the patients. In the cultures, *C. bantiana* and *Mycobacterium abscessus* were recovered from both patients. After treatment with voriconazole, clarithromycin, and amikacin, the symptoms gradually subsided in one patient. In the other patient, there were reminiscent abscesses, and *Mycobacterium abscessus* was recovered in cultures despite prolonged antibiotic therapy, although the individual eventually recovered. *C. bantiana* is a dematiaceous fungus from the environment with worldwide distribution. It is better known as a causal agent of disseminated disease and brain abscesses mainly in immunosuppressed patients, although sporadically it has been reported to cause damage in immunocompetent individuals.

Damage to Infrastructure

Damage to infrastructure usually results from the crowding of survivors into temporary shelters, as well as damage to water systems, electricity, and sanitation. It should also be mentioned that the destruction of roads and airports can prevent the arrival of all humanitarian aid on time.

The event that happened in Colombo, Sri Lanka, represents the danger of fungal infections when infrastructure is damaged. An outbreak of *Aspergillus fumigatus* meningitis occurred in 2005 (Gunaratne et al. 2007), between June 21 and July 17, 2005, after the administration of anesthesia in previously healthy women who underwent cesarean section. Approximately 11 days later, they developed fever, headache, and stiff neck. Three patients died. Fungi culture from cerebrospinal fluid (CSF) or postmortem brain tissue resulted in *A. fumigatus*. As the suspicion of an outbreak began, exhaustive microbiological investigations began. Disposable syringes, intravenous and spinal needles, cannulas, and ampules with anesthetics were collected. *A. fumigatus* was found in syringes from three different brands. When the places where the material was stored were reviewed, they found an inadequate space for tsunami donations: leaks in the ceiling, dust, and humidity. Although the origin of the contamination was not found, the most likely explanation is that the storage facilities could be involved given the poor storage conditions in which they were found after the tsunami in 2004. Because of the massive influx of donations, they were still in operation. *Aspergillus* meningitis has a high mortality rate; the survival of two of the heavily treated patients shows the importance of specific diagnostic tests to bring early recognition and administration of appropriate treatment. Removal and incineration of all unused material controlled the outbreak.

9.5 Conclusion

Very little has been studied about global warming and the impact on fungi, but it is believed that it could have an important effect on these microorganisms. A warmer ambient temperature will surely change the distribution of susceptible and heat-tolerant species favoring those that are more heat-tolerant and creating the conditions for environmental fungi to spread and come into close contact with humans. On the other hand, the rise in temperature in the world will lead to more drastic and frequent meteorological conditions. Fungal diseases occur after natural disasters.

These disasters damage human lives, infrastructure, the environment, and human shelter and water and food sources. It is not difficult to imagine the degree of harm to humans in the future caused by this climate-changed world.

> Future generations stand to inherit a greatly spoiled world. Our children and grandchildren should not have to pay the cost of our generation's irresponsibility. Today's ecological crisis, especially climate change, threatens the very future of the human family. This is no exaggeration.—Pope Francis, 2018

References

Allworth AM. Tsunami lung: a necrotising pneuminia in survivors of the Asian tsunami. Med J Aust. 2005;182:364. https://doi.org/10.5694/j.1326-5377.2005.tb06738.x.

Bajpai VK, Khan I, Shukla S, Kumar P, Rather IA, Park YH, et al. Invasive fungal infections and their epidemiology: measures in the clinical scenario. Biotechnol Bioprocess Eng. 2019;24:436–44.

Barrozo LV, Mendes RP, Marques SA, Benard G, Silva ME, Bagagli E. Climate and acute / subacute paracoccidioidomycosis in a hyper-endemic area in Brazil. Int J Epidemiol. 2009;38:1642–9. https://doi.org/10.1093/ije/dyp207.

Benedict K, Park BJ. Invasive fungal infections after natural disasters. Emerg Infect Dis. 2014;20:349–55. https://doi.org/10.3201/eid2003.131230.

Biegańska MJ. Two fundamentals of mammalian defense in fungal infections: endothermy and innate antifungal immunity. Pol J Vet Sci. 2014;17:555–67. https://doi.org/10.2478/pjvs-2014-0084.

Bongomin F, Gago S, Oladele RO, Denning DW. Global and multi-national prevalence of fungal diseases-estimate precisión. J Fungi. 2017;3(4):–57. https://doi.org/10.3390/jof3040057.

Bromuro C, Romano M, Chiani P, Berti F, Tontini M, Proietti D, Mori E, Torosanticci A, Costantino P, Rappuoli R, Cassone A. Beta-glucan-CRM197 conjugates as candidate antifungal vaccines. Vaccine. 2010;28:2615–23. https://doi.org/10.1016/j.vaccine.2010.01.012.

Convention of the Elimination of all Forms of Discrimination Against Women. Adopted by the United Nations General Assembly, New York, December 18, 1979.

Denning DW. Minimizing fungal disease deaths will allow the UNAIDS target of reducing annual AIDS deaths below 500,000 by 2020 to be realized. Philos Trans R SocBiol Sci. 2016;371(1709):20150468. https://doi.org/10.1098/rstb.2015.0468.71.

Elizondo-Zertuche M, de J Treviño-Rangel R, Robledo-Leal E, Luna-Rodriguez CE, Martínez-Fierro ML, Rodríguez-Sanchez IP, et al. Molecular identification and in vitro antifungal susceptibility of Scedosporium complex isolates from high-human-activity sites in Mexico. Mycologia. 2017;109:874–81. https://doi.org/10.1080/00275514.2017.1416260.

Enoch DA, Ludlam HA, Brown NM. Invasive fungal infections: a review of epidemiology and management options. J Med Microbiol. 2006;55:809–18. https://doi.org/10.1099/jmm.0.46548-0.

Floret N, Viel JF, Mauny F, Hoen B, Piarroux R. Negligible risk for epidemics after geophysical disasters. Emerg Infect Dis. 2006;12:543–7. https://doi.org/10.3201/eid1204.051569.

Flynn NM, Hoeprich PD, Kawachi MM, Lee KK, Lawrence RM, Goldstein E, et al. An unusual outbreak of windborne coccidioidomycosis. N Engl J Med. 1979;301:358–61. https://doi.org/10.1056/NEJM197908163010705.

Global Action Fund for Fungal Infections (GAFFI). Priority fungal infections. 2017. http://www.gaffi.org/media/fact-sheets/. Accessed 14 Aug 2017.

Gunaratne PS, Wijeyaratne CN, Seneviratne HR. Aspergillus meningitis in Sri Lanka-a post –tsunami effect? N Engl J Med. 2007;356:754–6. https://doi.org/10.1056/NEJMc062547.

Hawksworth DL. The magnitude of fungal diversity: the 1.5 million species estimate revisited. Mycol Res. 2001;105:1422–32.
Hawksworth DL, Lucking R. Fungal diversity revisited: 2.2 to 3.8 million species. Microbiol Spectr. 2017; https://doi.org/10.1128/microbiolspec.FUNK-0052-2016.
Hawksworth DL, Rossman AY. Where are all the undescribed fungi? Phytopathology. 1997;87:888–91. https://doi.org/10.1094/PHYTO.1997.87.9.888.
Hillenius WJ, Ruben JA. The evolution of endothermy in terrestrail vertebrates: who? When? Why? Physiol Biochem Zool. 2004;77:1019–42. https://doi.org/10.1086/425185.
Hoffman PF, Kaufman AJ, Halverson GP, Schrag DP. A neoproterozoic snowball earth. Science. 1998;281:1342–6. https://doi.org/10.1126/science.281.5381.1342.
Intergovernmental Panel on climate change. Climate change 2007: synthesis report. Geneva: Intergovernmental Panel on Climate Change; 2007.
International Covenant on Civil and Political Rights. Adopted by the United Nations General Assembly, New York, December 16, 1966.
International Covenant on Economic, Social and Cultural Rights. Adopted by the United Nations General Assembly, New York, December 16, 1966.
Kawakami Y, Tagami T, Kusakabe T, Kido N, Kawaguchi T, Omura M, et al. Disseminates aspergillosis associated with tsunami lung. Respir Care. 2012;57:1674–8. https://doi.org/10.4187/respcare.01701.
Keim ME. Chapter 7. Health-related impacts of tsunami disasters. Southampton. ISSN 1755–8336 (On-line): WIT Press; 2013. https://doi.org/10.2495/978-1-84564-770-4/007.
Marquez LM, Redman RS, Rodriguez RJ, Roossinck MJ. A virus in a fungus in a plant: three –way symbiosis required for thermal tolerance. Science. 2007;315:513–5. https://doi.org/10.1126/science.1136237.
McMichael C. Climate change-related migration and infectious disease. Virulence. 2015;6:548–53. https://doi.org/10.1080/21505594.2015.1021539.
Nakamura Y, Utsumi Y, Suzuki N, Nakajima Y, Murata O, Sasaki N, et al. Multiple Scedosporium apiospermum abscesses in a woman survivor of a tsunami in northeastern Japan: a case report. J Med Case Rep. 2011;5:526–30. https://doi.org/10.1186/1752-1947-5-526.
Noji EK. The public health consequences of disasters. Prehosp Disaster Med. 2000;15:147–57.
Petrini B, Farnebo F, Hedblad M-A, Appelgren. Concomitant late soft tissue infections by Cladophialophora bantiana and Mycobacterium abscessus following tsunami injuries. Med Mycol. 2006;44:89–92. https://doi.org/10.1080/13693780500294949.
Quaedvlieg W, Binder M, Groenewald JZ, Summerbell BA, Carnegie AJ, Burgess TI, et al. Introducing the consolidated species concept to resolve species in the Teratosphaericaeae. Persoonia. 2014;33:1–40. https://doi.org/10.3767/003158514X681981.
Richards RA. The species problem: a philosophical analysis. Cambridge: Cambridge University Press; 2010. https://doi.org/10.1017/CBO9780511762222.
Robert VA, Casadevall A. Vertebrate endothermy restricti most fungi as potential pathogen. J Infect Dis. 2009;200:1623–6. https://doi.org/10.1086/644642.
Rockstrom J, Steffen W, Noone K, Persson A, Stuart-Chapin F 3rd, Lambin EF, et al. A safe operating space for humanity. Nature. 2009;461:472–5. https://doi.org/10.1038/461472a.
Ruben J. The evolution of endothermy in mammals and birds: from physiology to fossils. Annu Rev Physiol. 1995;57:69–95. https://doi.org/10.1146/annurev.ph.57.030195.000441.
Schneider E, Haijch RA, Spiegel RA, Jibson RW, Harp EL, Marshall GA, et al. A coccidioidomycosis outbreak following the Northridge, Calif, earthquake. JAMA. 1997;277:904–8.
Taylor LH, Latham SM, Woolhouse MEJ. Risk factors for human disease emergence. Philos Trans R Soc Lond Ser B Biol Sci. 2001;356:983–9. https://doi.org/10.1098/rstb.2001.0889.
The Universal Declaration of Human Rights. Adopted by the United Nations General Assembly, Paris, December 10, 1948.
Vincent JL, Rello J, Marshall J, Silva E, Anzueto A, Martin CD, Moreno R, et al. International study of the prevalence and outcomes of infection in intensive care units. JAMA. 2009;302(21):2323–2329. https://doi.org/10.1001/jama.2009.1754.

Vivek BK, Khan I, Shukla S, Kumar P, Rather I, Park Y, et al. Invasive Fungal Infections and Their Epidemiology: Measures in the Clinical Scenario. Biotechnol Bioproc E. 2019;24;436–444. https://doi.org/10.1007/s12257-018-0477-0.

Watts N, Adger WN, Agnolucci P, Blackstock J, ByassP CW, et al. Health and climate change: policy responses to protect public health. Lancet. 2015;386:1861–914. https://doi.org/10.1016/S0140-6736(15)60854-6.

Wheeler N, Watts N. Climate change: from science to practice. Curr Environ Health Rep. 2018;5:170–8. https://doi.org/10.1007/s40572-018-0187-y.

World Health Organization. WHO definitions: emergencies (cited 2021 Jun 12). http://www.who.int/hac/about/definitions/en/index.html

World Health Organization. Tuberculosis. 2021. htpp://www.who.int/mediacentre/factsheets/fs104/en/.

Yutaka K, Tagami T, Kusakabe T, Kido N, Kawaguchi T, Omura M, et al. Disseminated aspergillosis associated with tsunami lung. Respir. Care. 2012;57(10):1674–8. https://doi.org/10.4187/respcare.01701.

Chapter 10
Opportunistic Yeast Infections and Climate Change: The Emergence of *Candida auris*

María Guadalupe Frías-De-León, Eduardo García-Salazar, María del Rocío Reyes-Montes, Esperanza Duarte-Escalante, and Gustavo Acosta-Altamirano

Contents

Abbreviations

CDC	Centers for Disease Control and Prevention
ITS	Internal transcribed spacer
NCEI	National Centers for Environmental Information
SNPs	Single nucleotide polymorphisms

M. G. Frías-De-León (✉)
Unit Research, Hospital Regional de Alta Especialidad de Ixtapaluca, Mexico State, Mexico

E. García-Salazar
Unidad de Investigación, Hospital Regional de Alta Especialidad de Ixtapaluca, Estado de México, Mexico

Master of Health Sciences, Escuela Superior de Medicina, Instituto Politécnico Nacional, CDMX, Mexico City, Mexico

M. d. R. Reyes-Montes · E. Duarte-Escalante
Microbiology and Parasitology Department, School of Medicine, Universidad Nacional Autónoma de México, Mexico City, Mexico
e-mail: remoa@unam.mx; dupe@unam.mx

G. Acosta-Altamirano
Unidad de Investigación, Hospital Regional de Alta Especialidad de Ixtapaluca, Ixtapaluca, Estado de México, Mexico

M. G. Frías-De-León et al. (eds.), *The Impact of Climate Change on Fungal Diseases*, Fungal Biology, https://doi.org/10.1007/978-3-030-89664-5_10

10.1 Introduction

In recent years, the frequency of opportunistic fungal infections has increased, particularly in some hospital areas like intensive care units (Gnat et al. 2021). This increase is associated with the growing susceptible population, which includes immunocompromised patients with serious diseases or immunosuppressive pharmacological treatments; patients receiving parenteral nutrition, broad-spectrum antibiotics, or organ transplants; as well as patients who have cancer or diabetes (de Oliveira Santos et al. 2018; Garbee et al. 2017; Gnat et al. 2021). The most common opportunistic fungal infection is candidiasis, caused by yeasts of the genus *Candida* which includes more than 200 species, but only about 10% are recognized as pathogenic to humans and some animals (Gnat et al. 2021; Pal et al. 2015). Some *Candida* species are part of the mycobiota of humans' skin and gastrointestinal tract (de Oliveira Santos et al. 2018). However, under certain immunosuppression conditions, they can become pathogenic and cause disease ranging from a superficial infection affecting the skin or mucous membranes to a systemic infection that can spread and invade internal organs (Pappas et al. 2018). Superficial candidiasis is the most common clinical form, usually presented chronically and recurrently, without endangering the lives of those who have it (Reyes-Montes et al. 2017). On the other hand, the invasive form is recognized as one of the main causes of morbidity and mortality in the hospital environment (Reyes-Montes et al. 2017; Spampinato and Leonardi 2013). The most frequent etiological agent of candidiasis, both superficial and invasive, is *Candida albicans*, which is the species with the highest pathogenicity within the genus (Borman et al. 2016). However, in the last two decades, the isolation of non-*albicans* species like *C. glabrata*, *C. parapsilosis*, *C. tropicalis*, *C. guilliermondii*, and *C. krusei* has been reported more frequently around the world (Bassetti et al. 2019; Garbee et al. 2017; Martins et al. 2014; Reyes-Montes et al. 2017). This change in etiology is relevant because the non-*albicans* species tend to have low susceptibility to antifungals, complicating the therapeutic management of patients (Gonzalez-Lara and Ostrosky-Zeichner 2020). It is important to mention that the number of pathogenic species within the genus *Candida* has increased with the discovery of new species, some of which are phenotypically indistinguishable from already known species due to a close phylogenetic relationship. Such is the case of *C. nivariensis* and *C. bracarensis*, which are related to *C. glabrata* (Alcoba-Flórez et al. 2005; Correia et al. 2006), as well as *C. metapsilosis* and *C. orthopsilosis* that are associated with *C. parapsilosis* (Bertini et al. 2013; Tavanti et al. 2005). Also, a little over a decade ago, a new clinically relevant species was reported, called *C. auris*, whose incidence increased rapidly worldwide, with a significant mortality rate, higher than 60%, and associated with the yeast tendency to show resistance to multiple types of antifungals (Ademe and Girma 2020; Satoh et al. 2009; Spivak and Hanson 2018; Tsay et al. 2017; Uppuluri 2020). In addition, other important features that *C. auris* possesses and that have made it a threat to global public health are the ability to contaminate and persist in hospital environments and spread to patients and healthcare personnel, causing colonization and outbreaks of invasive infections (Ademe and Girma 2020; Spivak and Hanson 2018; Uppuluri 2020). However, one of the aspects that draw more attention to this new pathogen is the cause that led to its sudden and apparently synchronized appearance on three continents (Lockhart et al. 2018). Although the emergence of *C. auris* can be associated with multiple factors, climate change seems to be a key factor (Casadevall et al. 2019, 2021).

This chapter presents the general aspects of *C. auris* and the disease caused by this fungus, as well as the evidence reflecting the possible participation of climate change in its appearance.

10.2 *C. auris:* An Emerging Yeast with Distinctive Features

In 2009, a yeast fungus isolated from the discharge of the ear canal of a 70-year-old woman suffering from otitis media was reported in Tokyo, Japan (Satoh et al. 2009). This yeast was described as a new species within the genus *Candida* and was called *C. auris* (Satoh et al. 2009). The identification of *C. auris* as a new species was based on both its genotypic and phenotypic characteristics.

At the genotypic level, the analysis of the sequences obtained from the internal transcribed spacer (ITS) and the D1/D2 domain of the (28S) rDNA gene established *C. auris* as a distinct species in the Metschnikowiaceae family within the clade *Candida/Clavispora*, showing a close phylogenetic relationship with the *C. haemulonii* species complex (Satoh et al. 2009). The *C. haemulonii* complex is composed of *C. haemulonii* sensu stricto, *C. haemulonii* var. *vulnera*, *C. duobushaemulonii*, and *C. pseudohaemulonii*, as well as the yeasts *C. ruelliae*, *C. vulturna*, and *C. heveicola*, which are isolated from flowers and plant material (Cendejas-Bueno et al. 2012; Jackson et al. 2019; Satoh et al. 2009).

At the phenotypic level, *C. auris* possesses unique features compared to other *Candida* species, including those of the *C. haemulonii* complex:

1. *Thermotolerance*. *C. auris* optimal growth, both in its aggregate form (i.e., cells where budding occurs without releasing daughter cells) and non-aggregate form (i.e., dissociated cells), is observed between 37 and 40 °C. It can even grow at 42 °C, although slowly, while species such as *C. albicans*, *C. haemulonii*, *C. pseudohaemulonii*, or *C. heveicola* do not tolerate this temperature.
2. *Osmotolerance*. *C. auris* tolerance to high salt concentrations (10% NaCl) is another distinctive feature. It has been observed that salinity induces the formation of elongated cellular morphologies, similar to the pseudohyphae of other *Candida* species, suggesting that the morphological transition may be adaptive under stressful conditions (Wang et al. 2018).

   Likewise, thermotolerance and osmotolerance are characteristics that can contribute to the persistence and survival of *C. auris* on biotic and abiotic surfaces over long periods (Kean et al. 2018; Welsh et al. 2017). It is known that *C. auris* survives several weeks on human skin and environmental surfaces and may even tolerate exposure to some commonly used disinfectants (Du et al. 2020; Uppuluri 2020).

3. *Carbon source*. *C. auris* can assimilate glucose, sucrose, maltose, D-trehalose, D-raffinose, D-melezitose, D-mannitol, sorbitol, citrate, inulin, starch, ribitol, and galactitol as carbon sources. However, it has little ability to assimilate galactose, l-sorbose, cellobiose, and l-arabinose, and the assimilation of N-acetylglucosamine, gluconate, and succinate varies depending on the geographic origin. For example, isolates from Japan or Korea do not assimilate N-acetylglucosamine (Iguchi et al. 2019).
4. *Morphological phenotypes*. Some of the *C. auris* isolates are unicellular yeasts, while others form large aggregates, simulating pseudohyphae, since stem and daughter cells do not dissociate, which can be caused by a defect in cell division (Borman et al. 2016; Du et al. 2020). There is evidence that *C. auris* isolates can form true hyphae in specific circumstances, such as the presence of 10% NaCl in culture media at 37 and 42 °C (Wang et al. 2018). Besides, the factors that induce filamentation in *C. albicans*, such as serum, N-acetylglucosamine, and high CO_2 levels, do not have the same effect on *C. auris* (Wang et al. 2018).
5. *Biofilm formation*. *C. auris* forms biofilms on surfaces, although these are weaker than the *C. albicans* biofilms. However, in the same way as with other *Candida* spp., *C. auris* cells in the biofilm tend to show greater resistance to antifungals than planktonic cells. It has been observed that biofilm formation varies between *C. auris* isolates. The isolates with a non-aggregate growth develop more robust biofilms than the isolates with aggregate growth (Singh et al. 2019).
6. *Antifungal resistance*. Another distinctive feature of *C. auris* is its intrinsic resistance to one or several types of antifungals (azoles, polyenes, and echinocandins) used for candidiasis treatment (Chybowska et al. 2020). Low susceptibility to triazoles (itraconazole, voriconazole, and isavuconazole) has been reported. Some isolates exhibit in vitro susceptibility to fluconazole, but in vivo treatment with this antifungal has failed (Vallabhaneni et al. 2016). *C. auris* susceptibility to polyenes, such as amphotericin B, is variable. It has been noted that between 10% and 35% of isolates are resistant to amphotericin B (Ostrowsky et al. 2020; Sarma et al. 2013). Concerning echinocandins, micafungin has shown a fungicidal effect against *C. auris* (Dudiuk et al. 2019), but caspofungin and anidulafungin have shown only a fungistatic effect in vitro, despite acting as fungicides with other *Candida* species (Kordalewska et al. 2018). Due to this variable resistance among isolates, it has not yet been possible to define optimal treatment regimens against *C. auris*. It is worth mentioning that the antifungal multiresistance that *C. auris* exhibits has also been observed in some phylogenetically related species, such as *C. haemulonii* (Ben-Ami et al. 2017; Kim et al. 2009). Therefore, their resistance mechanisms are also similar. The main resistance mechanisms to azoles and echinocandins are the Y132F mutations in ERG11 (gene encoding for lanosterol 14α-demethylase) and S639P in FKS1 (gene encoding for 1,3 β-D-glucan synthase), respectively (Chow et al. 2020).
7. *Virulence factors*. *C. auris* can adhere to the tissue surface and invade the host cell by producing lytic enzymes as proteinase, aspartyl protease, hemolysins,

lipases, and phospholipase (Chybowska et al. 2020). The level of activity of *C. auris* lytic enzymes may vary between isolates depending on their geographical origin; however, the phospholipase activity in *C. auris* is comparable to that of *C. albicans* (Friedman and Schwartz 2019; Jeffery-Smith et al. 2017; Navalkele et al. 2017; Spivak and Hanson 2018). The ability to form biofilms is another pathogenicity factor of *C. auris*, and it has been shown that isolates with a non-aggregate growth tend to produce more biofilms than those with aggregate growth (Sherry et al. 2017). Still, the biomass in *C. auris* biofilms is reduced compared to that of *C. albicans* biofilms (Jeffery-Smith et al. 2017). The form of growth is probably another virulence factor of *C. auris*. It has been observed that isolates with in vitro non-aggregate growth exhibit a higher or similar virulence to that of *C. albicans* compared to isolates with aggregate growth (Ben-Ami et al. 2017; Borman et al. 2016; Chybowska et al. 2020; Sherry et al. 2017). But it is still unknown whether aggregate-forming isolates cause less spread in human infections or if the aggregates protect the fungus against the host's immune response, the antifungal effects, and the detergents used to clean hospital environments (Borman et al. 2016; Spivak and Hanson 2018).

8. *Complex identification. C. auris* identification is difficult when conventional methods for *Candida* spp. and other clinically relevant yeasts are used. For instance, in chromogenic culture media, *C. auris* cells can form light pink to beige or even white colonies, preventing species identification (Bidaud et al. 2018). Automated systems, based on biochemical tests, often provide erroneous or confusing results for identifying *C. auris* isolates because these systems have not yet included the assimilation profiles for carbohydrates and other compounds corresponding to *C. auris*. Therefore, it is impossible to differentiate *C. auris* phenotypically from other *Candida* species or other yeasts (ElBaradei 2020; Kathuria et al. 2015; Mizusawa et al. 2017). For this reason, to alert users, the Centers for Disease Control and Prevention (CDC) of the United States published information on the *Candida* species and other yeasts that can be mistaken for *C. auris* when using the Vitek-MS® (bioMérieux, Marcy-l'Étoile, France), Vitek-2® (bioMérieux), Vitek 2YST® (bioMérieux), MicroScan (Beckman Coulter, Pasadena, USA), BD Phoenix® (Becton, Dickinson and Company, Franklin Lakes, USA), API 20C AUX® (bioMérieux), and RapID Yeast Plus® systems (Innovative Diagnostic Systems, Saint Eustache, Canada) (Table 10.1).

Due to the complexity to identify *C. auris* at the phenotypic level, the sequencing of the ITS or the D1/D2 domains of the (28S) rRNA gene is currently considered the gold standard methods for correctly identifying *C. auris* and differentiating it from other species, particularly those that conform the *C. haemulonii* complex. Likewise, it is possible to identify *C. auris* adequately through the protein profile obtained by MALDI-TOF MS (Kathuria et al. 2015; Kim et al. 2009; Kordalewska and Perlin 2019; Mahmoudi et al. 2019).

Table 10.1 Misidentification of *C. auris* through the most commonly used automated systems for clinically relevant yeasts based on biochemical tests

Yeast misidentified as *Candida auris*	Identification method							
	MicroScan©	Vitek-MS©	Vitek 2©	Vitek 2YST©	BD Phoenix system©	API 20C AUX©	API 32C©	RapID Yeast Plus©
Candida albicans	X	X						
Candida famata	X		X					
Candida guilliermondii	X							
Candida lusitaniae	X	X						
Candida parapsilosis	X							X
Candida catenulata	X				X			
Candida tropicalis	X							
Candida haemulonii		X	X	X	X			
Candida sake						X	X	
Candida duobushaemulonii				X				
Candida intermedia							X	
Candida spp.				X				
Rhodotorula glutinis				X		X		
Saccharomyces kluyveri							X	

Information obtained from the Centers for Disease Control and Prevention. Identification of *Candida auris*. Available online: https://www.cdc.gov/fungal/candida-auris/identifification.html?CDC_AA_refVal=https%3A%2F%2Fwww.cdc.gov%2Ffungal%2Fcandida-auris%2Frecommendations.html (accessed on 18 May 2021)

10.3 Infections Caused by *C. auris*

Since 2009, when the first case of infection (otitis media) by *C. auris* was reported in a geriatric patient (Satoh et al. 2009), this yeast has been associated with a variety of both superficial and invasive infections (Lockhart et al. 2018). However, invasive infections have been more frequent since most isolates have been recovered from sterile body fluids, such as blood, cerebrospinal fluid, sputum, urine, bile, and soft tissues, among others (Calvo et al. 2016; Du et al. 2020; Iguchi et al. 2019; Jeffery-Smith et al. 2017; Kathuria et al. 2015; Lee et al. 2011; Morales-Lopez et al. 2017). Among the invasive infections caused by *C. auris*, candidemia is the most observed one, affecting newborn to elderly patients. The estimated mortality rate is between

30% and 60%, mostly due to the yeast resistance to different antifungals (Spivak and Hanson 2018). Risk factors for candidemia caused by *C. auris* are the same as for any *Candida* spp. infection. They include hospitalization in critical areas, broad-spectrum antibiotic therapies or invasive medical procedures, advanced age, diabetes mellitus, recent surgery, chronic kidney disease, and a history of central venous catheter use, among others (Calvo et al. 2016; Du et al. 2020; Lockhart et al. 2018; Spivak and Hanson 2018). In the same way as with other invasive infections by *Candida* spp., infections caused by *C. auris* are usually acquired in the hospital. They can even occur several weeks after the patient's admission to the hospital, suggesting a source of exogenous infection associated with the poor practice of infection control recommendations (Lockhart et al. 2018). Nosocomial infections by *C. auris* usually occur with a clonal distribution of incident cases, demonstrating horizontal transmission between susceptible patients (Hata et al. 2020). Several studies have described infectious disease outbreaks and individuals colonized with *C. auris*, where the fungus was colonizing the skin (hands, armpit, groin) of patients and contacts of patients and health workers and numerous objects from the hospital environment. Such items include bed railings, manual bed controllers, chairs, bed trays and medical equipment, cell phones, floors, door handles, alcohol gel dispensers, sinks, urinals, and mop buckets (Escandón et al. 2018a, b; Tsay et al. 2017). These findings demonstrated that the hospital environment of a colonized or infected patient could lead to the transmission of *C. auris* and subsequent colonization or infection of other patients. A European hospital reported that more than 50% of infections caused by *C. auris* were preceded by patient colonization (Ruiz-Gaitan et al. 2018), which highlights the need to prevent colonization and identify colonized individuals. Therefore, it is recommended that all hospitalized patients with infection or colonization with *C. auris* are treated using both standard precautions and contact precautions (Hata et al. 2020).

10.4 Epidemiological Expansion of *C. auris*

Although it was in 2009, in Japan, the first time that *C. auris* was described as a new species causing infection in humans, the first known case was retrospectively identified by sequencing a bloodstream isolate obtained from a pediatric patient in South Korea in 1996, which had been erroneously described as *C. haemulonii* and *Rhodotorula glutinis* using the Vitek 2YST and API 20C systems, respectively (Lee et al. 2011). There are also records of seven *C. auris* isolates obtained in 1997 from patients with otic infection in Japan, misidentified as *C. sake* and *Saccharomyces kluyveri* with the API ID32C system (Iguchi et al. 2019). In 2008, a *C. auris* isolate from Pakistan that had not been appropriately typified was also retrospectively reclassified (Lockhart et al. 2018). It should be noted that these are the only reports of *C. auris* in the period between 1996 and 2009. Before 1996 there are no other reports, which indicates that the prevalence of this pathogen was low (Lockhart et al. 2018). However, as of 2009, cases of infections by *C. auris* increased steadily and in a short time in several countries. Furthermore, in 2015, the CDC reported outbreaks in India, South Africa, and Venezuela while handling an outbreak in Pakistan (Calvo et al. 2016; Chowdhary et al. 2013, 2014; Lockhart et al. 2018; Magobo et al. 2014). This sudden simultaneous outburst on three continents, Asia, Africa, and South America, created uncertainty on how the yeast had spread to these sites.

To better understand the epidemiology of *C. auris* and to define whether its emergence occurred independently in several countries or if it was caused by the spread of a single outbreak, international collaboration was established to perform the whole-genome sequencing (WGS) of isolates from Pakistan, India, South Africa, and Venezuela (Lockhart et al. 2018). The genome sequencing showed single nucleotide polymorphisms (SNPs) among the isolates, with which phylogenetic analyses were performed that revealed a phylogeographic structure composed of five genetically distinct clades: clade I, South Asia (India/Pakistan); clade II, East Asia (Japan); clade III, South Africa; clade IV, South America (Venezuela) (Chow et al. 2020; Lockhart et al. 2018); and clade V, Iran (Chow et al. 2019). It has been observed that clade IV has a more consistent phylogeographic structure, while clades I, II, and III are mixed. For example, isolates from different clades have been found in the United Kingdom, the United States, Kenya, and Germany (Borman et al. 2017; Chow et al. 2018, 2020; Hamprecht et al. 2019) (Fig. 10.1). It has been postulated that people exposed to *C. auris* who travel internationally, such as healthcare workers, may have been involved in the spread and phylogeographic mixing of the pathogen (Chow et al. 2018). On the other hand, clade V is composed of a single isolate obtained in Iran from a patient with otomycosis who reported never having traveled outside the country. This suggests that *C. auris* may have been introduced in that geographic region a long time ago, but it was not previously reported due to identification issues (Chow et al. 2019). To the present date, all analyzed isolates

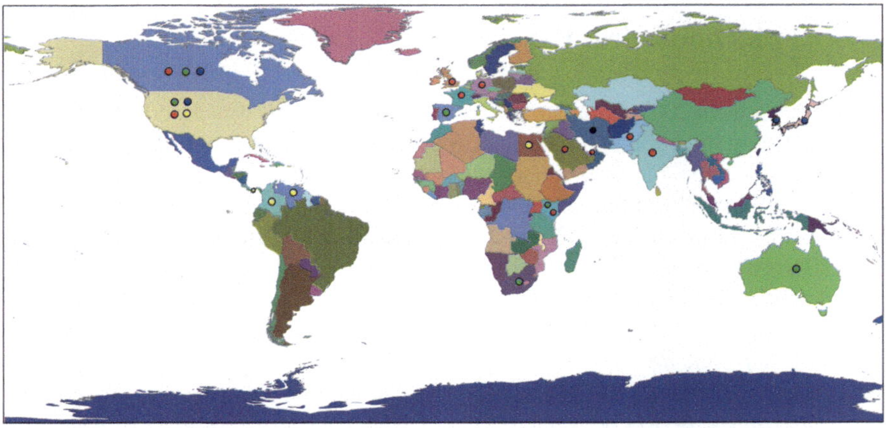

Fig. 10.1 Worldwide distribution of the five *Candida auris* phylogenetic clades: clade I, red; clade II, blue; clade III, green; clade IV, yellow; clade V, black

from different geographical origins have been integrated into any of the I–IV clades. However, no other isolate from clade V has been reported. It is relevant to mention that there is great diversity between clades, with differences of more than 200,000 SNPs, while within each clade, the differences are less than 70 SNPs (Chow et al. 2019; Lockhart et al. 2018; Muñoz et al. 2018). Clades I, III, and IV are related to invasive infections and exhibit more resistance to the three main types of antifungals (azoles, polyenes, and echinocandins). On the other hand, clades II and V present a noninvasive clinical phenotype linked to otic infections and low antifungal resistance (Chow et al. 2020; Sekizuka et al. 2019; Welsh et al. 2019). Thus, *C. auris* population structure demonstrates that the emergence of different clonal yeast populations occurred independently and almost simultaneously on three continents, Asia, Africa, and the Americas, ruling out the idea of a single-source spread (Chow et al. 2020; Lockhart et al. 2018). It also indicates that clonal isolates are distributed over long distances within countries and continents.

Currently, *C. auris* has been reported as the causative agent of infections or nosocomial outbreaks in more than 40 countries, including the United Arab Emirates, Australia, Austria, Belgium, Bangladesh, Canada, Switzerland, Chile, China, Colombia, Costa Rica, Germany, Egypt, Spain, France, the United Kingdom, Greece, India, Iran, Israel, Italy, Japan, Kenya, Kuwait, Lebanon, Malaysia, Mexico, the Netherlands, Norway, Oman, Pakistan, Panama, Peru, Poland, Russia, Saudi Arabia, Singapore, South Africa, South Korea, Sudan, Thailand, Turkey, the United States, and Venezuela (Fig. 10.2) (Ademe and Girma 2020; Allaw et al. 2021; Ayala-Gaytán et al. 2021; Chowdhary et al. 2017; Du et al. 2020; Hata et al. 2020; Kurt

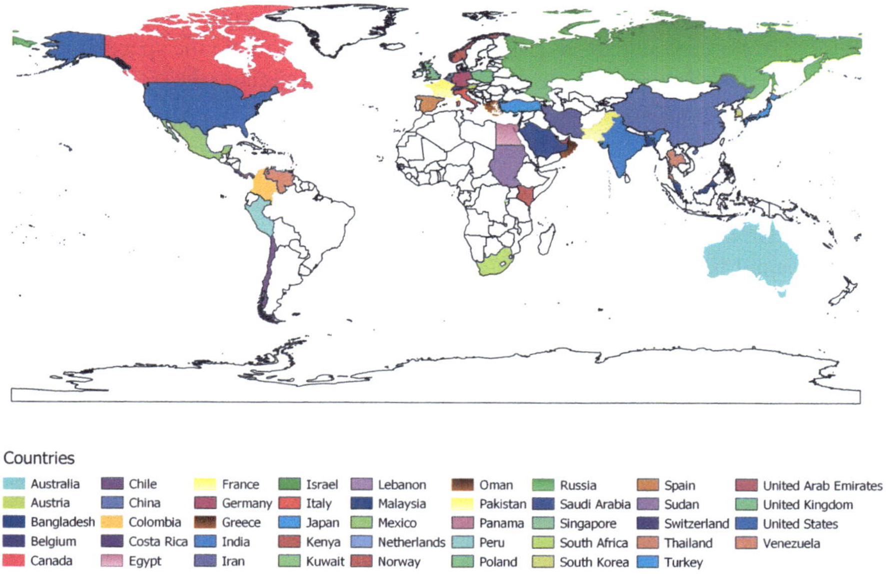

Fig. 10.2 Countries that have reported *Candida auris* isolates during the period 2009–2021

et al. 2021; Villanueva-Lozano et al. 2021). However, the distribution of *C. auris* might be even broader than what is currently known because, possibly, the yeast has not been correctly identified in some countries.

The causes of *C. auris* global expansion have not been clearly defined yet. It is speculated that selective pressures exerted by the increasing use of antifungals or the environment caused the recent adaptation of some characteristic features (osmotolerance, thermotolerance, antifungal resistance, cellular aggregation) to allow its expansion into new niches (Clancy and Nguyen 2017; Satoh et al. 2009).

10.5 The Origin of *C. auris*

The almost simultaneous way *C. auris* emerged in Asia, Africa, and America and its expanding speed to several countries generated uncertainty about its origin. It was initially suggested that the contributing factor to the emergence of *C. auris* was the excessive and indiscriminate use of antifungals, particularly azoles in the agricultural sector (Chowdhary et al. 2017; Cortegiani et al. 2019; Martin et al. 2020). However, while this may partially explain the emergence of new pathogenic microorganisms, it does not justify how the yeast expanded. Therefore, other hypotheses have been formulated proposing other factors may be involved, like some elements related to healthcare, changes in the behavior of both humans and the fungus, and climate change (Casadevall et al. 2019; Jackson et al. 2019; Misseri et al. 2019). Evidently, the spread of *C. auris* in healthcare settings is favored by its ability to colonize the human skin and contaminate surfaces persistently, its tolerance to many disinfecting agents used in hospitals, and the growing demand for intensive care unit services for diverse reasons. However, this does not explain the speed at which *C. auris* isolates, especially from clades I–III, spread throughout the world. Another hypothesis on the origin and expansion of *C. auris* poses that this fungus has always been present, with all its pathogenic potential, in some specific biogeographical niche (e.g., a plant, an insect, water) (Jackson et al. 2019). Although its ecological niche is currently unknown, it is assumed that humans had little contact with it (Steffen et al. 2015; Jackson et al. 2019). Nonetheless, human activities altering the environmental balance in ecosystems, such as deforestation and the expansion of agricultural land, modified or invaded this niche leading to greater contact between man and yeast. Consequently, the likelihood of fungal colonization in humans increased. Humans then introduced the fungus to the healthcare environment, which subsequently spread to vulnerable populations in different regions through international travel (Jackson et al. 2019; Steffen et al. 2015). It is thought that if water is *C. auris* habitat, then some fish farming activities, such as the practice of stabilizing pond microbiota adding yeast probiotics, increased the density of rare fungal communities, including *C. auris* (Hai 2015; Jackson et al. 2019). At the same time, the abundance of the fungus facilitated its contact with humans. Another hypothetical idea is that *C. auris* has been part of the mycobiota of relatively isolated human populations and that international travels increased contact with the rest of people, thus introducing the fungus to different world regions (Jackson et al. 2019). It has also been suggested that *C. auris* was a commensal yeast in the ear canal or some other human organ. The excessive use of antibiotics possibly generated a selective pressure that led to antifungal resistance and fungal adaptation to colonize other body parts, such as the skin, causing infection (Jackson et al. 2019). Finally, one of the most recent theories about the emergence of *C. auris* involves climate change (Casadevall et al. 2019, Casadevall et al. 2021; Misseri et al. 2019). As the average global temperature increases, a reduction in the gap between the environmental temperature and the body temperature of mammals is expected, causing new pathogenic fungi to appear (Garcia-Solache and Casadevall 2010). All these hypotheses provide information that will allow us to determine the possibly multifactorial origin of *C. auris*, which will lead to establishing the most appropriate measures to prevent the spread of this pathogen.

10.6 The Emergency of *C. auris* and Climate Change

None of the hypotheses put forward to explain the independent or simultaneous emergence of *C. auris* on three continents with genetically distinct clades have been yet proven. It was recently postulated that global warming is a crucial factor, though not the only, in the emergence of *C. auris* (Casadevall et al. 2019). Mammals are typically known to present resistance to invasive fungal diseases through their immune system's advanced mechanisms and high basal temperatures, which create a zone of thermal restriction (Casadevall 2012). However, it has been predicted that, with the imminent climate change, global warming will cause a reduction in the gradient between the environmental and basal temperatures of mammals, conferring fungi the ability to break through the mammalian thermal restriction zone and consequently greater pathogenicity (Garcia-Solache and Casadevall 2010; Casadevall et al. 2021). In addition, the ambient temperature increase can exert selective pressure on some fungi to acquire greater temperature tolerance (de Crecy et al. 2009; Robert et al. 2015). It has been observed that urban fungi tend to develop more tolerance to high soil and air temperatures compared to rural fungi (McClean et al. 2005). Therefore, urbanization may play a relevant role in the thermal adaptation of fungi. Based on the ability of various fungi to adapt to warmer temperatures, it was hypothesized that global warming would bring new fungal diseases (Garcia-Solache and Casadevall 2010) and new fungal pathogens, such as *C. auris* (Casadevall et al. 2019). There is evidence that proves the influence of global warming on the emergence of *C. auris* induced by anthropogenic activities. Evidence supporting this hypothesis is the higher thermotolerance of *C. auris*, which was demonstrated by growing in a wide temperature range from 25 up to 42 °C. Closely related species of the *C. haemulonii* complex only grow up to 37 °C, except for some *C. duobushaemulonii* isolates which can grow up to 40 °C (Casadevall et al. 2019). It is fitting to mention that, interestingly, the only species that exhibited a similar tolerance to temperature to *C. auris* was *C. ruelliae*. The latter is a new

Table 10.2 Thermotolerance in *C. auris* and phylogenetically related species

Species	Temperature (°C)							Source
	25	30	35	37	40	42	45	
Candida auris	X	X	X	X	X	X		Human ear
Candida haemulonis	X	X	X	X				Foot ulcer
Candida duobushaemulonii	X	X	X	X	X			Foot ulcer
Candida vulturna	X	X	X	X				Flowers
Candida ruelliae	X	X	X	X	X	X		Flowers
Candida heveicola	X	X	X	X	X			Soil

The table includes data from Casadevall et al. (2019)

environmental species isolated, in 2008, from flowers of the genus *Ruellia* within the family Acanthaceae, which is genetically related to the *C. haemulonii* species complex, sharing more than 80% of the genome (Casadevall et al. 2019; Puja and Prasad 2008) (Table 10.2).

On the other hand, *C. auris* was first isolated from the human ear, a cold body region, suggesting that the fungus possibly went through an adaptation period to higher temperatures before being associated with diseases (Kim et al. 2019). During such a period, it began to colonize larger areas like the skin. Furthermore, the fact that *C. auris* colonize cold body regions, coupled with its inability to grow in the absence of free oxygen, suggests that this fungus had an environmental origin. Several factors, including the overexpression of the 90 kDa heat-shock protein (Hsp90), favored not only its peculiar thermal tolerance but also its virulence, resistance to multiple antifungals, and tolerance to high salt concentrations (Kim et al. 2019). *C. auris* probably existed as a saprophyte in wetlands in the environment, where temperature fluctuations caused by global warming prompt it to adapt to warmer temperatures and higher salinity (Casadevall et al. 2019; Osland et al. 2018). Consequently, the community or biomass of *C. auris* yeasts increased in these ecological niches. An event that supports the idea that *C. auris* emerged from wetlands is that the geographic regions where the first isolations were reported included wetland ecosystems affected by global warming (Casadevall et al. 2019; Osland et al. 2018). Also, the idea that wetlands are *C. auris* ecological niche was recently reinforced when the yeast was isolated from samples taken from a sandy beach and a marsh (Arora et al. 2021). These were virgin areas inhabited by seabirds with no human activity, located around the Andaman Islands in the Indian Ocean. At the genetic level, these environmental isolates have a similar genome to that of the South Asian isolates; that is, they pertain to clade I (Arora et al. 2021). For *C. auris* to turn from a wetland fungus to a human pathogen, it is believed that some of the genes associated with virulence were transferred by plasmids from other *Candida* species that share the ecological niche and have intrinsic pathogenic potential, for example, *C. albicans* (Chatterjee et al. 2015; Stone et al. 2012). Interestingly, of the two *C. auris* isolates recovered from marine ecosystems, one was isolated along with *C. parapsilosis* yeasts and presented multiresistance to the antifungals. The other isolate was obtained along with *Trichosporon asahii* and resulted in being multisensitive and less temperature tolerant, as it grew more slowly at 37 and 42 °C than the multiresistant environmental isolate (Arora et al. 2021). It is speculated that another way by which *C. auris* may have acquired antifungal resistance is through mutations induced by increased ultraviolet radiation associated with climate change (Casadevall et al. 2019). It is also believed that birds with a body temperature between 41 and 42 °C could act as reservoirs from which the yeast leaped to humans and became infectious (Casadevall et al. 2019). For example, seabirds, such as seagulls, can act as reservoirs for the indirect transmission of multidrug-resistant *Candida* species to human hosts, as reported with *C. glabrata* (Al-Yasiri et al. 2016; Casadevall et al. 2019). However, this leap from animal to human may not be the only event involved in its evolution as a human pathogen, as it is feasible that global warming also affected the host (Casadevall et al. 2019).

Another fact supporting the theory that global warming is involved in the emergence of *C. auris* as a pathogen is that the first identified cases of infection occurred between 1996 and 2009. Before 1996, no other isolates of this species have been found, although extensive collections of clinical yeast isolates have been retrospectively analyzed (Chowdhary et al. 2017). This evidence is relevant because the National Centers for Environmental Information (NCEI) have classified 2009 as one of the ten hottest years in global history. It is also worth mentioning that 1998, a year close to the first records of *C. auris* in South Korea and Japan, is within the list of years when the global temperature rose. Incidentally, as of 2009, cases of infection by *C. auris* began to be reported more frequently in different parts of the world, and in this period, eight of the ten hottest years were recorded. In descending order, these years are 2019, 2015, 2017, 2018, 2014, 2010, 2013, and 2009. Hence,

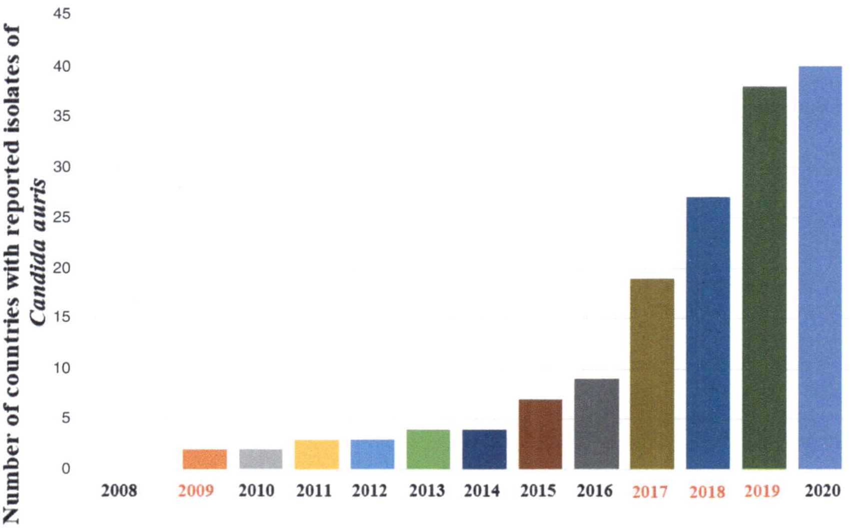

Fig. 10.3 Relationship between the temperature increase on Earth and *Candida auris* expansion in the world. The three hottest years in history are in red

there is a coincidence between the increased temperature on Earth and *C. auris* expansion worldwide (Fig. 10.3).

However, although there is sufficient evidence of the global warming involvement in the emergence of *C. auris* as a human pathogen, this variable does not explain the spontaneous appearance of four *C. auris* clades in geographically distant regions. Therefore, to have a better understanding of the origin and ecology of this fungus, as well as the interaction with humans, it is essential to consider other epidemiological variables, such as migrations and international travels, which may have contributed to its dispersion, persistence, and acquisition of antifungal resistance (Casadevall et al. 2019, 2021; Misseri et al. 2019).

10.7 Conclusion

Global warming is one of the main components of climate change, associated with anthropogenic activities, which significantly impact infectious diseases since it alters biodiversity and favors transmitting new pathogens to humans. To the present date, none of the hypotheses about the origin of *C. auris* has been verified. However, several pieces of evidence have been gathered showing that global warming is a key factor in the emergence of this yeast, which currently represents a global public health problem. Therefore, governments must take relevant actions to impact the causes that contribute to climate change and prevent the appearance of new pathogens.

Acknowledgments M.G.F.-D.-L. thanks the Consejo Nacional de Ciencia y Tecnología (CONACyT) for the support to the realization of this chapter. E.G.S thanks the Master's Program in Health Sciences of Escuela Superior de Medicina, Instituto Politécnico Nacional, and CONACyT for a scholarship (reference no. 1025718). The authors extend a special appreciation to Biol. Paola Betancourt Cisneros for the design of the figures.

References

Ademe M, Girma F. Candida auris: from multidrug resistance to pan-resistant strains. Infect Drug Resist. 2020;13:1287–94. https://doi.org/10.2147/IDR.S249864.

Alcoba-Flórez J, Méndez-Alvarez S, Cano J, Guarro J, Pérez-Roth E, del Pilar AM. Phenotypic and molecular characterization of Candida nivariensis sp. nov., a possible new opportunistic fungus. J Clin Microbiol. 2005;43(8):4107–11. https://doi.org/10.1128/JCM.43.8.4107-4111.2005.

Allaw F, Kara Zahreddine N, Ibrahim A, Tannous J, Taleb H, Bizri AR, et al. First Candida auris outbreak during a COVID-19 pandemic in a Tertiary-Care Center in Lebanon. Pathogens. 2021;10(2):157. https://doi.org/10.3390/pathogens10020157.

Al-Yasiri MH, Normand AC, L'Ollivier C, Lachaud L, Bourgeois N, Rebaudet S, et al. Opportunistic fungal pathogen Candida glabrata circulates between humans and yellow-legged gulls. Sci Rep. 2016;6:36157. https://doi.org/10.1038/srep36157.

Arora P, Singh P, Wang Y, Yadav A, Pawar K, Singh A, et al. Environmental isolation of Candida auris from the coastal wetlands of Andaman Islands, India. mBio. 2021;12:e03181–20. https://doi.org/10.1128/mBio.03181-20.

Ayala-Gaytán JJ, Montoya AM, Martínez-Resendez MF, Guajardo-Lara CE, de Treviño-Rangel JR, Salazar-Cavazos L, et al. First case of Candida auris isolated from the bloodstream of a Mexican patient with serious gastrointestinal complications from severe endometriosis. Infection. 2021;49(3):523–5. https://doi.org/10.1007/s15010-020-01525-1.

Bassetti M, Giacobbe DR, Vena A, Wolff M. Diagnosis and treatment of candidemia in the intensive care unit. Semin Respir Crit Care Med. 2019;40(4):524–39. https://doi.org/10.1055/s-0039-1693704.

Ben-Ami R, Berman J, Novikov A, et al. Multidrug-resistant Candida haemulonii and C. auris, Tel Aviv, Israel. Emerg Infect Dis. 2017;23(2):195–203. https://doi.org/10.3201/eid2302.161486.

Bertini A, de Bernardis F, Hensgens LA, Sandini S, Senesi S, Tavanti A. Comparison of Candida parapsilosis. Candida orthopsilosis, and Candida metapsilosis adhesive properties and pathogenicity. Int J Med Microbiol. 2013;303:98–103. https://doi.org/10.1016/j.ijmm.2012.12.006.

Bidaud AL, Chowdhary A, Dannaoui E. Candida auris: an emerging drug resistant yeast—a mini-review. J Mycol Med. 2018;28:568–73. https://doi.org/10.1016/j.mycmed.2018.06.007.

Borman AM, Szekely A, Johnson EM. Comparative pathogenicity of United Kingdom isolates of the emerging pathogen Candida auris and other key pathogenic Candida species. mSphere. 2016;1(4):e00189–16. https://doi.org/10.1128/mSphere.00189-16.

Borman AM, Szekely A, Johnson EM. Isolates of the emerging pathogen Candida auris present in the UK have several geographic origins. Med Mycol. 2017;55:563–7. https://doi.org/10.1093/mmy/myw147.

Calvo B, Melo AS, Perozo-Mena A, Hernandez M, Francisco EC, Hagen F, et al. First report of Candida auris in America: clinical and microbiological aspects of 18 episodes of candidemia. J Infect. 2016;73(4):369–74. https://doi.org/10.1016/j.jinf.2016.07.008.

Casadevall A. Fungi and the rise of mammals. PLoS Pathog. 2012;8:e1002808. https://doi.org/10.1371/journal.ppat.1002808.

Casadevall A, Kontoyiannis DP, Robert V. On the emergence of Candida auris: climate change, azoles, swamps, and birds. MBio. 2019;10:e01397-19. https://doi.org/10.1128/mBio.01397-19.

Casadevall A, Kontoyiannis DP, Robert V. Environmental Candida auris and the global warming emergence hypothesis. MBio. 2021;12:e00360-21. https://doi.org/10.1128/mBio.00360-21.

Cendejas-Bueno E, Kolecka A, Alastruey-Izquierdo A, Theelen B, Groenewald M, Kostrzewa M, et al. Reclassification of the Candida haemulonii complex as Candida haemulonii (C. haemulonii group I), C. duobushaemulonii sp. nov. (C. haemulonii group II), and C. haemulonii var. vulnera var. nov.: three multiresistant human pathogenic yeasts. J Clin Microbiol. 2012;50(11):3641–51. https://doi.org/10.1128/JCM.02248-12.

Chatterjee S, Alampalli SV, Nageshan RK, Chettiar ST, Joshi S, Tatu US. Draft genome of a commonly misdiagnosed multidrug resistant pathogen Candida auris. BMC Genomics. 2015;16:686. https://doi.org/10.1186/s12864-015-1863-z.

Chow NA, de Groot T, Badali H, Abastabar M, Chiller TM, Meis JF. Potential fifth clade of Candida auris, Iran, 2018. Emerg Infect Dis. 2019;25(9):1780–1. https://doi.org/10.3201/eid2509.190686.

Chow NA, Gade L, Tsay SV, Forsberg K, Greenko JA, Southwick KL, et al. Multiple introductions and subsequent transmission of multidrug-resistant *Candida auris* in the USA: a molecular epidemiological survey. Lancet Infect Dis. 2018;18:1377–84. https://doi.org/10.1016/S1473-3099(18)30597-8.

Chow NA, Muñoz JF, Gade L, Berkow EL, Li X, Welsh RM, Forsberg K, Lockhart SR, et al. Tracing the evolutionary history and global expansion of Candida auris using population genomic analyses. MBio. 2020;11:e03364–19. https://doi.org/10.1128/mBio.03364-19.

Chowdhary A, Anil Kumar V, Sharma C, Prakash A, Agarwal K, Babu R, et al. Multidrug-resistant endemic clonal strain of Candida auris in India. Eur J Clin Microbiol Infect Dis. 2014;33:919–26. https://doi.org/10.1007/s10096-013-2027-1.

Chowdhary A, Sharma C, Duggal S, Agarwal K, Prakash A, Singh PK, et al. New clonal strain of Candida auris, Delhi. India Emerg Infect Dis. 2013;19:1670–3. https://doi.org/10.3201/eid1910.130393.

Chowdhary A, Sharma C, Meis JF. Candida auris: a rapidly emerging cause of hospital-acquired multidrug-resistant fungal infections globally. PLoS Pathog. 2017;13:e1006290. https://doi.org/10.1371/journal.ppat.1006290.

Chybowska AD, Childers DS, Farrer RA. Nine things genomics can tell us about Candida auris. Front Genet. 2020;11:351. https://doi.org/10.3389/fgene.2020.00351.

Clancy CJ, Nguyen MH. Emergence of Candida auris: an international call to arms. Clin Infect Dis. 2017;64(2):141–3. https://doi.org/10.1093/cid/ciw696.

Correia A, Sampaio P, James S, Pais C. Candida bracarensis sp. nov., a novel anamorphic yeast species phenotypically similar to Candida glabrata. Int J Syst Evol Microbiol. 2006;56(Pt 1):313–7. https://doi.org/10.1099/ijs.0.64076-0.

Cortegiani A, Misseri G, Giarratano A, Bassetti M, Eyre D. The global challenge of Candida auris in the intensive care unit. Crit Care. 2019;23:150. https://doi.org/10.1186/s13054-019-2449-y.

de Crecy E, Jaronski S, Lyons B, Lyons TJ, Keyhani NO. Directed evolution of a filamentous fungus for thermotolerance. BMC Biotechnol. 2009;9:74. https://doi.org/10.1186/1472-6750-9-74.

de Oliveira Santos GC, Vasconcelos CC, Lopes AJO, de Sousa Cartágenes MDS, Filho AKDB, do Nascimento FRF, et al. Candida infections and therapeutic strategies: mechanisms of action for traditional and alternative agents. Front Microbiol. 2018;9:1351. https://doi.org/10.3389/fmicb.2018.01351.

Du H, Bing J, Hu T, Ennis CL, Nobile CJ, Huang G. Candida auris: epidemiology, biology, antifungal resistance, and virulence. PLoS Pathog. 2020;16(10):e1008921. https://doi.org/10.1371/journal.ppat.1008921.

Dudiuk C, Berrio I, Leonardelli F, Morales-Lopez S, Theill L, Macedo D, et al. Antifungal activity and killing kinetics of anidulafungin, caspofungin and amphotericin B against Candida auris. J Antimicrob Chemother. 2019;74(8):2295–302. https://doi.org/10.1093/jac/dkz178.

ElBaradei A. A decade after the emergence of Candida auris: what do we know? Eur J Clin Microbiol Infect Dis. 2020;39(9):1617–27. https://doi.org/10.1007/s10096-020-03886-9.

Escandón P, Cáceres DH, Espinosa-Bode A, Rivera S, Armstrong P, Vallabhaneni S, et al. Notes from the field: surveillance for Candida auris—Colombia, September 2016–May 2017. MMWR Morb Mortal Wkly Rep. 2018a;67:459–60. https://doi.org/10.15585/mmwr.mm6715a6.

Escandón P, Chow NA, Caceres DH, Gade L, Berkow EL, Armstrong P, et al. Molecular epidemiology of Candida auris in Colombia reveals a highly-related, country-wide colonization with regional patterns in amphotericin B resistance. Clin Infect Dis. 2018b;68(1):15–21. https://doi.org/10.1093/cid/ciy411.

Friedman DZP, Schwartz IS. Emerging fungal infections: new patients, new patterns, and new pathogens. J Fungi. 2019;5:67. https://doi.org/10.3390/jof5030067.

Garbee DD, Pierce SS, Manning J. Opportunistic fungal infections in critical care units. Crit Care Nurs Clin North Am. 2017;29(1):67–79. https://doi.org/10.1016/j.cnc.2016.09.011.

Garcia-Solache MA, Casadevall A. Global warming will bring new fungal diseases for mammals. MBio. 2010;1:e00061–10. https://doi.org/10.1128/mBio.00061-10.

Gnat S, Łagowski D, Nowakiewicz A, Dyląg M. A global view on fungal infections in humans and animals: opportunistic infections and microsporidioses. J Appl Microbiol. 2021; https://doi.org/10.1111/jam.15032.

Gonzalez-Lara MF, Ostrosky-Zeichner L. Invasive candidiasis. Semin Respir Crit Care Med. 2020;41(1):3–12. https://doi.org/10.1055/s-0040-1701215.

Hai NV. The use of probiotics in aquaculture. J Appl Microbiol. 2015;119:917–35. https://doi.org/10.1111/jam.12886.

Hamprecht A, Barber AE, Mellinghoff SC, Thelen P, Walther G, Yu Y, et al. Candida auris in Germany and previous exposure to foreign healthcare. Emerg Infect Dis. 2019;25:1763–5. https://doi.org/10.3201/eid2509.190262.

Hata DJ, Humphries R, Lockhart SR. College of American Pathologists Microbiology Committee. *Candida auris*: an emerging yeast pathogen posing distinct challenges for laboratory diagnostics, treatment, and infection prevention. Arch Pathol Lab Med. 2020;144(1):107–14. https://doi.org/10.5858/arpa.2018-0508-RA.

Iguchi S, Itakura Y, Yoshida A, Kamada K, Mizushima R, Arai Y, et al. Candida auris: a pathogen difficult to identify, treat, and eradicate and its characteristics in Japanese strains. J Infect Chemother. 2019;25:743–9. https://doi.org/10.1016/j.jiac.2019.05.034.

Jackson BR, Chow N, Forsberg K, Litvintseva AP, Lockhart SR, Welsh R, et al. On the origins of a species: what might explain the rise of Candida auris? J Fungi. 2019;5:58. https://doi.org/10.3390/jof5030058.

Jeffery-Smith A, Taori SK, Schelenz S, Jeffery K, Johnson EM, Borman A. Candida auris: a review of the literature. Clin Microbiol Rev. 2017;31(1):e00029–17. https://doi.org/10.1128/CMR.00029-17.

Kathuria S, Singh PK, Sharma C, Prakash A, Masih A, Kumar A, et al. Multidrug-resistant Candida auris misidentified as Candida haemulonii: characterization by matrix-assisted laser desorption ionization-time of flight mass spectrometry and DNA sequencing and its antifungal susceptibility profile variability by Vitek 2, CLSI broth microdilution, and etest method. J Clin Microbiol. 2015;53(6):1823–30. https://doi.org/10.1128/JCM.00367-15.

Kean R, Sherry L, Townsend E, McKloud E, Short B, Akinbobola A, et al. Surface disinfection challenges for Candida auris: an in-vitro study. J Hosp Infect. 2018;98(4):433–6. https://doi.org/10.1016/j.jhin.2017.11.015.

Kim M, Shin J, Sung H, Lee K, Kim EC, Ryoo N, et al. Candida haemulonii and closely related species at 5 university hospitals in Korea: identification, antifungal susceptibility, and clinical features. Clin Infect Dis. 2009;48(6):e57e61. https://doi.org/10.1086/597108.

Kim SH, Iyer KR, Pardeshi L, Muñoz JF, Robbins N, Cuomo CA. Genetic analysis of Candida auris implicates Hsp90 in morphogenesis and azole tolerance and Cdr1 in azole resistance. MBio. 2019;10:e02529–18. https://doi.org/10.1128/mBio.02529-18.

Kordalewska M, Lee A, Park S, Berrio I, Chowdhary A, Zhao Y, et al. Understanding echinocandin resistance in the emerging pathogen Candida auris. Antimicrob Agents Chemother. 2018;62(6):e00238–18. https://doi.org/10.1128/AAC.00238-18.

Kordalewska M, Perlin D. Molecular diagnostics in the times of surveillance for Candida auris. J Fungi. 2019;5(3):77. https://doi.org/10.3390/jof5030077.

Kurt AF, Kuskucu MA, Balkan II, Baris A, Yazgan Z, Serife Oz A, et al. Candida auris fungemia and a local spread taken under control with infection control measures: first report from Turkey. Indian J Med Microbiol. 2021;39(2):228–30. https://doi.org/10.1016/j.ijmmb.2021.03.007.

Lee WG, Shin JH, Uh Y, Kang MG, Kim SH, Park KH, et al. First three reported cases of nosocomial fungemia caused by Candida auris. J Clin Microbiol. 2011;49:3139–42. https://doi.org/10.1128/JCM.00319-11.

Lockhart SR, Etienne KA, Vallabhaneni S, Farooqi J, Chowdhary A, Govender NP, et al. Simultaneous emergence of multidrug-resistant Candida auris on 3 continents confirmed by whole-genome sequencing and epidemiological analyses. Clin Infect Dis. 2018;67(6):987. https://doi.org/10.1093/cid/ciw691.

Magobo RE, Corcoran C, Seetharam S, Govender NP. *Candida auris*-associated candidemia, South Africa. Emerg Infect Dis. 2014;20(7):1250–1. https://doi.org/10.3201/eid2007.131765.

Mahmoudi S, Kuchak SA, Gharehbolagh SA, Mirhendi H, Makimura K. Methods for identification of Candida auris , the yeast of global public health concern: a review. J Mycol Med. 2019;29(2):174–9. https://doi.org/10.1016/j.mycmed.2019.04.004.

Martin R, Litvintseva AP, Cuomo CA. Tracing the evolutionary history and global expansion of Candida auris using population genomic analyses. MBio. 2020;11:e03364–19. https://doi.org/10.1128/mBio.03364-19.

Martins N, Ferreira ICFR, Barros L, Silva S, Henriques M. Candidiasis: predisposing factors, prevention, diagnosis and alternative treatment. Mycopathologia. 2014;177(5–6):223–40. https://doi.org/10.1007/s11046-014-9749-1.

McClean MA, Angilletta MJ Jr, Williams KS. If you can't stand the heat, stay out of the city: thermal reaction norms of chitinolytic fungi in an urban heat island. J Therm Biol. 2005;20:384–91. https://doi.org/10.1016/j.jtherbio.2005.03.002.

Misseri G, Ippolito M, Cortegiani A. Global warming "heating up" the ICU through Candida auris infections: the climate changes theory. Crit Care. 2019;23:416. https://doi.org/10.1186/s13054-019-2702-4.

Mizusawa M, Miller H, Green R, Lee R, Durante M, Perkins R, et al. Can multidrug-resistant Candida auris be reliably identified in clinical microbiology laboratories? J Clin Microbiol. 2017;55:638–40. https://doi.org/10.1128/JCM.02202-16.

Morales-Lopez SE, Parra-Giraldo CM, Ceballos-Garzon A, Martinez HP, Rodriguez GJ, Alvarez-Moreno CA, et al. Invasive infections with multidrug-resistant yeast Candida auris. Colombia Emerg Infect Dis. 2017;23:162–4. https://doi.org/10.3201/eid2301.161497.

Muñoz JF, Gade L, Chow NA, Loparev VN, Juieng P, Berkow EL, et al. Genomic insights into multidrugresistance, mating and virulence in *Candida auris* and related emerging species. Nat Commun. 2018;9:5346. https://doi.org/10.1038/s41467-018-07779-6.

Navalkele BD, Revankar S, Chandrasekar P. Candida auris: a worrisome, globally emerging pathogen. Expert Rev Anti-Infect Ther. 2017;15(9):819–27. https://doi.org/10.1080/14787210.2017.1364992.

Osland MJ, Gabler CA, Grace JB, Day RH, McCoy ML, McLeod JL, et al. Climate and plant controls on soil organic matter in coastal wetlands. Glob Chang Biol. 2018;24:5361–79. https://doi.org/10.1111/gcb.14376.

Ostrowsky B, Greenko J, Adams E, Quinn M, O'Brien B, Chaturvedi V, et al. Candida auris isolates resistant to three classes of antifungal medications—New York, 2019. MMWR Morb Mortal Wkly Rep. 2020;69:6–9. https://doi.org/10.15585/mmwr.mm6901a2.

Pal M, Gebrezgabher W, Samajpati N, Manna AK. Growing role of non-Candida albicans species in clinical disorders of human and animals. J Mycopathol Res. 2015;53:41–8.

Pappas P, Lionakis M, Arendrup M, Ostrosky-Zeichner L, Kullberg BJ. Invasive candidiasis. Nat Rev Dis Primers. 2018;4:18026. https://doi.org/10.1038/nrdp.2018.26.

Puja S, Prasad GS. Candida ruelliae sp. nov., a novel yeast species isolated from flowers of Ruellia sp. (Acanthaceae). FEMS Yeast Res. 2008;8(4):660–6. https://doi.org/10.1111/j.1567-1364.2008.00372.x.

Reyes-Montes MDR, Duarte-Escalante E, Martínez-Herrera E, Acosta-Altamirano G, Frías-De León MG. Current status of the etiology of candidiasis in Mexico. Rev Iberoam Micol. 2017;34(4):203–10. https://doi.org/10.1016/j.riam.2017.05.001.

Robert V, Cardinali G, Casadevall A. Distribution and impact of yeast thermal tolerance permissive for mammalian infection. BMC Biol. 2015;13:18. https://doi.org/10.1186/s12915-015-0127-3.

Ruiz-Gaitan A, Moret AM, Tasias-Pitarch M, Aleixandre-López AI, Martínez-Morel H, Calabuig E, et al. An outbreak due to Candida auris with prolonged colonisation and candidaemia in a tertiary care European hospital. Mycoses. 2018;61(7):498–505. https://doi.org/10.1111/myc.12781.

Sarma S, Kumar N, Sharma S, Govil D, Ali T, Mehta Y, et al. Candidemia caused by amphotericin B and fluconazole resistant Candida auris. Indian J Med Microbiol. 2013;31:90–1. https://doi.org/10.4103/0255-0857.108746.

Satoh K, Makimura K, Hasumi Y, Nishiyama Y, Uchida K, Yamaguchi H. Candida auris sp. nov., a novel ascomycetous yeast isolated from the external ear canal of an inpatient in a Japanese hospital. Microbiol Immunol. 2009;53(1):41–4. https://doi.org/10.1111/j.1348-0421.2008.00083.x.

Sekizuka T, Iguchi S, Umeyama T, Inamine Y, Makimura K, Kuroda M, et al. Clade II *Candida auris* possess genomic structural variations related to an ancestral strain. PLoS One. 2019;14(10):e0223433. https://doi.org/10.1371/journal.pone.0223433.

Sherry L, Ramage G, Kean R, Borman A, Johnson EM, Richardson MD, et al. Biofilm-forming capability of highly virulent, multidrug-resistant Candida auris. Emerg Infect Dis. 2017;23:328–31. https://doi.org/10.3201/eid2302.161320.

Singh R, Kaur M, Chakrabarti A, Shankarnarayan SA, Rudramurthy SM. Biofilm formation by Candida auris isolated from colonising sites and candidemia cases. Mycoses. 2019;62(8):706–9. https://doi.org/10.1111/myc.12947.

Spampinato C, Leonardi D. Candida infections, causes, targets, and resistance mechanisms: traditional and alternative antifungal agents. Biomed Res Int. 2013;2013:204237. https://doi.org/10.1155/2013/204237.

Spivak ES, Hanson KE. Candida auris: an emerging fungal pathogen. J Clin Microbiol. 2018;56:e01588–17. https://doi.org/10.1128/JCM.01588-17.

Steffen W, Broadgate W, Deutsch L, Gaffney O, Ludwig C. The trajectory of the Anthropocene: the great acceleration. Anthropocene Rev. 2015;2:81–98. https://doi.org/10.1177/2053019614564785.

Stone W, Jones BL, Wilsenach J, Botha A. External ecological niche for Candida albicans within reducing, oxygen-limited zones of wetlands. Appl Environ Microbiol. 2012;78:2443–5. https://doi.org/10.1128/AEM.06343-11.

Tavanti A, Davidson AD, Gow NA, Maiden MC, Odds FC. Candida orthopsilosis and Candida metapsilosis spp. nov. to replace Candida parapsilosis groups II and III. J Clin Microbiol. 2005;43:284–92. https://doi.org/10.1128/JCM.43.1.284-292.2005.

Tsay S, Welsh RM, Adams EH, Chow NA, Gade L, Berkow EL, et al. Notes from the field: ongoing transmission of Candida auris in health care facilities—United States, June 2016–May 2017. MMWR Morb Mortal Wkly Rep. 2017;66:514–5. https://doi.org/10.15585/mmwr.mm6619a7.

Uppuluri P. Candida auris biofilm colonization on skin niche conditions. mSphere. 2020;5(1):e00972–19. https://doi.org/10.1128/mSphere.00972-19.

Vallabhaneni S, Kallen A, Tsay S, Chow N, Welsh R, Kerins J, et al. Investigation of the first seven reported cases of Candida auris, a globally emerging invasive, multidrug-resistant fungus—United States, May 2013–August 2016. MMWR Morb Mortal Wkly Rep. 2016;65:1234–7. https://doi.org/10.15585/mmwr.mm6544e1.

Villanueva-Lozano H, Treviño-Rangel RJ, González GM, Ramírez-Elizondo MT, Lara-Medrano R, Aleman-Bocanegra MC, et al. Outbreak of Candida auris infection in a COVID-19 hospital in Mexico. Clin Microbiol Infect. 2021.;S1198-743X(20)30790-4; https://doi.org/10.1016/j.cmi.2020.12.030.

Wang X, Bing J, Zheng Q, Zhang F, Liu J, Yue H, et al. The first isolate of Candida auris in China: clinical and biological aspects. Emerg Microbes Infect. 2018;7:93. https://doi.org/10.1038/s41426-018-0095-0.

Welsh RM, Bentz ML, Shams A, Houston H, Lyons A, Rose LJ, et al. Survival, persistence, and isolation of the emerging multidrug-resistant pathogenic yeast Candida auris on a plastic health care surface. J Clin Microbiol. 2017;55:2996–3005. https://doi.org/10.1128/JCM.00921-17.

Welsh RM, Sexton DJ, Forsberg K, Vallabhaneni S, Litvintseva A. Insights into the unique nature of the East Asian clade of the emerging pathogenic yeast *Candida auris*. J Clin Microbiol. 2019;57:e00007–19. https://doi.org/10.1128/JCM.00007-19.

Chapter 11
Climate Change and Global Distribution of Cryptococcosis

Edith Sánchez Paredes, Yurixhi Paola Aguilera León, Eidi Victoria Alvarado-Ramírez, Laura Rosio Castañón Olivares, and Magnolia Lizbeth Martínez Aguilar

Contents

Abbreviations

AFLP Amplified fragment length polymorphism
FAO Food and Agriculture Organization
FP Fingerprinting

E. Sánchez Paredes (✉) · Y. P. Aguilera León · E. V. Alvarado-Ramírez · L. R. Castañón Olivares · M. L. Martínez Aguilar
Departamento de Microbiología y Parasitología, Facultad de Medicina, Universidad Nacional Autónoma de México, CDMX, Mexico City, Mexico

M. G. Frías-De-León et al. (eds.), *The Impact of Climate Change on Fungal Diseases*, Fungal Biology, https://doi.org/10.1007/978-3-030-89664-5_11

RFLP Restriction fragment length polymorphism
UV Ultraviolet
UVR Ultraviolet Radiation
WHO World Health Organization

11.1 Introduction: General Characteristics of Cryptococcosis and Etiology

Cryptococcosis is a systemic mycosis caused by yeasts of the *C. neoformans/C. gattii* complexes. It is a globally distributed disease which impact increased with the AIDS epidemic. Before that, it was considered an infrequent pathology. Today, the number of cases is rising in some countries and decreasing in others, the latter due to antiretroviral therapy and the disease's established association with immunocompromised patients. Less frequently, it has also been reported in immunocompetent patients (Arenas 2011; Chen et al. 2014; Maziarz and Perfect 2016).

According to the World Health Organization (WHO) in 2020, 37.6 million people suffered from HIV/AIDS globally (World Health Organization 2020, available at: https://www.who.int/news-room/fact-sheets/detail/hiv-aids), being cryptococcosis the most frequent mycosis found in AIDS patients. It is estimated that 80–90% of global cryptococcosis cases are attributed to this group of patients, and 5–10% to individuals with a CD4+ count below 100 cells/mm^3 (Maziarz and Perfect 2016). In studies conducted by Rajasingham and collaborators, the number of cryptococcal meningitis incidents was re-estimated to be 223,100 cases, with 181,100 yearly deaths in 2014. Although this study represents a great leap forward in determining the number of cryptococcal meningitis incidents worldwide, these figures could still underestimate the actual number of cases (Bongomin et al. 2017; Rajasingham et al. 2017).

Cryptococcosis is more frequent in men than in women, with a ratio of 4:1 in the age range from 20 to 50 years old. *C. neoformans* species are responsible for the majority of cryptococcosis cases in immunosuppressed patients, while *C. gattii* species are associated with infections of individuals without immunologic alterations. The principal infection mechanism is airborne; the infection then spreads from the lungs into different organs and tissues, principally toward the central nervous system. Clinical diagnosis is difficult, since clinical manifestations depend on the characteristics of the studied population group, e.g., age, immunologic state, and site of infection (Arenas 2011; Maziarz and Perfect 2016).

Species of the *C. neoformans/C. gattii* complex live like saprophytes in the environment. Yeast from *C. neoformans* species has been isolated from soil, trees, water, arthropods, and avian and mammal excrement, while those of *C. gattii* species have been collected from soil and trees (Cogliati 2013). Generally, the reproduction is asexual through gemmation giving rise to blastoconidia; however, in adverse conditions where there is little access to nutrients, the reproduction could be sexual, producing filaments, basidia, and basidiospores (Velagapudi et al. 2009). In the

Table 11.1 Concordance between molecular typifications used for *C. neoformans*/*C. gattii*. The eight genotype groups are highlighted, seven of which are recognized as species

	Technique and author		
Serotype	FP o RFLP (Meyer)	AFLP (Boekhout)	Groups recognized as species
A	VNI	AFLP1	1 *C. neoformans*
A	VNII	AFLP1a/AFLP1b	
AD	VNIII (VNI x VNIV)	AFLP3	Hybrid
D	VNIV	AFLP2	2 *C. deneoformans*
B	VGI	AFLP4a/AFLP4b	3 *C. gattii*
B	VGII	AFLP6	4 *C. deuterogattii*
C	VGIII	AFLP5a/AFLP5b/AFLP5c	5 *C. bacillisporus*
B	VGIV	AFLP10	6 *C. decagattii*
C	VGIV	AFLP7	7 *C. tetragattii*
DB	VNIV x VGI	AFLP8	Hybrid
AB	VNI x VGI	AFLP9	Hybrid
AB	VNI x VGII	AFLP11	Hybrid

FP fingerprinting, *RFLP* restriction fragment length polymorphism, *AFLP* amplified fragment length polymorphism

laboratory, it has been corroborated that tiny blastoconidia and basidiospores are released into the air and can infect humans and other animals through the respiratory tract (Maziarz and Perfect 2016). Nevertheless, in natural conditions, the sexual forms have not been observed, and therefore, at the moment, the yeasts are considered to be the infectious forms.

Since 2015, seven species have been recognized as belonging to the *C. neoformans*/*C. gattii* complex (Table 11.1) (Hagen et al. 2015).

Based on the composition of the capsule, four principal serotypes have been described: A, B, C, and D. There are also three hybrid serotypes, a result of the cross between the two principal serotypes (Samarasinghe and Xu 2018).

Four molecular types have been identified for *C. neoformans:* VNI, VNII, VNIII, and VNIV (Meyer et al. 2009), while for *C. gattii*, five types have been identified: VGI, VGII, VGIII, VGIV, and VGV (Farrer et al. 2019; Meyer et al. 2009).

Although the hosts most susceptible to cryptococcosis are immunosuppressed patients, the genotypes belonging to the molecular types VGI and VGII can also infect immunocompetent individuals, acting as primary pathogens (Chen et al. 2014; Diaz 2020).

11.2 Climate Change, Global Warming, and Geographic Distribution

Climate change has existed since the beginning of history. The changes have often been gradual, but sometimes abrupt, and have had many different causes like changes in orbital parameters, variations in solar radiation, periods of intense volcanic activity, biotic processes, or meteorite impacts. Today, global warming is a currently ongoing climate change that implies an increase in temperature in the air and water surface of our planet, caused predominantly by human activities.

Global warming was identified decades ago and has triggered changes in physical systems, leading scientists to predict that the current and future ecological implications will have innumerable consequences for various sectors of the global population (Rosatti 2017).

Some hypotheses suggest that the behavior of certain pathogenic microorganisms, among them fungi, will be altered by global warming. Some will disappear, but others will be able to adapt to the rising temperature levels developing thermotolerance, thus increasing their potential as infectious agents (Raffa et al. 2012).

Another worrisome effect of these environmental variations are natural disasters that directly correlate with the appearance of infectious diseases, both from bacteria and fungi, posing various challenges to public healthcare (Benedict and Park 2014; Raffa et al. 2012). Although the role of global warming in the growth, distribution, and dispersion mechanisms of fungi and other microorganisms is not yet well established, some mycosis like coccidioidomycosis (Cavicchioli et al. 2019) and cryptococcosis have been categorized as emergent infections due to environmental change (Raffa et al. 2012).

C. neoformans species are globally spread and there are more isolations of them, both clinical and natural, compared to *C. gattii* species. The ratio of isolations varies between continents; the VNI molecular variant appears to be the most common worldwide except in Australia and Papua New Guinea, where the most common variant is VGI (Cogliati 2013). This is consistent with clinical data indicating that more than 90% of cryptococcal infections are caused by *C. neoformans* species (LIFE 2021).

Among the species of the *Cryptococcus* complex, *C. gattii* species has seen more significant changes in global distribution due to climate change, suggesting that its ecological niche has extended geographically both in Europe and in other parts of the world (Cogliati 2021). A gradual expansion of its niche from its areas of origin has been revealed from 1980 to 2009, followed by an important increase of cases in the last decade (2010–2019). However, *C. gattii* species were not discovered until 1970 in an isolation from an African child (Gatti and Eeckels 1970), and it was not until 1984 that, from a small sample, Kwon-Chung and Bennett concluded that the prevalence of *C. gattii* species were restricted to regions with tropical or subtropical climate. Lastly in 2010, Springer and Chaturvedi (2010) reviewed the existing literature about cases related to cryptococcosis caused by *C. gattii* species between 1948 and 2008. They found that the distribution of *C. gattii* species also extended to temperate regions and that its geographic extension was thus significantly wider than previously believed, arguing that previous studies on the identification of the fungi and their distribution had been insufficient. In line with this analysis, the number of cases of cryptococcosis caused by *C. gattii* species in the coming decade could increase, not only due to global warming but also for the ease of adaptation of these yeasts to new and different climates.

Based on clinical and natural isolations of *C. gattii* species in various parts of the world, it was assumed that these fungi were limited to tropical and subtropical climates (Chang and Chen 2015). Recently, there have been reports of cases where *C. gattii* species have infected immunocompetent hosts in temperate regions. It has been suggested that *C. gattii* species could reach more temperate regions via importation of agricultural products (including trees and livestock), propagation of infectious particles on the wind, and transportation through humans or other animals, or even due to natural phenomenon such as tsunamis, hurricanes, or earthquakes. However, the establishment, geographic expansion, and successful survival of *C. gattii* species in these territories have been directly attributed to changes in its ecological niches caused by global warming. The rising number of cases of cryptococcosis and isolation of *C. gattii* species in temperate regions is a cause for real concern, since certain molecular types (VGI and VGII), of particularly high virulence, can infect immunocompetent individuals (Cogliati 2021).

11.3 The Impact of Environmental Factors on the Growth and Development of *Cryptococcus* in Nature

11.3.1 Biotic Factors

Climate global changes affect nonliving and living components of the ecosystem. The nonliving components (abiotic factors) such as water, sunlight, temperature, oxygen, soil, and air interact with the living components (biotic factors). As both components are correlated, they will continue to be modified as the climate changes.

Biotic drivers include interactions with competitors, consumers, mutualists, and facilitators that influence the movement and establishment or extinction of populations (Morris et al. 2020). In recent years, it has been notable the increasing and/or emergence of pathogens as a result of climate change (El-Sayed and Kamel 2020). However, the relationship between *Cryptococcus* spp. with other microorganisms in the environment is not well understood. It is hypothesized that the establishment of ecological relationships with other microorganisms, birds, and certain species of trees is involved in the evolutionary processes of the yeast including the development of its pathogenesis.

Interactions with Plants

Ecology studies of the *C. neoformans* and *C. gattii* complexes have shown that these two groups can develop in distinct natural environments. *C. gattii* species have been isolated from different vegetal substrates like wood, dry leaves, and flowers and specifically from trunk cavities, in over 50 species of angiosperm and gymnosperm trees (Chaturvedi and Chaturvedi 2011; Schmertmann et al. 2019; Springer and Chaturvedi 2010).

Even then, the relationship between *C. gattii* species and eucalyptus trees is not exclusive (Fortes et al. 2001); in fact, isolations of *C. gattii* species in eucalyptus outside Australia are quite rare (Springer and Chaturvedi 2010). Moreover, various authors report isolations of these yeasts in endemic or common tree species in locations where samples have been taken (Acheson et al. 2018), e.g., on the island of Vancouver, Canada (Kidd et al. 2007; MacDougall et al. 2007), in India (Randhawa et al. 2006), in the Mediterranean Sea (Cogliati et al. 2016), in Brazil (Dos Santos et al. 2019), in Colombia (Angarita et al. 2019), and in other Latin American countries (Firacative et al. 2018).

C. gattii species can be isolated from forest areas, an ecological niche that might be suffering changes due to the effects of global warming. Deforestation due to soil exploitation leads to evapotranspiration reduction, increasing the flow of warm air from the soil into the atmosphere, triggering a feedback loop where reduced humidity causes less precipitation which increases the ambient temperature (Williams and Newbold 2019). We hypothesize that all these variables can affect the establishment of the yeast, forcing them to adapt or modify its survival factors in order to resist the new environmental conditions.

Similarly, *C. neoformans* species have also been isolated from bark and leaves from various species of trees (Angarita et al. 2019; Girish et al. 2010; Reimão et al. 2007). This suggests that this group of species could also see changes in its geographic distribution, as discussed above, for *C. gattii* species. However, these isolations are primarily associated with natural sources rich in nitrogen, such as soil contaminated with avian excrements (González et al. 2010; Reiss et al. 2012).

Interactions with Birds

A clear environmental connection is observed between *C. neoformans* species and various avian groups such as Passeriformes, Anseriformes, Accipitriformes, Apterygiformes, Psittaciformes, and Columbiformes (Johnston et al. 2016). Due to repeated isolations of these yeasts, it was documented how they can develop abundantly in places with great accumulation of *guano*, particularly in nests protected from direct sunlight (Malik et al. 2003).

The effects of regional climate changes cause variation in the migratory patterns and routes of certain avian species (Delmore et al. 2020). We assume that new migratory routes will directly impact the reproduction and natural selection, leading to the extinction or excessive growth of bird populations, inducing changes in the distribution of *C. neoformans* species; therefore, the proliferation of birds (with probable establishment of temporary nests along the routes) also suggests successful establishment of *Cryptococcus* spp. in other geographic areas (Amirrajab et al. 2016) where previously the yeast did not seem to be prevalent.

Although *C. neoformans* species have been found attached to feet and beaks of birds, the birds are not susceptible to the fungus, leading to the notion that the gastrointestinal tract of birds such as doves may have fungicidal activity (Mayer and Kronstad 2019). Together with the high corporal temperature (40–42 °C) of doves, these factors probably allow their macrophages to efficiently act against the yeasts (Johnston et al. 2016). More research is needed in this area to further clarify the role and interaction of avian microbiota in cryptococcosis.

Interaction with Other Microorganisms in Natural Habitats

It is evident that *C. neoformans* and *C. gattii* species coexist and interact with hundreds of microscopic organisms that share the same living spaces, particularly in tree cavities. These cavities are injuries or holes in the internal bark of the trees, originally caused by animals or fallen branches (Wiebe et al. 2020). They allow the formation of microhabitats where communities of saproxylic insects cohabit with various microorganisms (Sánchez-Galván et al. 2017). In vitro studies with *C. neoformans* species and ambiental bacteria like *Bacillus safensis* and *Acinetobacter baumannii* have shown that these bacteria can induce the formation of titan cells (cryptococcal cells of huge size and clinical relevance since they are refractory to phagocytosis) and the production of melanin, which can confer an advantage to the yeast in surviving adverse environmental conditions and also contribute to its pathogenicity (Mayer and Kronstad 2019). Also, *C. neoformans* species are unable to grow on cellulose or lignin, and the joint activities of other microorganisms may be required in order to release this source of carbon (Watkins et al. 2017).

11.3.2 Abiotic Factors

In addition to interactions with different organisms, other environmental elements from regions where these yeasts were isolated can also influence their development. There are hypotheses that seek to explain how the principal virulence factors of *Cryptococcus* spp. and other fungi developed through adaptation and evolutionary modifications needed to survive in new environments, and how, curiously enough, these adaptations have also turned them into a pathogenic organism (Perfect 2006).

Temperature

There is evidence that mammal resistance to fungal infections is due from both the development of their immune system and their base corporal temperature. In the case of *Cryptococcus* spp., the acquisition of thermotolerance presumably increased its potential to become a pathogen. One such case is *Cryptococcus laurentii*, which normally does not grow at 37 °C, but where thermotolerant strains are increasingly associated with diseases in severely immunosuppressed patients (Castro-Lainez et al. 2019).

The *C. neoformans/C. gattii* complex within the *Tremellales* family can grow in temperatures up to 37 °C (Perfect 2006). Thermotolerance has been a big step in the evolution of pathogenesis of these species, given that the ability to tolerate high environmental temperatures could be related to a greater capacity to grow in mammal-like temperatures (Garcia and Casadevall 2010).

Wind

In 2017, Acheson et al. (2018) suggested that deforestation might be related to a higher quantity of basidiospores of *Cryptococcus* in the air. The basidiospores produced by the sexual phase of *C. neoformans* (*Filobasidiella neoformans*) and *C. gattii* (*Filobasidiella bacillisporus*) (Kwon-Chung 1976) are apparently dispersed by the wind, similar to what happens with the spores of other basidiomycetes. The sexual reproduction of these species, as well as their mechanism of dispersion, has only been observed in laboratory conditions. Nonetheless, it is probable that this process also occurs in nature, where the basidiospores together with the desiccated yeasts would function like infective propagules for cryptococcosis when inhaled and deposited in the alveoli of the host (Springer et al. 2013; Velagapudi et al. 2009).

Given its potential as a pathogen, knowing the quantity of spores in the air would be of great importance to establish a relationship between the infection and the size of the inhaled inoculum. Studies performed on the coasts of Vancouver showed that the quantity of basidiospores of *C. gattii* species in the air was higher in the warmest and driest months and lower during wintertime after precipitation (Kidd et al. 2007).

Increased tolerance to oxidative stress, high temperatures, chemical damage, and desiccation are important factors to consider as resistance structures of *Cryptococcus* spp., given that in some ways they affect the germination, viability, and dispersion of the spores of these species (Botts and Hull 2010).

11.4 Exotic Species and Cryptococcosis Outbreaks

Twentieth-century outbreaks of cryptococcosis caused by *C. gattii* species have been related, one way or another, with exotic species not normally present in the area of the outbreak. In 1998, Baró et al. (1998) described various outbreaks of cryptococcosis caused by *C. gattii* species in the Spanish province of Cáceres. Herds of goats affected by this disease displayed symptoms of pulmonary cryptococcosis with neurologic impact, causing the death of hundreds of animals. In the affected grazing fields, it was observed the presence of eucalyptus trees, an exotic species that was introduced to these areas in the 1950s intended for reforestation; however, even after sampling plants and soil in these zones, no yeast capsules were detected.

According to the Food and Agriculture Organization (FAO) of the United Nations, eucalyptus was, in 1981 (http://www.fao.org/3/ac459s/ac459s.pdf), expected to be one of the most commonly used trees used for reforestation. Perhaps due to this, great numbers of eucalyptus trees were distributed and replanted in countries like India, Brazil, Spain, and even Canada; these trees, originally from Australia, were replanted in new habitats and exposed to new living creatures. In 2007, after the outbreak of cryptococcosis on the island of Vancouver, Kidd et al. (2007) intended to evaluate the relationship between *Cryptococcus* spp. and eucalyptus trees planted in the forests of Vancouver. After sampling a great quantity and variety of vegetal species, it was only possible to isolate *C. gattii* species in 10% of the samples in over ten species of trees in this study, insufficient to prove a relationship between the two.

In 2006, various authors, among them Torres-Rodríguez et al. (2006), listed various types of mammals, birds, and reptiles where disease-causing species of both *C. neoformans* and *C. gattii* had been isolated. The forced introduction of exotic animals in cities, in many cases as a result of illegal hunting, long transportations, and inadequate conditions, may influence their health and immune system. Moreover, the diminution or destruction of their natural habitats results in forced migrations and adaptations for many species, including the microbiota which needs to adapt to new hosts and new environments.

11.5 Virulence and Survival Factors of *Cryptococcus* Species Facing Climate Change

By its very survival factors, this encapsulated fungus can induce disease in different hosts. Among these factors are the capsule, melanin production, and various enzymes that degrade many types of proteins. It is beyond question that they confer *Cryptococcus* spp. the capacity to survive, both outside and inside of infected individuals. Trees, soil, nests and bird droppings, seawater, fruit juice (Gatti and Eeckels 1970), cerebrospinal fluid, blood, the skin, and the bone marrow are just a few examples of the various environments in which it has been possible to isolate *Cryptococcus*. These yeasts are able to survive arid, temperate, and warm environments, elevated concentrations of salt, acidic pH, suspension in liquid mediums, and acid rain and even tolerate the toxicity of heavy metals. All these evolutionary successes are climate change adaptations or survival factors of *Cryptococcus* (Fig. 11.1).

11.5.1 The Capsule

One of the most important pathogenic factors of this yeast is the polysaccharide capsule that surrounds it (Fig. 11.2). The capsule consists mainly of xylose, mannose, and glucuronide acid, a quite particular combination that distinguishes it from other clinically interesting fungi. Pioneering studies of *Cryptococcus* spp. have shown that

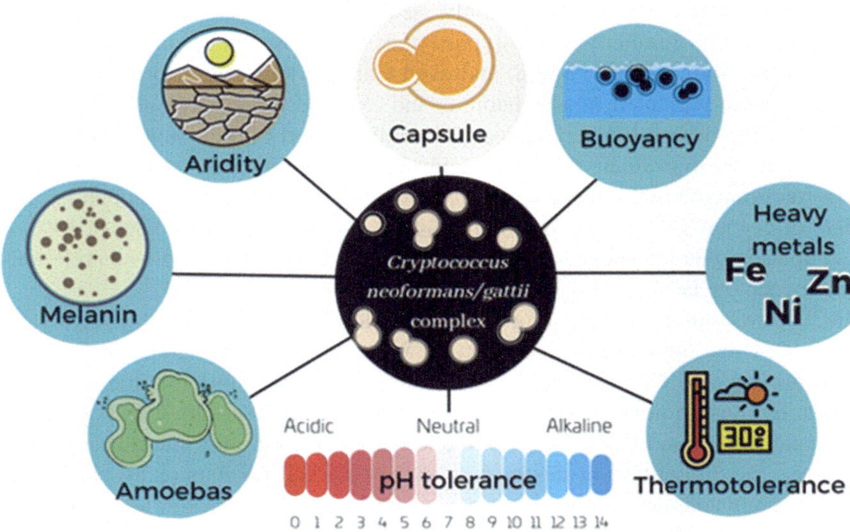

Fig. 11.1 Climate change survival factors of *Cryptococcus* spp.: The capsule, the production of melanin, the capacity to float in liquids (buoyancy), the tolerance of different pH, the assimilation of heavy metals, survival in arid climates, and defending itself against natural predators are some of the factors that allow *Cryptococcus* spp. to adapt to and survive changes in the environment

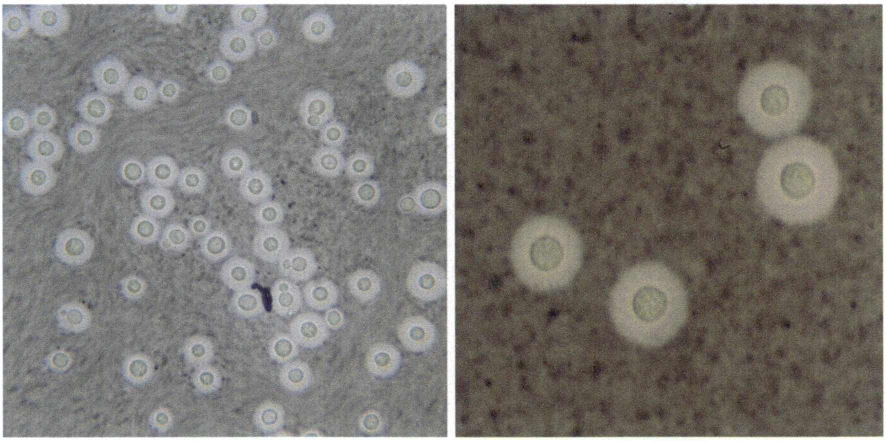

Fig. 11.2 Yeast capsules of *C. gattii* species observed with Indian ink contrast

strains without the genes that intervene in the synthesis of the capsule became avirulent and incapable of infecting laboratory mice (Chang and Kwon-Chung 1994).

Casadevall et al. (2019) describe a marked difference in capsule size when *Cryptococcus* grows in environments with or without oxidative stress, with larger sizes when under such stress.

In nature, the capsule of the *Cryptococcus* complex protects it from desiccation and changes in osmolarity and temperature.

The capsule is also related to buoyancy or suspension in liquids. Numerous species of *Cryptococcus* spp. have adapted and live in seawater. The constituents of the *C. gattii* and *C. neoformans* complexes possess an intrinsic mechanism to support cellular buoyancy, specifically its capacity to increase the capsule production, which diminishes the cellular density and supports buoyancy in salt water.

In 2018, Vij et al. (2018) evaluated the density of yeast-like cells with or without capsule and found differences in the strains' densities. When the capsule was removed, by chemical or mechanical methods, it increased the *C. neoformans* cell density and reduced buoyancy. This survival factor is of great importance since it allows the cryptococcal cells to float in liquids, facilitating its transport. It has been shown that the strains of *C. gattii* species originating from the Pacific Northwest of the USA survive for at least 1 year in ocean water. Environmental studies performed immediately after the Vancouver outbreak in 1999 reported growth of *Cryptococcus* spp. in various oceanic samples taken from the island's coastline. Furthermore, dozens of infected cetaceans were found along the northwestern Pacific coastline, including that of Vancouver.

Infected pinnipedian animals have been documented in the region, which suggests an ongoing contamination at great scale in the marine environment. Pathologically, nearly all marine mammal infections have been pulmonary, suggesting infection via inhalation. As marine mammals breathe at the surface of the water, it is possible that *C. gattii* survives at the sea surface microlayer, as described for other *Cryptococcus* species, and is subsequently inhaled (Engelthaler and Casadevall 2019).

In the natural environment, the cryptococcal capsule avoids phagocytosis of the yeast from ambiental amoebas (Bunting et al. 1979). In 2021, Fu et al. (2021) did phenotype and genetic characterization of *C. neoformans* species colonies co-incubated with amoebas for 1 month. Yeasts that survived being exposed to this predator were more virulent. Although the exact mechanisms involved in this evolutionary step are still unknown, it is evident that the forces of natural selection are at work for these organisms.

Similarly, but related to pathogenesis, the presence of the capsule confers protection from phagocytosis from the macrophages of an infected host's immune system.

11.5.2 Melanin Production

C. neoformans/C. gattii complex produces a spotted, dark, almost black pigment called melanin, an important virulence factor (Fig. 11.3). Fungal melanins play multiple biological functions including photoprotection, energy harvest, and thermoregulation by readily absorbing and transducing electromagnetic radiation. Fungal melanins also function in free radical and metal binding; protection against

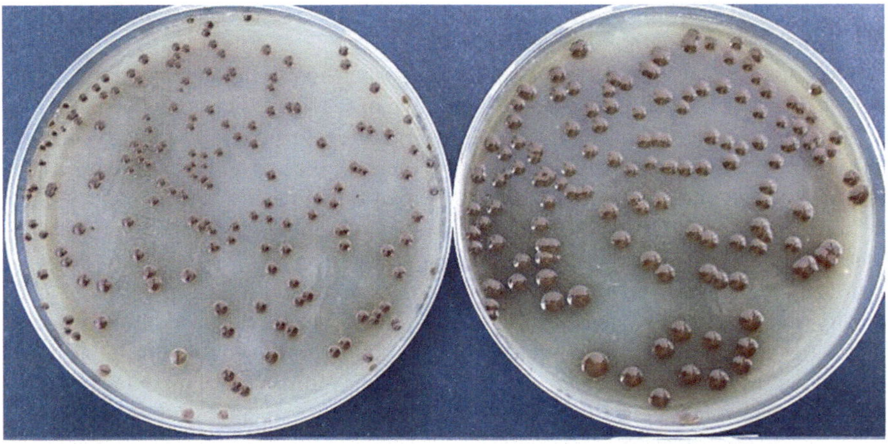

Fig. 11.3 Melanized colonies of *C. gattii* (left) and *C. neoformans* (right) on Niger seed agar

dehydration, chemical, and mechanical stressors; and fungal development and conidiation (Cordero and Casadevall 2017).

Global warming seems to favor melanin-rich species. As ultraviolet radiation penetrates the atmosphere more easily, it is expected that individuals with high concentration of this component will be able to resist the effects of climate change, giving *Cryptococcus* spp. an adaptability tool to endure UV radiation (Roulin 2014).

It is also worth considering that increasing melanin production might also take place due to changes in soil composition due to contamination and acid rain, which modify the concentration of heavy metals in the soil. In studies by Wang and Li (2019), it was shown that melanin production in bacteria of the *Streptomyces* genus was enhanced in the presence of metals like iron (Fe) and nickel (Ni). In 1996, Fogarty and Tobin (1996) showed the protective role of the melanin polymer in a diverse group of dematiaceous fungi when exposed to heavy metals.

On the other hand, high temperatures also regulate melanin synthesis. In 1992, Coyne and Al-Harthi (1992) showed that *Vibrio cholerae*, a bacterium highly tolerant to salt, promotes melanin synthesis at temperatures above 30 °C and in hyperosmotic environments.

The increasing concentration of melanin is relevant due to its implication in the pathogenesis of infections. Specifically for *Cryptococcus* spp., a faster melanization increases virulence in experimental and human cryptococcosis, highlighting the importance of melanin production in the prevalence of the disease (van Rhijn and Bromley 2021).

11.5.3 pH Tolerance

The *Cryptococcus* yeasts tolerate different ambiental pH levels. As previously mentioned, *C. neoformans* species have been isolated from various types of bird excrement, most commonly from domestic doves (*Columba livia*). Global warming will impact the diversity of many species but probably not the domestic doves. They adapt very well to cities, eat a wide variety of foods, and survive in countries with different temperatures all over the world. *C. neoformans* frequently appear in domestic doves' excrement, which also serves as a geographic propagation mechanism as the birds release their feces into the environment in different locations.

Regarding the pH, by releasing their feces to the ground, domestic pigeons reintroduce cryptococcal cells into the environment (Soltani et al. 2013). The bird excrements reduce the pH of the soil due to the presence of uric acid in the excretions (Chae et al. 2012). The cryptococcal cells form capsules to protect themselves from these acid conditions. Thus, it is not surprising to find marked capsule formation in yeast isolations from environmental samples. The rate of growth is affected by changes in pH, and the ability of yeast to adapt to these changes is fundamental for survival in both the environment and inside its hosts.

Though cryptococcal cells often spread by hematogenous dissemination where the blood pH is neutral, they have also been discovered surviving inside phagosomes and in bird intestines, where pH is generally acidic. Apparently, this acidic environment does not inhibit the growth of these cells, suggesting that the *C. neoformans* and *C. gattii* complexes are well adapted to variations in pH.

11.5.4 Capacity to Survive Exposure to Natural Predators

In 2019, Naranjo and Gabaldón (2019) identified amoebas as natural predators of various microorganisms, among them *Cryptococcus* spp. yeasts. In nature, both are found in environments with humid soil. The protozoans phagocytose the yeasts, which appear to be using the capsule as a defense mechanism. It seems that this strategy of survival has served *Cryptococcus* spp. well, since, by protecting itself from wild amoeba phagocytosis, it has also learned to evade phagocytosis by macrophages of its host, as previously evidenced by the study by Fu et al. (2021).

Climate change has caused, and surely will continue to cause, the migration of living creatures to new habitats. This is also true for microorganisms. The increased temperature in rivers and lakes will lead to changes in the ambiental microbiota, among others for the free-living amoebas that can cause infrequent but lethal encephalitis in humans. In 2019, authors like Cooper et al. (2019) suggested that this rare disease caused by the *Naegleria fowleri* amoeba is spreading to new geographies due to an increase in ambient temperature.

Even today, the complete set of factors involved in the outbreak of cryptococcosis by the *C. gattii* species in Vancouver in 1999 is unknown. This event highlighted a series of questions regarding how the microorganism passed from the environment to the host, particularly since it was an unexpected and inexplicable event where both animals and immunocompetent humans were infected. Various theories have been proposed to help explain the evolutionary leap taken by the yeast during this outbreak of cryptococcosis. In 2019, Engelthaler and Casadevall (2019) proposed that *C. gattii* species arrived to this cold-temperature habitat nearly 100 years earlier, when a tsunami flooded the coasts of Vancouver in the early twentieth century, an event that may have helped *C. gattii* species establish itself in this region. How come the outbreak of cryptococcosis occurred nearly a century later and how did the yeast pass from environment to host? This is an important and still unresolved question.

11.6 Factors that Increase Host Susceptibility to Cryptococcosis

11.6.1 Malnutrition

Climate change impacts the production, quality, and quantity of food. As an effect of this, in 2016 an estimated 529,000 deaths were associated with malnutrition (Springmann et al. 2016). In 2050, it is predicted a 3.2% reduction in the food availability per capita, with particular decreases in the consumption of fruit, vegetables, and red meat.

Of these deaths associated with malnutrition, a great majority are estimated to be caused by complications associated with infections. A poor diet low in micronutrients restricts the proper development of lymphoid organs like the spleen, thymus, lymph nodes, and bone marrow, all of which are essential to the development and differentiation of the immune system cells, both innate and adaptive, generating cells with low immunogenic activity. Cells from the complement pathway, antigen-presenting cells, macrophages, and different subtypes of T- and B-lymphocytes are diminished and altered during nutritional deficits (França et al. 2009). There are few precedents indicating that malnutrition is a factor of predisposition for developing cryptococcosis. However, in malnourished patients who also have HIV, there is an increased predisposition to suffer opportunistic infections, as has been documented for cryptococcosis (Enwonwu 2006). Given the projected future restrictions in the availability of basic foods, the long-term outlook suggests an increased number of cases of cryptococcosis in this group of individuals.

11.6.2 Dehydration

Water quantity and quality have diminished globally due to rising temperature, change in precipitation, increased contamination, inability to purify bodies of water, and the reduced dissolution of oxygen, to mention a few factors. This scarcity impacts the water consumed daily, as well as the water used to clean and disinfect foods, to clean surfaces, and for daily hygiene (UNESCO, UN-Water 2020).

The lack of water and hygienic measures brings with them various diseases, both organic and infectious. Being well hydrated keeps our organisms functioning properly, while dehydration causes renal, cardiovascular, neurological, gastrointestinal, metabolic, and respiratory diseases. In the lungs in particular, a reduction in bodily fluids causes changes in the production of surfactant, impacting normal cell functions and preventing an adequate gas exchange. This results in constricted bronchi and airway obstruction, facilitating infectious processes (El-Sharkawy et al. 2015).

11.6.3 Stress

Global warming is an ongoing event that connotes the deterioration of the planet, something that concerns all humanity. There are myriad sources of information, from scientific publications to tabloids. This constant deluge of information together with the actual impact of climate change on society (increases in migration, extreme natural events, and food and water shortage, among others) has raised stress and depression levels, principally in the very young or very old populations, and individuals living in developing countries. Specifically, an increase in cases of posttraumatic stress has been seen in subjects that have survived natural disasters, alongside a similar increase in the number of hurricanes, fires, tornadoes, floods, and even heat waves (Cianconi et al. 2020).

There is scarce information regarding the possible connection between stress and fungal infections. Some authors have suggested that women with chronic stress suffer increased vulvovaginal infections by *Candida albicans*, related to high levels of cortisol, though the molecular mechanism relating these two conditions has not been deeply investigated (Ehrström et al. 2005; Moshfeghy et al. 2020). For other fungi of medical importance like *Cryptococcus*, there are no direct antecedents; however, it is plausible that a similar mechanism may exist.

11.6.4 Ultraviolet Light

Ultraviolet (UV) light from solar radiation is vital to life on earth. The atmospheric "ozone layer" filters ultraviolet radiation (UVR) and only those called UV-A can penetrate the layer. Due to the depletion of atmospheric ozone, increased cloud cover, increased CO_2 levels, increased temperature, and changed circulatory patterns (Williamson et al. 2014), it has been estimated that by 2100 (Bais et al. 2011) UVR will increase in the tropics and decrease in the poles.

This increase in UVR affects human health, changing vitamin D levels which are regulated by solar exposition as well as particular characteristics of the immune defense system.

It has been demonstrated that the immune defense system is suppressed if exposed to high levels of UVR, in both humans and animals, principally altering Th1 and Treg lymphocytes and antigen-producing cells (Sleijffers et al. 2002; Swaminathan et al. 2014). These precedents allow us to suggest that immunity to the *Cryptococcus* complex might possibly also be affected by the increased exposure to UVR.

The effects of UVR in depleting the innate and adaptive immune system response to both cryptococcosis itself and HIV, which is one of its major risk factors, will have repercussions in both the number of cases and in a more dire clinical picture for infected individuals, predicting a poor prognosis for both the disease itself and its associated comorbidities, among them HIV.

11.7 Conclusion

The adaptive capabilities of the *Cryptococcus* complex when faced with global warming are proven beyond doubt. This adaptation has been explained in this chapter, analyzing the intrinsic physiological characteristics of these yeasts, which allow them to survive in different environments even in adverse or hostile conditions. The ecology and geographical distribution of *Cryptococcus* spp. have changed during these one and a half centuries of global warming, but the following events must be kept in mind:

- The etiologic agents have varied from 1894 to 2021, not from being new or emergent species, but because the tools and techniques of identification and typification have constantly improved. As a consequence, the ecology, physiology, and virulence factors of each of the seven species known today are distinct, explaining the various epidemiological and clinical differences encountered in patients during the current period of global warming.
- The distribution of *Cryptococcus* spp. and cases of cryptococcosis can also vary with human activity and habits. The most recent proof of this can be seen with the appearance of AIDS on a global scale, which concomitantly raised the morbidity and geographic distribution of cryptococcosis.
- Literature shows that the geographic distribution of cryptococcosis is not tied to a source of infection; that is, in various cases of cryptococcosis caused by a particular species, this very same species has not been possible to isolate from the surrounding environment – an inexplicable mystery.

To infect mammals, *Cryptococcus* spp. need to adapt to, among other things, drastic changes in temperature and extremely acidic environments, and must also develop alternative means of respiration in different concentrations of CO_2. We can predict that with the ability to adapt to two different environments—inside and outside the host—the etiologic agents of cryptococcosis will be able to adapt to the projected changes in climate over the coming 30 years, with hot and dry summers, winters with heavy downpours of rain, and mountains without snow cover. In such environments, the prediction is that in 30 years from now, the distribution of cryptococcosis on our planet will be even wider than it is today. This slow, ongoing event requires commitment and action from governments, health authorities, and personnel to consistently and efficiently track the epidemiology of the *C. neoformans/C. gattii* complex, and to take adequate measures enabling them to rapidly detect and contain future outbreaks of cryptococcosis.

Acknowledgments The authors extend a special appreciation to Aurora Erandi Chico González and Nadia Joyce Dominguez Aviles for the design of the figure "Survival factors of *Cryptococcus* spp." They also extend a special thanks to Jonas Wallenius for his review and translation of the manuscript.

References

Acheson E, Galanis E, Bartlett K, Mak S, Klinkenberg B. Searching for clues for eighteen years: deciphering the ecological determinants of Cryptococcus gattii on Vancouver Island, British Columbia. Med Mycol. 2018;56(2):129–44. https://doi.org/10.1093/mmy/myx037.

Amirrajab N, Haghani I, Rasuli M, Shokohi T. Migratory birds as a potential reservoirs of Cryptococcus neoformans. Int J Environ Res. 2016;10(3):459–64. https://doi.org/10.22059/IJER.2016.58765.

Angarita A, Cárdenas D, Parra-Giraldo C, Diaz-Carvajal C, Escandon-Hernandez P. Recuperación de Cryptococcus neoformans y C. gattii ambientales y su asociación con aislados clínicos en Cúcuta, Colombia. Rev MVZ Córdoba. 2019;24(1):7137–44. https://doi.org/10.21897/rmvz.1601.

Arenas R. Dermatología Atlas, diagnóstico y tratamiento. 5th ed. México: Interamericana-McGraw-Hill; 2011. ISBN 13-978-970-10-7268-2

Bais A, Tourpali K, Kazantzidis A, Akiyoshi H, Bekki S, Braesicke P, et al. Projections of UV radiation changes in the 21st century: impact of ozone recovery and cloud effects Atmos. Chem Phys. 2011;11:7533–45. https://doi.org/10.5194/acp-11-7533-2011.

Baró T, Torres J, De Mendoza M, Morera Y, Alía C. First identification of autochthonous Cryptococcus neoformans var. gattii isolated from goats with predominantly severe pulmonary disease in Spain. J Clin Microbiol. 1998;36:458–61. https://doi.org/10.1128/JCM.36.2.458-461.1998.

Benedict K, Park B. Invasive fungal infections after natural disasters. Emerg Infect Dis. 2014;20(3):349–55. https://doi.org/10.3201/eid2003.131230.

Bongomin F, Gago S, Oladele RO, Denning DW. Global and multi-national prevalence of fungal diseases-estimate precision. J Fungi (Basel). 2017;3(4):57. https://doi.org/10.3390/jof3040057.

Botts M, Hull C. Dueling in the lung: how Cryptococcus spores race the host for survival. Curr Opin Microbiol. 2010;13(4):437–42. https://doi.org/10.1016/j.mib.2010.05.003.

Bunting L, Neilson J, Bulmer G. Cryptococcus neoformans: gastronomic delight of a soil ameba. Sabouraudia. 1979;17(3):225–32. https://doi.org/10.1080/00362177985380341.

Casadevall A, Coelho C, Cordero R, Dragotakes Q, Jung E, Vij R, et al. The capsule of Cryptococcus neoformans. Virulence. 2019;10(1):822–31. https://doi.org/10.1080/21505594.2018.1431087.

Castro-Lainez M, Deliz-Aguirre R, Antunez D, Cruz-Codina M, Cahuayme-Zuniga L, Vitale K, et al. Cryptococcus laurentii meningitis in a non-HIV patient. IDCases. 2019;18:e00612. https://doi.org/10.1016/j.idcr.2019.e00612.

Cavicchioli R, Ripple W, Timmis K, Azam F, Bakken L, Baylis M, et al. Scientists' warning to humanity: microorganisms and climate change. Nat Rev Microbiol. 2019;17(9):569–86. https://doi.org/10.1038/s41579-019-0222-5.

Chae H, Jang G, Kim N, Son H, Lee J, Kim S, et al. Classification of Cryptococcus neoformans and yeast-like fungus isolates from pigeon droppings by colony phenotyping and ITS genotyping and their seasonal variations in Korea. Avian Dis. 2012;56(1):58–64. https://doi.org/10.1637/9703-030711-Reg.1.

Chang C, Chen S. Colliding epidemics and the rise of cryptococcosis. J Fungi. 2015;2(1):1. https://doi.org/10.3390/jof2010001.

Chang Y, Kwon-Chung K. Complementation of a capsule-deficient mutation of Cryptococcus neoformans restores its virulence. Mol Cell Biol. 1994;14:4912–9. https://doi.org/10.1128/mcb.14.7.4912-4919.1994.

Chaturvedi V, Chaturvedi S. Cryptococcus gattii a resurgent fungal pathogen. Trends Microbiol. 2011;19(11):564–71. https://doi.org/10.1016/j.tim.2011.07.010.

Chen S, Meyer W, Sorrell T. Cryptococcus gattii infections. Clin Microbiol Rev. 2014;27(4):980–1024. https://doi.org/10.1128/CMR.00126-13.

Cianconi P, Betrò S, Janiri L. The impact of climate change on mental health: a systematic descriptive review. Front Psychiatry. 2020;6(11):74. https://doi.org/10.3389/fpsyt.2020.00074.

Cogliati M. Global molecular epidemiology of Cryptococcus neoformans and Cryptococcus gattii: an atlas of the molecular types. Scientifica (Cairo). 2013;2013:675213. https://doi.org/10.1155/2013/675213.

Cogliati M. Global warming impact on the expansion of the fundamental niche of Cryptococcus gattii VGI in Europe. Environ Microbiol Rep. 2021;13(3):375–83. https://doi.org/10.1111/1758-2229.12945.

Cogliati M, D'Amicis R, Zani A, Montagna MT, Caggiano G, De Giglio O, et al. Environmental distribution of Cryptococcus neoformans and C. gattii around the Mediterranean basin. FEMS Yeast Res. 2016;16(4):12. https://doi.org/10.1093/femsyr/fow086.

Cooper A, Aouthmany S, Shah K, Rega PP. Killer amoebas: primary amoebic meningoencephalitis in a changing climate. JAAPA. 2019;32(6):30–5. https://doi.org/10.1097/01.JAA.0000558238.99250.4a.

Cordero R, Casadevall A. Functions of fungal melanin beyond virulence. Fungal Biol Rev. 2017;31(2):99–112. https://doi.org/10.1016/j.fbr.2016.12.003.

Coyne V, Al-Harthi L. Induction of melanin biosynthesis in vibrio cholerae. Appl Environ Microbiol. 1992;58(9):2861–5. https://doi.org/10.1128/aem.58.9.2861-2865.1992.

Delmore K, Van Doren B, Conway G, Curk T, Garrido-Garduño T, Germain R, et al. Individual variability and versatility in an eco-evolutionary model of avian migration. Proc R Soc B. 2020;287(1938):20201339. https://doi.org/10.1098/rspb.2020.1339.

Diaz J. The disease ecology, epidemiology, clinical manifestations, and management of emerging Cryptococcus gattii complex infections. Wilderness Environ Med. 2020;31(1):101–9. https://doi.org/10.1016/j.wem.2019.10.004.

Dos Santos BA, Wanke B, Dos Santos LM, Freire A, da Silva JR, Rocha D, et al. Cryptococcus gattii VGII isolated from native forest and river in Northern Brazil. Braz J Microbiol. 2019;50(2):495–500. https://doi.org/10.1007/s42770-019-00066-6.

Ehrström S, Kornfeld D, Thuresson J, Rylander E. Signs of chronic stress in women with recurrent candida vulvovaginitis. Am J Obstet Gynecol. 2005;193(4):1376–81. https://doi.org/10.1016/j.ajog.2005.03.068.

El-Sayed A, Kamel M. Climatic changes and their role in emergence and re-emergence of diseases. Environ Sci Pollut Res Int. 2020;27(18):22336–52. https://doi.org/10.1007/s11356-020-08896-w.

El-Sharkawy A, Sahota O, Lobo D. Acute and chronic effects of hydration status on health. Nutr Rev. 2015;73(2):97–109. https://doi.org/10.1093/nutrit/nuv038.

Engelthaler D, Casadevall A. On the emergence of Cryptococcus gattii in the Pacific northwest: ballast tanks, Tsunamis, and Black Swans. mBio. 2019;10(5):e02193–19. https://doi.org/10.1128/mBio.02193-19.

Enwonwu C. Complex interactions between malnutrition, infection and immunity: relevance to HIV/AIDS infection. Niger J Clin Biomed Res. 2006;1(1):6–14.

Farrer R, Chang M, Davis M, van Dorp L, Yang D, Shea T, et al. A new lineage of Cryptococcus gattii (VGV) discovered in the central Zambezian Miombo Woodlands. MBio. 2019;10(6):e02306–19. https://doi.org/10.1128/mBio.02306-19.

Firacative C, Lizarazo J, Illnait M, Castañeda E, Latin American Cryptococcal Study Group. The status of cryptococcosis in Latin America. Mem Inst Oswaldo Cruz. 2018;113(7):e170554. https://doi.org/10.1590/0074-02760170554.

Fogarty R, Tobin J. Fungal melanins and their interactions with metals. Enzym Microb Technol. 1996;19(4):311–7. https://doi.org/10.1016/0141-0229(96)00002-6.

Food and Agriculture Organization (FAO). El eucalipto en la repoblación forestal. [Internet] FAO. 1981; [http://www.fao.org/3/ac459s/ac459s.pdf].

Fortes S, Lazéra M, Nishikawa M, Macedo R, Wanke B. First isolation of Cryptococcus neoformans var. gattii from a native jungle tree in the Brazilian Amazon rainforest. Mycoses. 2001;44(5):137–40. https://doi.org/10.1046/j.1439-0507.2001.00651.x.

França T, Ishikawa L, Zorzella S, Chiuso F, da Cunha M, Sartori A. Impact of malnutrition on immunity and infection. J Venom Anim Toxins Incl Trop Dis. 2009;15(3):374–90. https://doi.org/10.1590/S1678-91992009000300003.

Fu M, Liporagi L, Dos Santos SR, Júnior TJ, Perfect J, Cuomo C, Casadevall A. Amoeba predation of Cryptococcus neoformans results in pleiotropic changes to traits associated with virulence. MBio. 2021;12(2):e00567–21. https://doi.org/10.1128/mBio.00567-21.

Garcia M, Casadevall A. Global warming will bring new fungal diseases for mammals. MBio. 2010;1(1):e00061–10. https://doi.org/10.1128/mBio.00061-10.

Gatti F, Eeckels R. An atypical strain of Cryptococcus neoformans (Sanfelice) Vuillemin. Part 1. Description of the disease and of the strain. Ann Soc Belg Med Trop. 1970;50(6):689–93.

Girish C, Prabu D, Mitani H, Mikami Y, Menon T. Environmental isolation of Cryptococcus neoformans and Cryptococcus gattii from living trees in Guindy National Park, Chennai, South India. Mycoses. 2010;53(3):262–4. https://doi.org/10.1111/j.1439-0507.2009.01699.x.

González G, González J, Díaz M. Isolation of Cryptococcus neoformans in dry droppings of captive birds in Santiago, Chile. J Avian Med Surg. 2010;24(3):227–36. https://doi.org/10.1647/2009-003.1.

Hagen F, Khayhan K, Theelen B, Kolecka A, Polacheck I, Sionov E, et al. Recognition of seven species in the Cryptococcus gattii/Cryptococcus neoformans species complex. Fungal Genet Biol. 2015;78:16–48. https://doi.org/10.1016/j.fgb.2015.02.009.

Johnston S, Voelz K, May R. Cryptococcus neoformans thermotolerance to avian body temperature is sufficient for extracellular growth but not intracellular survival in macrophages. Sci Rep. 2016;6:20977. https://doi.org/10.1038/srep20977.

Kidd S, Bach P, Hingston A, Mak S, Chow Y, MacDougall L, et al. Cryptococcus gattii dispersal mechanisms, British Columbia. Canada Emerg Infect Dis. 2007;13(1):51–7. https://doi.org/10.3201/eid1301.060823.

Kwon-Chung K. A new species of Filobasidiella, the sexual state of Cryptococcus neoformans B and C serotypes. Mycology. 1976;68(4):942–6.

Kwon-Chung K, Bennett J. Epidemiologic differences between the two varieties of Cryptococcus neoformans. Am J Epidemiol. 1984;120(1):123–30. https://doi.org/10.1093/oxfordjournals.aje.a113861.

LIFE Fungal infections: Cryptococcus neoformans. Leading International fungal Education. [Internet]. United Kingdom [Consulted on 4 Jun 2021] [Life (life-worldwide.org)].

MacDougall L, Kidd S, Galanis E, Mak S, Leslie MJ, Cieslak PR, Kronstad JW, Morshed MG, Bartlett KH. Spread of Cryptococcus gattii in British Columbia, Canada, and detection in the Pacific northwest, USA. Emerg Infect Dis. 2007;13(1):42–50. https://doi.org/10.3201/eid1301.060827.

Malik R, Krockenberger M, Cross G, Doneley R, Madill DN, Black D, Love DN. Avian cryptococcosis. Med Mycol. 2003;41(2):115–24. https://doi.org/10.1080/mmy.41.2.115.124.

Mayer F, Kronstad J. The spectrum of interactions between Cryptococcus neoformans and bacteria. J Fungi. 2019;5(2):31. https://doi.org/10.3390/jof5020031.

Maziarz E, Perfect J. Cryptococcosis. Infect Dis Clin N Am. 2016;30(1):179–206. https://doi.org/10.1016/j.idc.2015.10.006.

Meyer W, Aanensen DM, Boekhout T, Cogliati M, Diaz MR, Esposto MC, et al. Consensus multilocus sequence typing scheme for Cryptococcus neoformans and Cryptococcus gattii. Med Mycol. 2009;47(6):561–70. https://doi.org/10.1080/13693780902953886.

Morris W, Ehrlén J, Dahlgren J, Loomis A, Louthan A. Biotic and anthropogenic forces rival climatic/abiotic factors in determining global plant population growth and fitness. Proc Natl Acad Sci U S A. 2020;117(2):1107–12. https://doi.org/10.1073/pnas.1918363117.

Moshfeghy Z, Tahari S, Janghorban R, Najib FS, Mani A, Sayadi M. Association of sexual function and psychological symptoms including depression, anxiety and stress in women with recurrent vulvovaginal candidiasis. J Turk Ger Gynecol Assoc. 2020;21(2):90–6. https://doi.org/10.4274/jtgga.galenos.2019.2019.0077.

Naranjo M, Gabaldón T. Fungal evolution: major ecological adaptations and evolutionary transitions. Biol Rev Camb Philos Soc. 2019;94(4):1443–76. https://doi.org/10.1111/brv.12510.

Perfect J. Cryptococcus neoformans: the yeast that likes it hot. FEMS Yeast Res. 2006;6(4):463–8. https://doi.org/10.1111/j.1567-1364.2006.00051.x.

Raffa R, Eltoukhy N, Raffa K. Implications of climate change (global warming) for the healthcare system. J Clin Pharm Ther. 2012;37(5):502–4. https://doi.org/10.1111/j.1365-2710.2012.01355.x.

Rajasingham R, Smith R, Park B, Jarvis J, Govender N, Chiller T, et al. Global burden of disease of HIV-associated cryptococcal meningitis: an updated analysis. Lancet Infect Dis. 2017;17(8):873–81. https://doi.org/10.1016/S1473-3099(17)30243-8.

Randhawa H, Kowshik T, Preeti Sinha K, Chowdhary A, Khan ZU, Yan Z, Xu J, Kumar A. Distribution of Cryptococcus gattii and Cryptococcus neoformans in decayed trunk wood of Syzygium cumini trees in North-Western India. Med Mycol. 2006;44(7):623–30. https://doi.org/10.1080/13693780600860946.

Reimão J, Drummond E, Terceti Mde S, Lyon J, Franco M, de Siqueira A. Isolation of Cryptococcus neoformans from hollows of living trees in the city of Alfenas, MG. Brazil Mycoses. 2007;50(4):261–4. https://doi.org/10.1111/j.1439-0507.2007.01374.x.

Reiss E, Shadomy H, Marshall G. Cryptococcosis. In: Reiss E, Shadomy H, Marshall G, editors. Fundamental medical mycology. 3rd ed. E.U.A: Willey Black Well; 2012. p. 311–5.

Rosatti A. Global warming and its health impact. Int J Occup Environ Med. 2017;8(1):7–20. https://doi.org/10.15171/ijoem.2017.963.

Roulin A. Melanin-based colour polymorphism responding to climate change. Glob Chang Biol. 2014;20:3344–50. https://doi.org/10.1111/gcb.12594.

Samarasinghe H, Xu J. Hybrids and hybridization in the Cryptococcus neoformans and Cryptococcus gattii species complexes. Infect Genet Evol. 2018;66:245–55. https://doi.org/10.1016/j.meegid.2018.10.011.

Sánchez-Galván I, Ferrer J, Galante E, Marcos M. Bacteria and Hoverflies (Diptera: Syrphidae) in tree hollows from the Iberian Mediterranean Forest. Environ Entomol. 2017;46(1):137–42. https://doi.org/10.1093/ee/nvw158.

Schmertmann L, Irinyi L, Malik R, Powell J, Meyer W, Krockenberger MB. The mycobiome of Australian tree hollows in relation to the Cryptococcus gattii and C. neoformans species complexes. Ecol Evol. 2019;9(17):9684–700. https://doi.org/10.1002/ece3.5498.

Sleijffers A, Garssen J, Van Loveren H. Ultraviolet radiation, resistance to infectious diseases, and vaccination responses. Methods. 2002;28(1):111–21. https://doi.org/10.1016/s1046-2023(02)00215-3.

Soltani M, Bayat M, Hashemi S, Zia M, Pestechian N. Isolation of Cryptococcus neoformans and other opportunistic fungi from pigeon droppings. J Res Med Sci. 2013;18(1):56–60.

Springer D, Chaturvedi V. Projecting global occurrence of Cryptococcus gattii. Emerg Infect Dis. 2010;16(1):14–20. https://doi.org/10.3201/eid1601.090369.

Springer D, Saini D, Byrnes E, Heitman J, Frothingham R. Development of an aerosol model of Cryptococcus reveals humidity as an important factor affecting the viability of Cryptococcus during aerosolization. PLoS One. 2013;8(7):e69804. https://doi.org/10.1371/journal.pone.0069804.

Springmann M, Mason-D'Croz D, Robinson S, Garnett T, Godfray HC, Gollin D, Rayner M, Ballon P, Scarborough P. Global and regional health effects of future food production under climate change: a modelling study. Lancet. 2016;387(10031):1937–46. https://doi.org/10.1016/S0140-6736(15)01156-3.

Swaminathan A, Lucas RM, Harley D, McMichael AJ. Will global climate change alter fundamental human immune reactivity: implications for child health? Children (Basel). 2014;1(3):403–23. https://doi.org/10.3390/children1030403.

Torres-Rodríguez J, Hermoso de Mendoza M, Alvarado-Ramírez E, Segura-Roca G. Cryptococcosis by Cryptococcus gattii in immunocompetent goats in Spain and review of the literature. Acta Sci Vet. 2006;34(3):245–53.

UNESCO, UN-Water. United Nations World Water Development ReportWater and climate change. Paris: UNESCO; 2020.

van Rhijn N, Bromley M. The consequences of our changing environment on life threatening and debilitating fungal diseases in humans. J Fungi (Basel). 2021;7(5):367. https://doi.org/10.3390/jof7050367.

Velagapudi R, Hsueh Y, Geunes-Boyer S, Wright J, Heitman J. Spores as infectious propagules of Cryptococcus neoformans. Infect Immun. 2009;77(10):4345. https://doi.org/10.1128/IAI.00542-09.

Vij R, Cordero R, Casadevall A. The buoyancy of Cryptococcus neoformans is affected by capsule size. mSphere. 2018;3(6):e00534–18. https://doi.org/10.1128/mSphere.00534-18.

Wang L, Li Y. Metal ions driven production, characterization and bioactivity of extracellular melanin from Streptomyces sp. ZL-24. Int J Biol Macromol. 2019;123:521–30. https://doi.org/10.1016/j.ijbiomac.2018.11.061.

Watkins R, King J, Johnston S. Nutritional requirements and their importance for virulence of pathogenic Cryptococcus species. Microorganisms. 2017;5(4):65. https://doi.org/10.3390/microorganisms5040065.

Wiebe K, Cockle K, Trzcinski M, Edworthy A, Martin K. Gaps and runs in nest cavity occupancy: cavity" destroyers" and" cleaners" affect reuse by secondary cavity nesting vertebrates. Front Ecol Evol. 2020;8:205.

Williams J, Newbold T. Local climatic changes affect biodiversity responses to land use: a review. Divers Distrib. 2019;26(1):76–92.

Williamson C, Zepp R, Lucas R, Madronich S, Austin A, Ballaré C, et al. Solar ultraviolet radiation in a changing climate. Nat Clim Change. 2014;4:434–41.

World Health Organization. Fact sheet on HIV/AIDS. [Internet]. November, 2020. https://www.who.int/news-room/fact-sheets/detail/hiv-aids.

Chapter 12
Climate Change, Hurricanes, and Fungal Diseases

Eduardo García-Salazar, Paola Betancourt-Cisneros, María Guadalupe Frías-De-León, Gustavo Acosta-Altamirano, Alma Rosa Sánchez-Conejo, and Faustino López Barrera

Contents

Eduardo García-Salazar and Paola Betancourt-Cisneros contributed equally to this work

---

E. García-Salazar (✉)
Unidad de Investigación, Hospital Regional de Alta Especialidad de Ixtapaluca, Estado de México, Mexico

Master of Health Sciences, Escuela Superior de Medicina, Instituto Politécnico Nacional, CDMX, Mexico City, Mexico

P. Betancourt-Cisneros · F. López Barrera
Unidad de Investigación en Sistemática Vegetal y Suelo. Facultad de Estudios Superiores Zaragoza, Universidad Nacional Autónoma de México, CDMX, Mexico City, Mexico
e-mail: lopezbf@unam.mx

M. G. Frías-De-León · G. Acosta-Altamirano
Unit Research, Hospital Regional de Alta Especialidad de Ixtapaluca, Mexico State, Mexico

A. R. Sánchez-Conejo
Dirección General, Hospital Regional de Alta Especialidad de Ixtapaluca, Ixtapaluca Estado de México, Mexico
e-mail: asanchez@hraei.gob.mx

M. G. Frías-De-León et al. (eds.), *The Impact of Climate Change on Fungal Diseases*, Fungal Biology, https://doi.org/10.1007/978-3-030-89664-5_12

Abbreviations

CO_2	Carbon dioxide
IPCC	Intergovernmental Panel on Climate Change
NCEI	National Centers for Environmental Information
NOAA	National Oceanic and Atmospheric Administration
WHO	World Health Organization

12.1 Introduction

The consumption of fossil fuels in anthropogenic activities has generated the release of large amounts of carbon dioxide (CO_2) and other greenhouse gases, causing the retention of more heat in the lower layers of the atmosphere and, therefore, altering the climate on our planet (WHO 2021). The National Oceanic and Atmospheric Administration (NOAA) has recorded that the warmest years globally have been after 1998. This increase in temperature on Earth has been gradual, contributing to extreme weather and climate events with greater frequency and intensity. These phenomena, including extreme peak temperatures, heat waves, windstorms, hurricanes, high precipitation levels with associated floods, and scarce precipitation with droughts and wildfires, have led to natural disasters that impact humans and wildlife (Van Aalst 2006). It has been estimated that natural disasters, whether geophysical (earthquakes, volcanic eruptions, and tsunamis), hydrometeorological (floods, hurricanes, and tornadoes), or geomorphological (landslides and avalanches), affect at least 160,000,000 people annually in their physical, biological, social, and economic environment, in addition to the consequences for public health (Fig. 12.1) (Noji 2000, 2005a, b).

Hurricanes are one of the meteorological phenomena that have allowed us to witness the effect of climate change (Woodward and Samet 2018). In addition to alterations of fungal ecology, major hurricanes can cause population displacement and unsanitary conditions that can increase the risk of fungal diseases (Kouadio et al. 2012; Noji 2005a). Post-hurricane mycoses can be acquired through different pathways (inhalation, traumatic inoculation, near drowning) and can occur in both immunocompromised and immunocompetent people (Kontoyiannis et al. 2019; Webb et al. 2013). Although mycoses are not the only or most critical health problem associated with hurricanes, health personnel in disaster areas must be aware of endemic and non-endemic fungal diseases to provide appropriate treatment promptly.

This chapter presents the close relationship between climate change, the occurrence of intense hurricanes, and post-hurricane fungal diseases.

Fig. 12.1 Destructive effect of hydrometeorological phenomena in Baja California Sur (BCS), Mexico, the state with the longest coastline in the country (Salazar Frías)

12.2 The Hurricanes

A hurricane, typhoon, cyclone, or willy-willy is a tropical storm formed by a warm, moist air mass with strong winds ≥74 mph (119 kilometers per hour) spiraling around a central area (Hidalgo and Baez 2019). Hurricanes generally occur during the hottest time of the year in the Atlantic and Pacific basins and the Indo/Australian, Australian/Pacific, and southwest Indian Ocean basins (Fig. 12.2).

Hurricanes are formed over the warm ocean waters, with a surface temperature of ≥26 °C providing the energy a storm needs to become a hurricane. When warm, moist air evaporates from the ocean's surface, it rises rapidly and encounters cold air, causing it to condense. Condensation turns into storm clouds, and when the introduction of moist, warmer air into the storm clouds persists, a circular wind pattern forms in the storm, called a hurricane, which spirals around the center of the storm (Hidalgo and Baez 2019). Hurricanes are classified into five categories, according to the Saffir-Simpson Hurricane Wind Scale: 1, winds of 74–95 mph; 2,

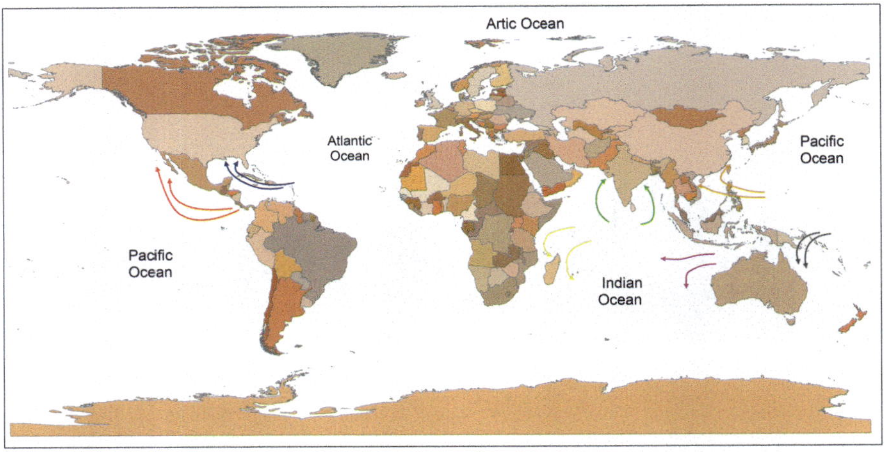

Fig. 12.2 Geographical locations of hurricane formation. Hurricanes generally form in the Atlantic (blue arrow) and the Pacific [Australia/Southwest (gray arrow) and Northwest (red and orange arrows)] basins, as well as the Indo/Australian (pink arrow) basin, and in the north (green arrow) and southwest (yellow arrow) of the Indian Ocean. The direction of the arrows indicates the typical path of hurricanes

winds of 96–110 mph; 3, winds of 111–129 mph; 4, winds of 130–156 mph; and 5, winds ≥156 mph, being categories 4 and 5 the most catastrophic. However, Category 1 hurricanes can also cause significant damage, depending on where they touch land (Simpson and Rielh 1981; www.nhc.noaa.gov/aboutsshs.shtml).

When a hurricane makes landfall, it pushes a wall of ocean water to the shore, called storm surge, reaching heights of more than 20 feet and covering the coast for hundreds of miles. Storm surges, along with the large waves and heavy rains characteristic of hurricanes, can cause substantial flooding along the coast, threatening the lives of those living in these areas (Waddell et al. 2021).

12.3 Climate Change and Hurricanes

In general terms, the warmer the sea temperatures, the more thermal energy is available and the greater the chance of hurricanes developing. Therefore, as humans continue to release greenhouse gases that warm the planet, the likelihood of hurricane activity increases.

Since 2012, the Intergovernmental Panel on Climate Change (IPCC 2012) concluded that the frequency of Category 4 or 5 hurricanes, according to the Saffir-Simpson scale, is likely to double with global warming driven by greenhouse gas emissions from anthropogenic activities (Bender et al. 2010). Global warming increases the ocean's temperature, causing alterations in species migration, melting glaciers, and, consequently, the rise in sea level (Mimura 2013). Rising sea levels lead to more destructive storm surges during hurricanes, causing greater vulnerability in the island and coastal communities. It has been estimated that in the following six decades, the number of people affected will increase almost eightfold (Neumann et al. 2015; Shultz et al. 2019).

Another factor contributing to changes in the frequency and intensity of hurricanes is the development of coastal cities, as this implies an increase in energy consumption based on fossil fuels and the emission of greenhouse gases, mainly CO_2 (Nordhaus 2010).

In 2014, it was reported that most anthropogenic warming, generated by the gases and aerosols emissions of human activities, occurred in the last four decades (Holland and Bruyère 2014). The prior was related to the increase in Category 4 or 5 hurricanes with concerning all hurricanes. Evidence of the latter is the increasing intensity of Atlantic hurricane seasons, which has changed from about 50 hurricanes on average per decade to 70 over the past two decades (Table 12.1).

More evidence on the wavering scenario of hurricanes enhanced by global warming and the sea level rise is the intense 2017 Atlantic basin hurricane season. This period stood out for the presence of ten hurricanes, four of them classified as Category ≥4, which had devastating effects on both material resources and human lives (Shultz et al. 2019). Interestingly, 2017 is the fourth warmest year on record, according to the National Centers for Environmental Information (NCEI).

The increase in the proportion of large-magnitude hurricanes due to global warming has been recorded in all ocean basins. However, it has been balanced with a similar decrease in Category 1 and Category 2 hurricanes (Holland and Bruyère 2014). Although it is complex to establish trends in the amount and intensity of hurricanes, NOAA data show an increase in the number of Atlantic hurricanes, particularly Category ≥3 hurricanes, from 1970 to 2020 (Fig. 12.3). Moreover, the year 2020 had a historical record of catastrophic hurricanes. According to NOAA, at least ten cyclones are expected to form in 2021, with five of them being potentially catastrophic.

It should be noted that the relationship between global warming and more intense hurricanes is not linear. Moreover, it is predicted that the increase in Category 4 or 5 hurricanes will not continue at the same rate with future global warming. After an

Table 12.1 Atlantic hurricanes recorded in recent decades

Decade	Total number of Atlantic hurricanes	Total number of category ≥4 hurricanes
1970–1979	49	8
1980–1989	52	10
1990–1999	63	13
2000–2009	74	23
2010–2019	72	18

Data obtained from the National Oceanic and Atmospheric Administration (http://www.aoml.noaa.gov/hrd/tcfq/E17.html)

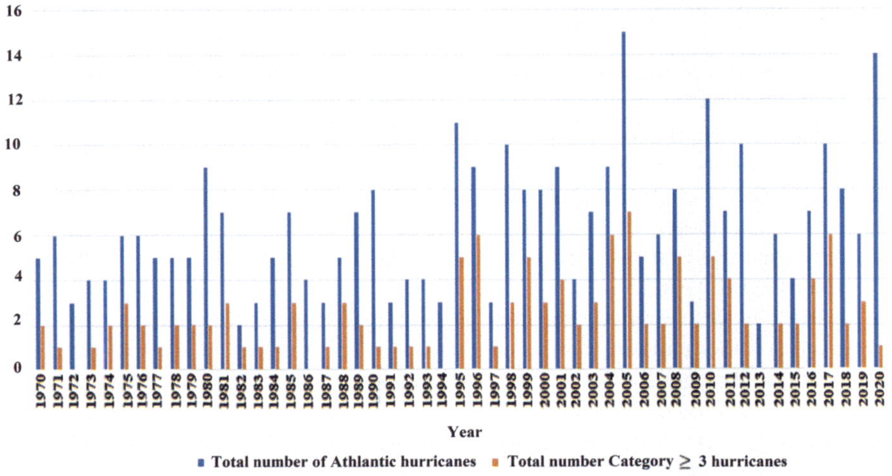

Fig. 12.3 Frequency of hurricanes in the period 1970–2020

initial climate increase, a level of saturation of intense hurricanes will be reached, and any further global warming will have little effect (Holland and Bruyère 2014). However, it has been projected that climate changes will continue to increase the formation of hurricanes during the twenty-first century (Shultz et al. 2019). Therefore, vulnerability to these hazards will also increase for populations in coastal areas.

12.4 Hurricanes and Fungal Diseases

More than one million species of fungi exist on Earth, but only a small percentage, 0.02%, cause disease in humans (Garcia-Solache and Casadevall 2010). Many of the pathogenic fungi are found in the environment. Some are ubiquitous, such as *Aspergillus* spp., while others have habitats restricted to specific climatic conditions, such as *Coccidioides* spp. (Kollath et al. 2019; Latgé and Chamilos 2019). The abundance and distribution of environmental fungi depend on factors such as temperature and humidity. Their presence in natural habitats can be altered due to environmental disturbances, which can be caused by human activities (excavations, constructions, among others) or by natural events of great magnitudes, such as hurricanes (Hernandez and Martinez 2018). The development of more dangerous hurricanes, with winds and storms of greater extents, can significantly impact at an economic and health level depending on their size, their intensity, and their location in terms of population density and proximity to the coast (Dinan 2017). An example of this is Hurricane Katrina, Category 5 on the Saffir-Simpson scale, which was recorded in the United States as one of the most devastating in history, causing the death of more than 1000 people and leaving damage estimated at $ 170,000,000,000 (Waddell et al. 2021). Within the health scope, hurricanes can lead to diseases, such as those caused by fungi (Benedict and Park 2014). While fungal diseases are not the main adverse health event after a hurricane, there is a growing evidence on the possibility of fungal proliferation and fungal diseases associated with these weather phenomena (Benedict and Park 2014). Large floods and strong winds during intense hurricanes can alter the growth, distribution, and dispersion of fungi in the environment (Garcia-Solache and Casadevall 2010; Benedict and Park 2014). Fungi can be displaced from their habitats, many of them with pathogenic potential, causing them to appear in areas where they were not common (Sood et al. 2018). Likewise, in areas where the fungus was already present, the fungal load may increase (Barbeau et al. 2010; Schieffelin et al. 2013; Sood et al. 2018). A higher, or a broader geographically widespread, fungal load facilitates contact between fungi and people affected by hurricanes, causing disease (Barbeau et al. 2010). In some cases, contact with fungi has not caused disease, but colonization by clinically relevant agents, such as *Syncephalastrum*, has been observed (Rao et al. 2007a). Furthermore, it should be noted that during a hurricane, like in any other natural disaster, the host susceptibility to pathogens can change as people can suffer traumatic injuries during evacuation or cleaning activities afterward, and these injuries can be the route of entry for fungi (Garcia-Solache and Casadevall 2010; Panackal et al. 2010). Other entry routes may be inhalation or aspiration of water in near-drowning (Kontoyiannis et al. 2019; Webb et al. 2013).

Among the diseases caused by fungal contamination associated with floods after major hurricanes are acute fungal infection of the upper respiratory tract (sinusitis), pneumonia, post-traumatic skin infection, endophthalmitis, keratitis, as well as infection of the central nervous system as a result of transient lung infection with hematogenous spread after massive aspiration of contaminated water (near-drowning), allergic rhinitis, conjunctivitis, asthma, hypersensitivity pneumonitis, and allergic bronchopulmonary aspergillosis (Brandt et al. 2006). Some of these diseases have been recorded in people exposed to fungal contamination in areas damaged by Category 3 to 5 hurricanes, particularly by Katrina, one of the deadliest hurricanes in the United States (Table 12.2).

However, it has not yet been possible to record an increase in post-hurricane fungal illnesses, which attributed to different factors: (1) people who had mild symptoms of infection did not seek medical attention or the diagnosis was not reported to surveillance systems; (2) people are forced to seek shelter in homes not affected by the hurricane; therefore, their exposure to the fungus is minimal, or they were diagnosed and treated in unaffected areas; and (3) people who exposed to the fungi were immunocompetent and the exposure time was limited or minimal (Barbeau et al. 2010).

Table 12.2 Fungal diseases associated with hurricanes

Hurricane (Name, year)	Hurricane category (Saffir-Simpson scale)	Fungal agent	Disease	Reference
Katrina, 2005	5	*Cladosporium* sp.	Lung infection	Rao et al. (2007b)
Katrina, 2005	5	*Blastomyces dermatitidis*	Systemic infection (central nervous system)	Szeder et al. (2007)
Ike, 2008	4	Unidentified agent causing chromoblastomycosis	Subcutaneous infection (soft tissue)	Riddel et al. (2011)
Katrina y Wilma, 2005	5	*Drechslera* sp., other unidentified fungi	Allergic sinusitis, invasive infection	Sridhar et al. (2012)
Katrina, 2005	5	*Coccidioides*	Disseminated and lung infection in non-endemic area	Schieffelin et al. (2013)
Katrina, 2005	5	*Aspergillus fumigatus*, *Mucor*	Invasive lung infection	Webb et al. (2013)
Sandy, 2012	3	*Aspergillus fumigatus*, *Aspergillus flavus*, *Aspergillus terreus*, *Aspergillus niger*, *Trichosporon asahii*, *Fusarium*, *Penicillium*, *Mucor*, *Curvularia*, *Bipolaris*, *Rhizopus*, *Trichoderma*, *Trichosporon*	Skin infection	Sood et al. (2018)
Harvey, 2017	4	*Aspergillus*, *Fusarium*, *Rhizopus*, *Syncephalastrum*, *Conidiobolus*	Invasive infection	Kontoyiannis et al. (2019)

12.5 Prevention of Post-Hurricane Fungal Diseases

After a hurricane, excess humidity and stagnant water promote the abundant growth of fungi in homes, representing a health risk. Therefore, it is necessary to remove the fungi and control the humidity to prevent the fungus from proliferating. The presence of the fungus poses a substantial health hazard for people suffering from asthma, allergies or other respiratory conditions, or immunosuppression (Chow et al. 2019). Thus, the most vulnerable people must not participate in post-hurricane cleanup activities. On the other hand, people involved in cleaning or rehabilitating areas affected by hurricanes should avoid areas where fungal contamination is evident, use personal protective equipment (respirator, boots, glasses, gloves) to prevent trauma, and keep hands, skin, and clothing clean and free of fungus-contaminated dust (Brandt et al. 2006; Chow et al. 2019; Liang and Messenger 2018). If anyone experiences a health problem after being exposed to fungi, medical attention must be provided immediately. Healthcare workers must be aware of the spectrum of possible fungal diseases, including unusual conditions, that may occur after a hurricane to establish a timely diagnosis and appropriate treatment.

Preventing fungal diseases after a hurricane disaster requires systems-based emergency management and public health strategies that address population displacement (Chow et al. 2019; Cummings et al. 2007). Evacuation of survivors from contaminated environments followed by access to safe shelter, potable water and food, and basic healthcare services are integral components of a coordinated response to hurricanes and other hydrometeorological disasters (Liang and Messenger 2018).

12.6 Conclusion

Climate change has contributed to increasingly intense hurricane seasons. Large-scale hurricanes can alter the growth, distribution, and dispersal of fungi in the environment and favor greater contact between fungi and people affected by hurricanes, leading to various fungal diseases. Therefore, it is necessary to implement multidisciplinary actions to limit the anthropogenic effect on climate change and establish and communicate measures for the prevention and control of fungal diseases in communities vulnerable to hurricanes.

Acknowledgments E.G.S thanks the Master's Program in Health Sciences of Escuela Superior de Medicina, Instituto Politécnico Nacional, and the Consejo Nacional de Ciencia y Tecnología (CONACyT) for a scholarship (reference no. 1025718). M.G.F.-D.-L., and G.A.-A. thank the CONACyT for the support to the realization of this chapter.

References

Barbeau DN, Grimsley LF, Whitte LE, El-Dahr JM, Lichtveld M. Mold exposure and health effects following hurricanes Katrina and Rita. Annu Rev Public Health. 2010;31:165–78. https://doi.org/10.1146/annurev.publhealth.012809.103643.

Bender MA, Knutson TR, Tuleya RE, Sirutis JJ, Vecchi GA, Garner ST, et al. Modeled impact of anthropogenic warming on the frequency of intense Atlantic hurricanes. Science. 2010;327:454–8. https://doi.org/10.1126/science.1180568.

Benedict K, Park BJ. Invasive fungal infections after natural disasters. Emerg Infect Dis. 2014;20(3):349–55. https://doi.org/10.3201/eid2003.131230.

Brandt M, Brown C, Burkhart J, Burton N, Cox-Ganser J, Damon S, et al. Mold prevention strategies and possible health effects in the aftermath of hurricanes and major floods. MMWR Recomm Rep. 2006;55(RR-8):1–27.

Chow NA, Toda M, Pennington AF, Anassi E, Atmar RL, Cox-Ganser JM, et al. Hurricane-associated mold exposures among patients at risk for invasive mold infections after hurricane Harvey – Houston, Texas, 2017. MMWR Morb Mortal Wkly Rep. 2019;68(21):469–73. https://doi.org/10.15585/mmwr.mm6821a1.

Cummings KJ, Cox-Ganser J, Riggs MA, et al. Respirator donning in post-hurricane New Orleans. Emerg Infect Dis. 2007;13(5):700–7. https://doi.org/10.3201/eid1305.061490.

Dinan T. Projected increases in hurricane damage in the United States: the role of climate change and coastal development. Ecol Econ. 2017;138:186–98. https://doi.org/10.1016/j.ecolecon.2017.03.034.

Garcia-Solache MA, Casadevall A. Global warming will bring new fungal diseases for mammals. MBio. 2010;1:e00061–10. https://doi.org/10.1128/mBio.00061-10.

Hernandez H, Martinez LR. Relationship of environmental disturbances and the infectious potential of fungi. Microbiology. 2018;164(3):233–41. https://doi.org/10.1099/mic.0.000620.

Hidalgo J, Baez AA. Natural Disasters. Crit Care Clin. 2019;35(4):591–607. https://doi.org/10.1016/j.ccc.2019.05.001.

Holland G, Bruyère CL. Recent intense hurricane response to global climate change. Clim Dyn. 2014;42:617–27. https://doi.org/10.1007/s00382-013-1713-0.

IPCC. Managing the risks of extreme events and disasters to advance climate change adaptation. 2012. https://www.ipcc.ch/report/managing-the-risks-of-extreme-events-and-disasters-to-advance-climate-change-adaptation. Accessed 27 Jan 2021.

Kollath DR, Miller KJ, Barker BM. The mysterious desert dwellers: Coccidioides immitis and Coccidioides posadasii, causative fungal agents of coccidioidomycosis. Virulence. 2019;10(1):222–33. https://doi.org/10.1080/21505594.2019.1589363.

Kontoyiannis DP, Shah EC, Wurster S, Aitken SL, Graviss L, Raad II, et al. Culture-documented invasive mold infections at MD Anderson Cancer Center in Houston, Texas, pre- and post-hurricane Harvey. Open Forum Infect Dis. 2019;6(4):ofz138. https://doi.org/10.1093/ofid/ofz138.

Kouadio IK, Aljunid S, Kamigaki T, Hammad K, Oshitani H. Infectious diseases following natural disasters: prevention and control measures. Expert Rev Anti-Infect Ther. 2012;10(1):95–104. https://doi.org/10.1586/eri.11.155.

Latgé JP, Chamilos G. Aspergillus fumigatus and Aspergillosis in 2019. Clin Microbiol Rev. 2019;33(1):e00140–18. https://doi.org/10.1128/CMR.00140-18.

Liang SY, Messenger N. Infectious diseases after hydrologic disasters. Emerg Med Clin North Am. 2018;36(4):835–51. https://doi.org/10.1016/j.emc.2018.07.002.

Mimura N. Sea-level rise caused by climate change and its implications for society. Proc Jpn Acad Ser B Phys Biol Sci. 2013;89(7):281–301. https://doi.org/10.2183/pjab.89.281.

Neumann J, Emanuel K, Ravela S, Ludwig L, Kirshen P, Bosma K, et al. Joint effects of storm surge and sea-level rise on US coasts: new economic estimates of impacts, adaptation, and benefits of mitigation policy. Clim Chang. 2015;129(1):337–49. https://doi.org/10.1007/s10584-014-1304-z.

Noji EK. The public health consequences of disasters. Prehosp Disaster Med. 2000;15:147–57.

Noji EK. Public health issues in disasters. Crit Care Med. 2005a;33:S29–33. https://doi.org/10.1097/01.CCM.0000151064.98207.9C.

Noji EK. Public health in the aftermath of disasters. BMJ. 2005b;330(7504):1379–81. https://doi.org/10.1136/bmj.330.7504.1379.

Nordhaus W. The economics of hurricanes and the implications of global warming. Clim Chang Econ. 2010;1(1):1–20. https://doi.org/10.1142/S2010007810000054.

Panackal AA, Li H, Kontoyiannis DP, Mori M, Perego CA, Boeckh M, et al. Geoclimatic influences on invasive aspergillosis after hematopoietic stem cell transplantation. Clin Infect Dis. 2010;50:1588–97. https://doi.org/10.1086/652761.

Rao CY, Kurukularatne C, Garcia-Diaz JB, Kemmerly SA, Reed D, Fridkin SK, et al. Implications of detecting the mold Syncephalastrum in clinical specimens of New Orleans residents after hurricanes Katrina and Rita. J Occup Environ Med. 2007a;49:411–6. https://doi.org/10.1097/JOM.0b013e31803b94f9.

Rao CY, Reed D, Kemmerly S, Morgan J, Fridkin SK. Assessing invasive mold infections, mold exposures and personal protective equipment use among immunocompromised New Orleans residents after hurricane Katrina. Society for Healthcare Epidemiology of America 17th Annual Scientific Meeting 2007; 2007b Apr 14–17; Baltimore. http://www.shea-online.org/Assets/ files/The_Environment.doc.pdf. Accessed 12 Feb 2021.

Riddel CE, Surovik JG, Chon SY, Wang WL, Cho-Vega JH, Cutlan JE, et al. Fungal foes: presentations of chromoblastomycosis post-hurricane Ike. Cutis. 2011;87:269–72.

Schieffelin JS, Torrellas M, Lartchenko S, Gill F, Garcia-Diaz J, McGoey R. How natural disasters change natural patterns: coccidioidomycosis imported to New Orleans. J La State Med Soc. 2013;165(3):145–9.

Shultz JM, Kossin JP, Marshall Shepherd J, Ransdell JM, Walshe R, Kelman I, et al. Risks, health consequences, and response challenges for small-island-based populations: observations from the 2017 Atlantic hurricane season. Disaster Med Public Health Prep. 2019;13(1):5–17. https://doi.org/10.1017/dmp.2018.28.

Simpson RH, Rielh H. The hurricane and its impact. Baton Rouge: Louisiana State University Press; 1981.

Sood G, Vaidya D, Dam L, Grubb LM, Zenilman J, Krout K, et al. A polymicrobial fungal outbreak in a regional burn center after hurricane Sandy. Am J Infect Control. 2018;46(9):1047–50. https://doi.org/10.1016/j.ajic.2018.01.011.

Sridhar J, Lam BL, Pasol J, Sternau L. Three cases of fungal disease with neuro-ophthalmic manifestations associated with post-hurricane environment. J Neuroophthalmol. 2012;32(3):197–201. https://doi.org/10.1097/WNO.0b013e3182268746.

Szeder V, Ortega-Gutierrez S, Frank M, Jaradeh SS. CNS blastomycosis in a young man working in fields after hurricane Katrina. Neurology. 2007;68(20):1746–7. https://doi.org/10.1212/01.wnl.0000265229.31844.45.

Van Aalst MK. The impacts of climate change on the risk of natural disasters. Disasters. 2006;30:5–18. https://doi.org/10.1111/j.1467-9523.2006.00303.x.

Waddell SL, Jayaweera DT, Mirsaeidi M, Beier JC, Kumar N. Perspectives on the health effects of hurricanes: a review and challenges. Int J Environ Res Public Health. 2021;18(5):2756. https://doi.org/10.3390/ijerph18052756.

Webb BJ, Blair JE, Kusne S, Scott RL, Steidley DE, Arabia FA, et al. Concurrent pulmonary aspergillus fumigatus and Mucor infection in a cardiac transplant recipient: a case report. Transplant Proc. 2013;45(2):792–7. https://doi.org/10.1016/j.transproceed.2012.03.056.

Woodward AJ, Samet JM. Climate change, hurricanes, and health. Am J Public Health. 2018;108(1):33–5. https://doi.org/10.2105/AJPH.2017.304197.

World Health Organization WHO Climate change and health. 2021. www.who.int/es/news-room/fact-sheets/detail/climate-change-and-health

Chapter 13
Climate Change and Allergies

Mirta Álvarez Castelló and Michel Almaguer Chávez

Contents

Abbreviations

ABPA	Allergic bronchopulmonary aspergillosis
ABPM	Allergic bronchopulmonary mycosis
ACOS	Asthma-COPD overlap syndrome
AFRS	Allergic fungal rhinosinusitis
CRS	Chronic rhinosinusitis
CRSsNP	CRS without nasal polyps
CRSwNP	CRS with nasal polyps
DMHS	Mold hypersensitivity syndrome
HP	Hypersensitivity pneumonitis

M. Álvarez Castelló (✉)
University Hospital "General Calixto García", Havana, Cuba
e-mail: mirtaac@infomed.sld.cu

M. Almaguer Chávez
Faculty of Biology, University of Havana, Havana, Cuba
e-mail: michelalm@fbio.uh.cu

M. G. Frías-De-León et al. (eds.), *The Impact of Climate Change on Fungal Diseases*, Fungal Biology, https://doi.org/10.1007/978-3-030-89664-5_13

13.1 Introduction

Climate change affects human health and represents a challenge for all countries (Osborne et al. 2015; Tummon et al. 2021; Villacís 2019). It is a global phenomenon with differential influences on populations, depending on their exposure, sensitivity, and adaptive capacity (Paavola 2017; Prescott 2013). These changes could affect ecosystems, populations, and all spheres of our lives and represent the biggest global health threat of the twenty-first century (Katelaris and Beggs 2018). Its main characteristics are the sea level rise, the global warming, the changes in the precipitations, and extreme weather events (Pendrey et al. 2020). These changes could cause several negative consequences depending on geographic region, socioeconomic status of people, and preexisting population vulnerabilities (Jhan et al. 2020). This chapter is about how these changes affect human health, focusing on fungal allergic diseases and conditions.

Allergic diseases have high prevalence around the world. Exposure to fungal allergens affects the human health causing allergies, irritation, and intoxications. The effects vary according to the nature and dose of exposure (Agache et al. 2019; Katelaris and Beggs 2018; Osborne et al. 2015; Tummon et al. 2021). The fungal components (enzymes, toxins, cell wall components, and proteins) interact with the immune system and cause diverse responses such as asthma, rhinitis, and rhinoconjunctivitis. Other significant diseases have been identified, like allergic bronchopulmonary aspergillosis, hypersensitivity pneumonitis, and allergic fungal sinusitis (Pfavayi et al. 2020).

Climate change affects the quantity, quality, and temporal and spatial dynamics of aeroallergens (Patella et al. 2018), and environmental conditions play an important role in the appearance of sensitization to these aeroallergens. Also the flooding associated with tropical hurricanes impacts on allergic diseases in these latitudes, causing a significant proliferation of mold growth in damp homes. Their involvement role as allergens has been demonstrated by many aerobiology studies (Levetin et al. 2016). It is a topic of research by allergologist, and not only as a risk factor for asthma but also their involvement in severe and fatal asthma.

13.2 Allergenic Fungi

Many studies described different fungal-related respiratory conditions, especially those related to asthma (Hulin et al. 2013; Reponen et al. 2011, 2012). Some investigations on environmental medicine have established a causal relationship between asthma and environmental fungi. Furthermore, these studies indicate that the prevalence of fungal-induced asthma in adults is currently increasing (Denning et al. 2014). In the adult population, the incidence of work-attributable asthma is around 30% (Karjalainen et al. 2001). Adult asthma is strongly associated with the level of exposure to fungi (Park et al. 2008). In Finland, environmental fungi became the main causative factor of occupational asthma (Denning et al. 2014).

There are more than 80 genera of fungi reported as causal agents of allergies; among these are *Aspergillus*, *Penicillium*, *Cladosporium*, *Alternaria*, *Helminthosporium*, *Epicoccum*, *Fusarium*, *Rhizopus*, and *Mucor* (Levetin et al. 2016). Of these fungal genera, *Aspergillus*, *Penicillium*, *Cladosporium*, and *Alternaria* stand out due to their greater clinical importance and the majority to identified allergens (Adhikari et al. 2004; Cafarchia et al. 2014; Kochar et al. 2014). The phyla *Zygomycota*, *Ascomycota*, and *Basidiomycota* contain most of the genus of fungi that produce allergens (Fig. 13.1) (Hibbett et al. 2007; Levetin et al. 2016; Soeria-Atmadja et al. 2010a, b).

The phylum *Zygomycota* contains allergenic saprobic genera, highlighting *Mucor* and *Rhizopus* (Hibbett et al. 2007; Levetin et al. 2016). Soeria-Atmadja et al. (2007) and Andersen et al. (2011) showed the species *Mucor racemosus* induces sensitization in some individuals, as well as symptoms of asthma and rhinitis. On the other hand, Donthi et al. (2011) stated that aplanospores of *Rhizopus* are allergenic and tend to disperse in hot and dry climates. *Rhizopus nigricans*, *R. stolonifer*, and *R. oryzae* possess several allergenic proteins associated with rhinitis, asthma, fungal sinusitis, hypersensitivity pneumonitis, and infection (Levetin et al. 2016; Maniwa et al. 2002; Sircar et al. 2012).

Ascomycota contains a large group of allergenic fungi that produce asexual spores (conidia) and sexual spores (ascospores) that are dispersed through the air (Fig. 13.2). Based on the DNA sequence, many genera of allergological interest are located within this phylum (Levetin et al. 2016). *Cladosporium* is a highly studied

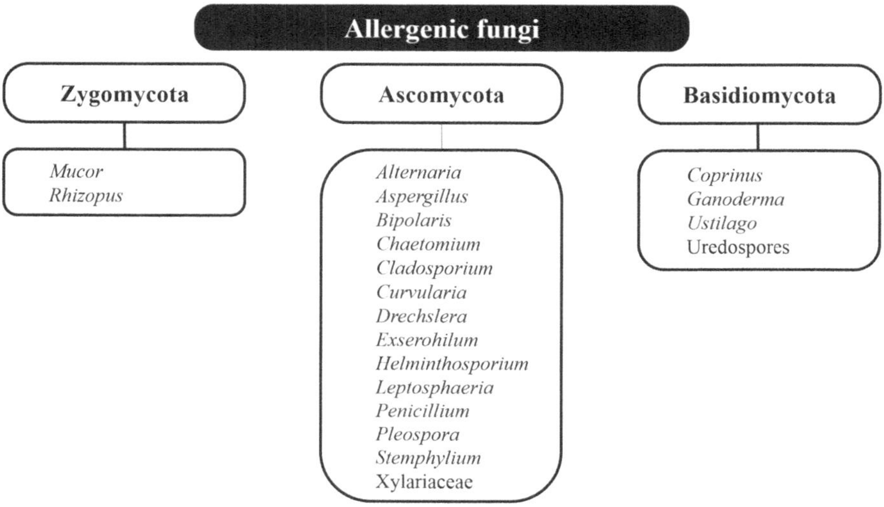

Fig. 13.1 Examples of allergenic fungi grouped in the three main phyla

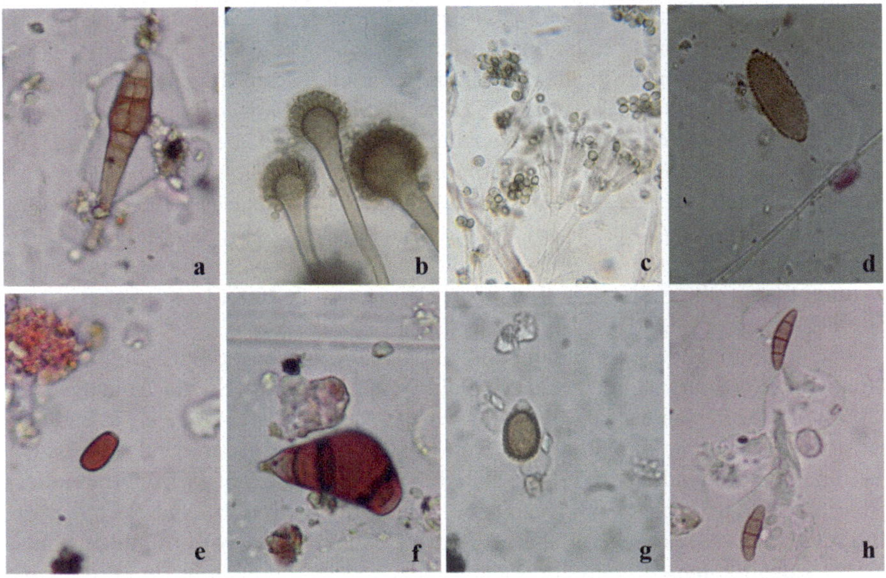

Fig. 13.2 Some airborne allergenic fungi. (**a**) *Alternaria* (spore, conidium), (**b**) *Aspergillus* (conidial head), (**c**) *Penicillium* (penicilla and conidia), (**d**) *Cladosporium* (spore, conidium), (**e**) *Coprinus* (spore, basidiospore), (**f**) *Curvularia* (spore, conidium), (**g**) *Ganoderma* (spore, basidiospore), (**h**) *Leptosphaeria* (spore, ascospore)

genus with a high presence in the air. The allergens Cla h 1 and Cla h 2 have been identified in *C. herbarum* and could affect sensitized individuals with allergic rhinitis and asthma (Breitenbach and Simon-Nobbe 2002; Raphoz et al. 2010). Multiple *Alternaria* allergens have also been characterized (Williams et al. 2016). Some authors have shown that there is a cross-reactivity between these genera, so that the high concentrations of *Cladosporium* enhance the immune response of those sensitive to *Alternaria* (Aira et al. 2012, 2013). The species *Alternaria alternata* and *Cladosporium herbarum* are probably the most important from the allergological point of view (Aira et al. 2012, 213).

Some related genera of the *Pleosporaceae* family, such as *Curvularia*, *Bipolaris*, *Drechslera*, *Helminthosporium*, and *Exserohilum*, are also related to allergic processes (Fryen et al. 1999; McAleer et al. 1981; Menezes et al. 1998). More than four IgE-binding proteins have been identified from *Bipolaris* extracts (Lim et al. 1995). Several species of *Helminthosporium* can cause allergic rhinitis and asthma (Weber 2006), and allergens from *Exserohilum rostratum* have been detected by Green et al. (2003). On the other hand, *Curvularia* shares IgE cross-reactivity with *Alternaria*, *Bipolaris*, and *Stemphylium* (Agarwal et al. 1982). Several *Stemphylium* allergens (Ste h 1 and Ste. b 1) appear to have a high cross-reaction with *Alternaria* Alt 1.63 to *S. herbarum* associated with allergic asthma and hypersensitivity pneumonitis (Gutierrez-Rodriguez et al. 2011; Lahoute et al. 1983).

Aspergillus is another ascomycetous genus of allergological importance, with a large number of species that produce small conidia (asexual spores), which allows their aspiration and penetration into the respiratory tract (Chacón 2012). *Aspergillus* can cause allergies to invasive infections and can affect the upper airways or lungs, respectively (Cruz et al. 2005; Singh and Pursell 2008). The most frequently identified species in clinical isolates is *Aspergillus fumigatus* followed by *A. flavus*, *A. niger*, *A. terreus*, and *A. nidulans*. These last four also play an important role in various human pathologies (Hedayati et al. 2007; Oliveira and Carmalho 2014). The most important species from the allergenic point of view is *A. fumigatus*, which tolerates a wide range of growth temperatures, between 12 and 52 °C (Oliveira and Carmalho 2014). In humans, it causes allergy symptoms (rhinoconjunctivitis and asthma), hypersensitivity pneumonitis or extrinsic allergic alveolitis and allergic bronchopulmonary aspergillosis, localized infections such as aspergilloma, or generalized infections such as invasive aspergillosis, although compared to other species of this genus generally in low concentrations in the air (Salim and Runco 2008; Zubeldia et al. 2012). The main allergen of *A. fumigatus*, Asp f1, is an 18 kD IgE-binding protein, which is a cytotoxin also produced by *A. restrictus* (Arruda et al. 1990; Ryan et al. 2001). In other investigations, specific IgE was found in 2% of 1800 children to *A. versicolor*, which produces the allergen Asp v 13, a subtilisin-like serine protease (Kolossa-Gehring et al. 2007; Shi et al. 2011).

Penicillium is another genus with species that produce small conidia, and it has been identified as a risk factor for asthma and allergies (Scott et al. 2004). Some species are major indoor fungi that are associated with respiratory atopic disorders. Investigation of the genus *Penicillium* allergens and its allergenic cross-reactivity among fungal allergens provides important information in clinical mold allergy (Shen et al. 2005).

Also the spores of *Fusarium* can be airborne, especially during humid weather, and can cause allergies and may cause also sinusitis and allergic bronchopulmonary mycosis. Several studies have shown the reactivity of skin tests to its extracts (Levetin et al. 2016).

Basidiospores (sexual spores of the phylum *Basidiomycota*) have been associated with increases in emergency department visits for asthma. This has been recognized as one of the strongest connections between asthma and the environment. In some reports, the role of *Ganoderma* spores in sensitization of 10–48% has been identified by skin prick tests (Jedryczka et al. 2015; Sadys et al. 2014). In the Caribbean, a greater sensitization to basidiospores than to conidia was detected in Puerto Rico (Rivera-Mariani et al. 2011). However, in this same country, the need to further evaluate the role of basidiomycetes in allergies was highlighted (Rivera-Mariani et al. 2012).

Several studies show the temporal dynamics of fungal spores in the external environment, in which differences in seasonal behavior for the different genera are described, related to the environmental factors characteristic of each region. It has been suggested that atmospheric fungal concentration tends to be higher in temperate and tropical regions than in high latitudes and deserts, where fungal colonization is limited by adverse climatic conditions (Barriga 2007). The concentration of spores in the air can vary depending on the type of vegetation, the local microenvironment, and human activity.

In countries with a tropical climate during most of the year, the temperature varies between 25 °C and 30 °C, and the relative humidity remains above 70%. These conditions favor fungal development and the formation and release of spores (Quintero et al. 2010).

Numerous authors reported that the genera *Cladosporium*, *Aspergillus*, and *Penicillium* are the most common in the air (Adhikari et al. 2004; Aira et al. 2013; Docampo et al. 2010; Levetin et al. 2016). Their concentration varies depending on biological factors (the daily sporulation rate and the availability of the substrate for the development of the mycelium), climatic (temperature, humidity, and precipitation) or physical (movement of the atmosphere, turbulence, and warming) (Sidel et al. 2015). *Cladosporium* spores are the most abundant in the exterior of most latitudes and in low latitudes of tropical climate; they tend to appear outside the summer period (Aira et al. 2013; Rocha et al. 2013).

Aspergillus and *Penicillium* spores are widely represented in the air (Docampo et al. 2010). *Aspergillus* releases large amounts of conidia into the air from the conidiophores that project from the mycelium, and due to their small size, they can remain suspended in the environment for a long period of time, increasing the probability of being inhaled (Fairs et al. 2013).

In tropical areas, the genus *Curvularia* contributes significantly to airborne spore concentration values with maximums between 4000 and 9000 spores/m^3, while *Alternaria* conidia appear in daily mean concentrations that can reach up to 150 spores/m^3 (Griffin 2007). Most air fungi studies around the world have been conducted mostly in urban areas. However, in agricultural areas, higher concentrations of fungal propagules are usually detected due to the abundant vegetation and decomposing organic matter and the agricultural activity itself.

The origin of airborne spores in indoor environments is (1) from outside and penetrates through ventilation systems and (2) released by fungi that grow in situ on various substrates (Shelton et al. 2002). Although there is no international standard that indicates when an environment is contaminated or not, some authors suggest that an environment can be considered contaminated if it has more than 1000 spores/m^3 (Borrego et al. 2010). The most abundant fungal genera indoors are *Cladosporium*, *Aspergillus*, and *Penicillium*. However, there are reports of other fungal genera such as *Curvularia*, *Alternaria*, *Fusarium*, *Acremonium*, and *Epicoccum* (Díaz et al. 2010a, b). The predominance of one or the other is given by the climatic region and the specific conditions of the premises, such as the presence of organic substances, the microclimatic conditions of temperature and relative humidity, ventilation, and the predominant microbiota in the outdoor air (Pyrri and Kapsanaki-Gotsi 2007).

13.3 Responses to Fungal Allergens

There is association between home dampness and respiratory symptoms in children and adults, with evidence for dose-response relationships (Cai et al. 2019; Mendell et al. 2011). Several studies show that early exposure to environments with dampness, visible mold, and moldy odor is associated with successive development of asthma (Baxi et al. 2016; D'Amato et al. 2015) and pediatric asthma emergency admissions (Tham et al. 2017). The prevalence of allergic rhinitis is three times that of asthma, and many patients with asthma have associated rhinitis. Allergic rhinitis and asthma are currently not considered two independent conditions (Navarro et al. 2017), but rather manifestations of a single syndrome: chronic respiratory allergy syndrome (Montes-Angel et al. 2020; Pawankar et al. 2011).

Fungi contribute to asthma by different mechanisms, but the role of fungal allergens has been incompletely studied (Agarwal and Gupta 2011). In the case of thermotolerant fungi, such as *Aspergillus*, colonization of numerous substrates allows a continuous source of fungal allergens, leading to persistent immune response, airway inflammation, and tissue damage. *Alternaria*, a temperature sensitive fungus, may cause disease exacerbations in response to extend exposure to fungal spores or fragments and fungal products. Several epidemiological studies worldwide have recognized an association between observed mold or water damage and asthma (Zhang et al. 2016).

With respect to severe asthma, many patients are frequently sensitized to multiple fungal species, documented by skin prick test or serum IgE test. To date, severe asthma with fungal sensitization is recognized as a phenotype. These patients had more bronchiectasis and fixed airflow obstruction than other asthmatic patients (Mota and Borrego 2016). Elevated spore levels in the air correlate with an increased number of hospitalizations and deaths from asthma (Sabino et al. 2019) as well as a greater risk of developing rhinitis (Fernández-Soto et al. 2018). An association between emergency visits for asthma and thunderstorms has been linked to high airborne fungal spore counts in Canada and the United Kingdom, mainly by *Alternaria* spores (Dales et al. 2003). Bioaerosols are relevant in damp indoor environments and include fungi, fungal spores, hyphae, as well as fungal fragments and other allergens (Zhang et al. 2016).

Hinson et al. (1952) described allergic bronchopulmonary aspergillosis (ABPA) in 1952 characterized by increased serum levels of *Aspergillus*-specific IgE and IgG antibodies, peripheral blood eosinophilia, immediate and/or delayed skin reactions to fungal antigen, and radiographic findings including pulmonary opacities, central bronchiectasis, and mucus plugs (Asano et al. 2018). The prevalence of ABPA in asthma and cystic fibrosis is about 13% and 9%, respectively (Agarwal et al. 2016 Jubin et al. 2010). Worldwide, more than four million people are affected by ABPA. Positive skin prick test to *Aspergillus* antigen is present in almost 25% asthmatics and 50% of cystic fibrosis patients, but ABPA is not that much prevalent (Santos et al. 2021). It was lately named allergic bronchopulmonary mycosis (ABPM) because other fungi were implicated. ABPM is characterized by fungal colonization of the airway and IgE sensitization to fungi, with recurrent and transient radiographic peripheral infiltrates, pulmonary eosinophilia, and bronchiectasis (Mota and Borrego 2016). *Candida albicans* is the most commonly reported associated fungus with *Bipolaris* species, *Schizophyllum commune*, and *Curvularia* species (Denning et al. 2014). *Penicillium*, *Drechslera*, *Torulopsis*, *Mucor*, *Fusarium*, *Cladosporium*, and *Saccharomyces* have also been identified (Cottin 2016).

Fungal antigens can cause hypersensitivity pneumonitis (HP), or extrinsic allergic alveolitis, and these can be found in the workplace, at home, and in recreational activities. HP is characterized by non-IgE-mediated inflammation of the parenchyma, alveoli, and terminal airways of the lung (Corry et al. 2019; Grammer 2019). HP is secondary to repeated and prolonged exposure to specific antigens in a susceptible host (Watts and Grammer 2019). *Penicillium* spp. and *Aspergillus* spp. are the fungi species most frequently linked with the onset of disease (Gomes et al. 2021). Diagnosis is established by clinical, radiological, and pathological data. Only a small proportion of exposed individuals develops this lung disease (Nogueira et al. 2019). HP is a complex clinical problem, and there are still many areas in the pathogenesis, diagnosis, and treatment, which require more knowledge (Vasakova et al. 2019).

Chronic rhinosinusitis (CRS) is a complex disease characterized by chronic inflammation of the paranasal sinuses, eosinophilic mucin, and IgE antibodies to fungi. There are two major subtypes based upon phenotypic appearance: CRS with nasal polyps (CRSwNP) and CRS without nasal polyps (CRSsNP) (Grayson et al. 2019). At this time, it is known as allergic fungal rhinosinusitis (AFRS). AFRS occurs in approximately 5–10% of individuals with CRS and approximately 50% of individuals with AFRS also have asthma (Bachert et al. 2021; Park et al. 2019). AFRS is associated with various fungi, including *Bipolaris*, *Curvularia*, *Cladosporium*, *Alternaria*, and *Aspergillus*, and is more severe than other forms of CRS. Thus, both upper and lower airway subtypes of asthma and CRS clearly involve increased immune sensitivity to fungi (Mota and Borrego 2016).

Fungi have been implicated in occupational diseases. Occupational exposure to indoor molds and dampness may cause mold hypersensitivity syndrome (DMHS), a complex multiorgan disorder and lung damage (Tuuminen and Lohi 2018). In Finland, researchers found that exposure to mold odor at work increased the risk of asthma-COPD overlap syndrome (ACOS), and this risk was stronger among current smokers, suggesting a synergistic effect between these exposures (Jaakkola et al. 2020).

Some studies reported the presence of fungi and fungal elements in archives and libraries and with an increase in the prevalence of fungal sensitization in librarian workers (Alvarez-Castelló et al. 2020; Molina-Veloso and Borrego-Alonso 2017). Exposure of workers to molds in an office of work building was associated with occupational asthma (Caillaud et al. 2018). Occupational mold allergy can also occur in farm workers, bakers, brewers, florists, carpenters, winemakers, and wallpaper hangers (Kerbelker 2019).

IgE-binding trials, and skin test that induce histamine release in basophils from sensitized patients, have demonstrated the importance of molds to induce

IgE-mediated reactions. Bronchial and nasal challenge tests with extracts of fungal spores or hyphae are used to induce rhinitis and asthma. A prerequisite for the development of allergic sensitization is that individuals are exposed to the allergen source in clinically relevant amounts (Twaroch et al. 2015).

In the near future, the rising sea levels, warming ocean temperatures, and an increase in the frequency and intensity of storms and hurricanes which will result in coastal flooding will have an effect on indoor mold levels. This resulting exposure to fungal allergens will adversely affect the respiratory health of humans in these environments (Poole et al. 2019).

There is a significant need to evaluate the impact of fungi on allergic diseases and general health. Even though there are many progresses in the knowledge of fungal allergens and the allergic immune responses, there are still many unanswered questions.

13.4 Fungal Allergy and its Relationship with Environmental Changes

The impact of climate change on human health is a complex phenomenon (Sarofim et al. 2016). The increase of atmospheric greenhouse gas (such as carbon dioxide, methane, and nitrogen dioxide) and the global warming are the major manifestations of climate change, and it has been suggested their relationship to mold allergies (Patella et al. 2018; Peter Mshelia et al. 2020). Therefore, the frequency of extreme weather events such as snowfall, heavy rain, storms, hurricanes, tornadoes, and big storms hail events will increase globally with differences in season and regions. One of these consequences is related to thunderstorm asthma in various geographical zones (Dales et al. 2003; Grinn-Gofroń et al. 2013). Many studies have shown the impacts on emergency visits and hospitalizations for exacerbations of asthma (Bannister et al. 2021; Baxi et al. 2016). Most of these events have occurred during the spring or early summer periods, when aeroallergen counts are usually highest (Xu et al. 2021). Asthma has estimated to affect 339 million people and it is the most common chronic disease in children (Olsson et al. 2021). Mortality rates vary widely, with more affected low- and middle-income countries and 2030 millions of people in the world do not have accessible affordable medical care (Bannister et al. 2021; Fletcher et al. 2020). The prevalence of allergic rhinitis is even higher; occurring in approximately 500 million people worldwide for this reason, allergic diseases are considered a major public health problem globally (Davies et al. 2021).

Nowadays, there is increased evidence of the effects of climate change on allergic diseases. Global warming is influencing the production of aeroallergens, allergenicity, seasonality, distribution, and atmospheric concentration of airborne fungi (D'Amato et al. 2015; Weinberger et al. 2016).

The mechanisms through air pollutants and climate that may influence allergic diseases are complex and not fully understood. Air pollutants can act as adjuvants and alter the immunogenicity of allergenic proteins, affecting atmospheric concentration and human exposure to aeroallergens, increasing health risks (Reinmuth-Selzle et al. 2017; Weinberger et al. 2016). It has been also reported that concomitant to exposure to aeroallergens occurs a co-exposure to air pollutants (Weinberger et al. 2016).

Fungi have effects on human health, and these occur in indoor and outdoor environments. The air quality inside buildings influences allergic diseases. Moreover, most people spend more than 90% of their time indoors. The indoor environment quality depends on the outdoor air and the presence of indoor air pollution sources (Baldacci et al. 2015; Davies et al. 2021). In the home or workplace, the key factor for fungal growth is humidity. Fungi are most often found in basements, kitchens, and bathrooms. The most common sources of moisture in buildings are plumbing, roof, and window leaks and flooding.

Aspergillus, *Penicillium*, *Alternaria*, and *Cladosporium* species have been identified at higher concentrations in homes of patients with asthma symptoms (Segura-Medina et al. 2019; Sharpe et al. 2015). Fungal exposure in work buildings is related with the incidence and exacerbations of occupational asthma. As industrialization and furnishing of homes and workplaces have progressed, so has exposure to indoor fungi.

There is an association between asthma exacerbation and mortality in adults with peak days of high concentrations of exposure to airborne fungi in the outdoor air (Batra et al. 2021). Air temperature, relative humidity, precipitation, wind speed, and direction have effects on the levels of airborne spores and their release, presence, composition, movement, and viability. *Alternaria* and *Cladosporium* have been the most considered in aerobiological studies due to their pathogenic and allergenic properties (Ijadpanahsaravi et al. 2021; Stamper et al. 2016).

Rain reduces the temperature in many regions of the world, and fungi can release their fungal spores and cause the highest concentrations in the atmosphere. In addition, dry spores can also be released during rain by mechanical shock or fast airflow when raindrops spread. Numerous aerobiological studies indicate that most airborne fungal spores are positively temperature-dependent. Consequently, an increase in temperature may cause an increase in spore counts of strongly temperature-dependent fungi, especially in the continental climate (Grinn-Gofroń et al. 2020).

Due to the ubiquity of air pollution both indoors and outdoors, it is realistic to assume that exposures to aeroallergens occur with co-exposure to one or more air pollutants. These pollutants can act as adjuvants and affect the immune system by inducing tissue damage and successively more uptake of allergens. These induce oxidative stress and activation of immune cells, by co-exposure with the allergen that favors Th2 responses, or by modifying allergens enhancing their allergic potential (Diaz-Sanchez et al. 2003).

Persistent exposure to particulate air pollution from motor vehicles has been implicated as one of the factors that is responsible for the observed increased prevalence of atopy, mainly in industrialized countries. There is an important association between ambient levels of motor vehicle traffic emissions and increased symptoms of asthma and rhinitis. Particulate toxic pollutants, and in particular diesel exhaust particles, can enhance allergic inflammation and induce the development of allergic immune responses (Diaz-Sanchez et al. 2003; Muñoz et al. 2019). Several epidemiological studies have reported that respiratory allergies and atopic dermatitis are associated with exposure to traffic-related air pollution.

Cladosporium spore counts have been positively correlated with average temperature and humidity and negatively correlated with precipitation; however, no significant trends were found for *Alternaria* or *Epicoccum*. Hurricane Katrina triggered an extensive mold growth in buildings. Higher mold spore concentrations were found indoors compared to outdoors, and much higher spore concentrations were found in flooded areas compared to non-flooded areas. There is a clear need of epidemiologic studies of allergic diseases and their relationship to aeroallergens and climatic factors (Reid and Gamble 2009).

Injuries in the airway mucosal and compromised mucociliary clearance produced by air pollution enable the penetration of inhaled allergens and access to the immune system cells (Glencross et al. 2020; Reinmuth-Selzle et al. 2017). So these events promote airway sensitization. Consequently, an enhanced immunoglobulin E-mediated response to aeroallergens and greater airway inflammation by air pollution could account for the increasing prevalence of allergic respiratory diseases in urban areas (D'Amato et al. 2003).

Air pollution has been associated with chronic lung disease, cancer, and cardiac disease. The ground-level ozone (O_3), has a direct impact on the human respiratory system, increasing frequency of asthma exacerbations, increasing susceptibility to infection and possible promotion of chronic obstructive pulmonary disease (Demain et al. 2021). Some studies indicate that fungi exposure is affected by CO_2. The sporulation of allergenic fungi is probable to be enlarged as atmospheric CO_2 concentration increases and is also likely to contribute to the increasing prevalence and severity of asthma and allergies (Demain et al. 2021). When *Alternaria* grows on leaves of timothy grass plants grown at elevated CO_2 concentrations, it produces nearly three times the number of spores and more than twice the total antigenic protein per plant than at lower CO_2 concentrations (Tu et al. 2015; Wolf et al. 2010). Outdoor exposure to *Aspergillus fumigatus* and $O_3/NO_2/O_3 + NO$ produces an exposure time-dependent changes of IgE-binding capacity, suggesting links to chemical modifications of allergen Asp f 1 such as nitration and deamidation (Lang-Yona et al. 2015; Ortega-Rosas et al. 2021; Reinmuth-Selzle et al. 2017).

13.5 Climate Change and Allergenic Fungi

Climate change can affect the dispersal, transport, and deposition of allergenic fungal spores. There are many meteorological processes that affect spore dynamics in the atmosphere. From the point of view of atmospheric transport, fungal spores are aerosols that are lighter than most airborne particles and can be transported to hundreds of kilometers or more, if weather conditions are favorable (Sofiev and Prank 2016). As mentioned earlier in Chap. 3, the concentration of airborne spores can increase in a wide range of temperatures as has been documented in countries around the world. In addition, the annual amounts of spores and their relationship with temperature were different between cities with climatic differences.

In particular, this environmental factor can influence the presence of certain allergenic spores in a season of the year (seasonality). This relationship is important because the increase in the concentration of aeroallergens is associated with increases in hospital admissions for asthma (Forkel et al. 2021; Pfavayi et al. 2020; Ziska 2016, 2020; Ziska et al. 2019). The study by Dales et al. (2003) in Canada concluded when the fungal spore count doubled, emergency admissions for pediatric asthma increased by more than 15%. This illustrates atmospheric allergens are a risk factor for asthma (exacerbations). The presence and high concentration of *Alternaria*, *Cladosporium*, and *Aspergillus* spores have been related to a higher prevalence of hospital admissions for asthma (Cecchi et al. 2010). Sensitivity to *Alternaria alternata* is a predictor of epidemic asthma in patients with seasonal asthma and is likely to be the important factor in storm-related asthma (Pulimood et al. 2007). Basidiospore and ascospore spores are also associated with early wheezing in children (Harley et al. 2009; Sheffield et al. 2011).

Fungi, in general, require high humidity and temperatures above a minimum physiological threshold. Therefore, anthropogenic climate change and warmer temperatures are almost certain to have a significant impact on the life cycles of fungi, including sporulation (Peden and Reed 2010). In addition, it is documented that electrical storms, floods, and increased pollutants can also have an impact (Ziska 2016). Barnes et al. (2013) reviewed the effects of climate change and the environment on respiratory and allergic diseases and emphasized that the approach of allergists and/or environmental physicians to global climate change should be as integrated and forward-looking as possible.

Climate change also affects allergen levels in homes, schools, and offices. People allergic to outdoor allergies will not only experience more symptoms, so will people with outdoor allergies. Chew and Saha (2016) analyzed the impacts of climate change on indoor allergens, taking into account several aspects that can change inside homes. In this sense, the emergence of microenvironments with high temperatures, generated by modern electronic equipment, must also be taken into account.

Fungal spores have been monitored quarterly at the Wieliczka Salt Mine, located in southern Poland, for 8 years. Concentration in the subterranean environment differs between seasons, which indicates that the biological material penetrates the interior by anthropic action, appears to be typical of public facilities, and is more stable than that of the outside air (Myszkowska et al. 2020).

Various adaptation measures have been proposed to reduce the impact of climate change on allergenic fungi in the interior of homes or other occupational premises. These include legislative and specific building design and construction measures to minimize the risks of leakage and surface flooding (D'Amato et al. 2020; D'Amato

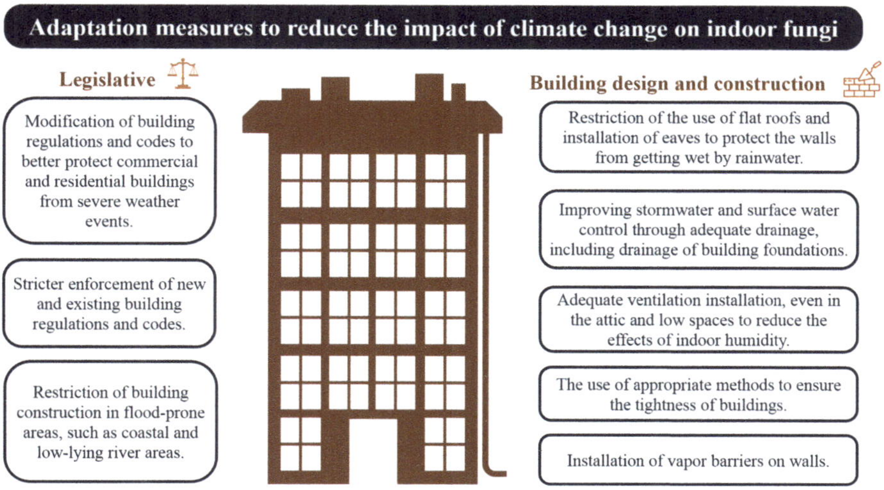

Fig. 13.3 Some adaptation measures to reduce the impact of climate change on indoor fungi

et al. 2015; De Marco et al. 2010; Eguiluz-Gracia et al. 2020; Gent et al. 2003; Islam et al. 2007; Mayaux et al. 2005; McConnell et al. 2002; Shea et al. 2008; Smith 2000). These measures are aimed at minimizing the presence of fungi inside buildings, due to their possible impact on allergies (Fig. 13.3), and although they focus mainly on the relationship with humidity, temperature is a fundamental point for the analysis.

Considerable research remains to be done on changes in the production, dispersal, and allergen content of fungal spores, which may be region and species specific. However, several authors have analyzed the impacts of climate change on aeroallergens, as well as on the interaction between air pollution and the exacerbation of allergic diseases (Bartra et al. 2007; Beggs 2004; Beggs and Bambrick 2005; Cecchi et al. 2010; D'Amato and Cecchi 2008; Reid and Gamble 2009; Shea et al. 2008).

Short-term forecasts indicate that although temperature can be a strong predictor of fungal spore concentrations in the air, the effect varies by fungus species and geography. In Denver, USA, an upward trend in *Cladosporium* spore counts was observed for 8 years that was significantly positively correlated with temperature (Katial et al. 1997). A similar result was found in another US city, Tulsa, by Troutt and Levetin (2001). For their part, Freye et al. (2001) observed higher counts of airborne fungi after a warmer El Niño event in New England.

In another zone, Sindt et al. (2016) verified whether in recent years there have been any changes in the spore content of atmospheric fungi, as well as whether these modifications can be linked to changes in the climate and whether the observed trend is uniform throughout the French territory or different from one climatic region to another. Depending on the context, the same heating can trigger more or less spore production in the air and alter the distribution of the pattern and time in one direction or the other. Longer periods of aerobiological monitoring and more sites that record fungal data are urgently required to determine the airborne fungi dynamics, increasing or decreasing levels of certain types of spores that could have an impact in public health.

Studies examining local temperature increases in two UK cities indicated that *Alternaria* spore counts increased from the early 1970s to the late 1990s and that the spore season started earlier and increased by duration (Corden et al. 2003; Corden and Millington 2001). Seasonal concentrations of *Alternaria* spores showed a clear upward trend during this period, beginning gradually from 1970 to 1991 and then increasing markedly from 1992 onward. *Alternaria* start dates showed that there was a trend toward an earlier seasonal onset (from late June around the beginning of the record to early June around the end of the record). Associated with this was a trend toward a longer *Alternaria* season. Corden and Millington (2001) related the trend for an earlier seasonal onset to an increase in accumulated winter and early spring temperatures in Derby.

Ogawa et al. (2012) raised the possible sun of the basidiomycete *Bjerkandera adusta* to increase the severity of allergic fungal cough through sensitization. They also analyzed their likely long-term transport from the Taklamakan and Gobi deserts, highlighting the effect of climate change on desertification. Sautour et al. (2009) in France had reported the frequent detection of this fungus in indoor and outdoor samples. It is the study by Ogawa et al. (2012) that raises the importance of knowledge of the dispersion of allergenic fungal spores to prevent sensitization. In addition, it leaves some questions about how to reduce fungal spores in the indoor environment and the selection of the type of ventilation. Integrated research, which would not be possible with an exclusively reactionary methodology, can contribute to the protection of human health against allergy to fungi enhanced by climate change (Ogawa et al. 2013).

In a study conducted on the allergenicity of *A. fumigatus*, it was found that high concentrations of carbon dioxide affect its ability to induce allergies (Campos and Mendivil 2006).

13.6 Conclusion

Climate change could transform the fungal exposure and allergies. Many factors contribute to the variation in fungal diversity and concentrations, as well as the built environment, the behavior of residents, and the fungal ecology. A multidisciplinary approach involving house/workplace design and management, environmental characterization, molecular categorization, and health professionals may be necessary to a better understanding of the role of fungi in the appearance of allergic diseases. It is important the interdisciplinarity in research on allergic diseases and their key relationship with the climate change. Reducing risk factors mainly environmental pollutants may improve nutrition and physical activity and increase tolerance. It is essential to increase global research on allergic diseases and its relation to aeroallergens as fungi. Reducing the burden of allergic diseases is crucial today and the near future for our health and well-being.

References

Adhikari A, Sen MM, Gupta-Bhattacharya S, Chanda S. Airborne viable, non-viable, and allergenic fungi in a rural agricultural area of India: a 2-year study at five outdoor sampling stations. Sci Total Environ. 2004;326(1):123–41. https://doi.org/10.1016/j.scitotenv.2003.12.007.

Agache I, Annesi-Maesano I, Bonertz A, Branca F, Cant A, Fras Z, et al. Prioritizing research challenges and funding for allergy and asthma and the need for translational research—the European Strategic Forum on Allergic Diseases. Allergy. 2019;74(11):2064–76. https://doi.org/10.1111/all.13856.

Agarwal MK, Jones RT, Yunginger JW. Shared allergenic and antigenic determinants in *Alternaria* and *Stemphylium* extracts. J Allergy Clin Immunol. 1982;70:437–44. https://doi.org/10.1016/0091-6749(82)90006-9.

Agarwal R, Gupta D. Severe asthma and fungi: current evidence. Med Mycol. 2011;49(1):S150–7. https://doi.org/10.3109/13693786.2010.504752.

Agarwal R, Sehgal IS, Dhooria S, Aggarwal AN. Developments in the diagnosis and treatment of allergic bronchopulmonary aspergillosis. Expert Rev Respir Med. 2016;10(12):1317–34. https://doi.org/10.1080/17476348.2016.1249853.

Aira MJ, Rodríguez-Rajo FJ, Fernández-González M, Seijo C, Elvira-Rendueles B, Abreu I, et al. Spatial and temporal distribution of *Alternaria* spores in the Iberian Peninsula atmosphere, and meteorological relationships: 1993–2009. Int J Biometeorol. 2013;57(2):265–74. https://doi.org/10.1007/s00484-012-0550-x.

Aira MJ, Rodríguez-Rajo FJ, Fernández-González M, Seijo C, Elvira-Rendueles B, Gutiérrez-Bustillo M. Cladosporium airborne spore incidence in the environmental quality of the Iberian Peninsula. Grana. 2012;51(4):293–304. https://doi.org/10.1080/00173134.2012.717636.

Andersen B, Frisvad JC, Sondergaard I, Rasmussen IS, Larsen LS. Associations between fungal species and water-damaged building materials. Appl Environ Microbiol. 2011;77:4180–8. https://doi.org/10.1128/AEM.02513-10.

Arruda LK, Platts-Mills TA, Fox JW, Chapman MD. Aspergillus fumigatus allergen I, a major IgE-binding protein, is a member of the mitogillin family of cytotoxins. J Exp Med. 1990;172:1529–32. https://doi.org/10.1084/jem.172.5.1529.

Asano K, Kamei K, Hebisawa A. Allergic bronchopulmonary mycosis–pathophysiology, histology, diagnosis, and treatment. Asia Pac Allergy. 2018;8(3):29–32. https://doi.org/10.5415/apallergy.2018.8.e24.

Bachert C, Bhattacharyya N, Desrosiers M, Khan AH. Reducing fungal exposure critical for treating rhinosinusitis with or without polyps [response to letter]. J Asthma Allergy. 2021;14:393–5. https://doi.org/10.2147/JAA.S314846.

Baldacci S, Maio S, Cerrai S, Sarno G, Baïz N, Simoni M, Study HEALS. Allergy and asthma: effects of the exposure to particulate matter and biological allergens. Respir Med. 2015;109(9):1089–104. https://doi.org/10.1016/j.rmed.2015.05.017.

Bannister T, Ebert EE, Williams T, Douglas P, Wain A, Carroll M, et al. A pilot forecasting system for epidemic thunderstorm asthma in Southeastern Australia. BAMS. 2021;102(2):399–420. https://doi.org/10.1175/BAMS-D-19-0140.1.

Barnes CS, Alexis NE, Bernstein JA, Cohn JR, Demain JG, Horner E, et al. Climate change and our environment: the effect on respiratory and allergic disease. J Allergy Clin Immunol: In Practice. 2013;1:137–41. https://doi.org/10.1016/j.jaip.2012.07.002.

Barriga AG. Microbios africanos de vacaciones en el Caribe. Rev Mex Patol Clin. 2007;54(4):168–76.

Bartra J, Mullol J, del Cuvillo A, Davila I, Ferrer M, Jauregui I. Air pollution and allergens. JIACI. 2007;2(Suppl):3–8.

Batra M, Vicendese D, Newbigin E, Lambert KA, Tang M, Abramson MJ, et al. The association between outdoor allergens–pollen, fungal spore season and high asthma admission days in children and adolescents. Int J Environ Health Res. 2021:1–10. https://doi.org/10.1080/09603123.2021.1885633.

Baxi SN, Portnoy JM, Larenas-Linnemann D, Phipatanakul W, Barnes C, Baxi S, et al. Exposure and health effects of fungi on humans. J Allergy Clin Immunol. 2016;4(3):396–404. https://doi.org/10.1016/j.jaip.2016.01.008.

Beggs PJ. Impacts of climate change on aeroallergens: past and future. Clin Exp Allergy. 2004;34(10):1507–13. https://doi.org/10.1111/j.1365-2222.2004.02061.x.

Beggs PJ, Bambrick HJ. Is the global rise of asthma an early impact of anthropogenic climate change? Environ Health Perspect. 2005;113(8):915–9. https://doi.org/10.1289/ehp.7724.

Borrego S, Guiamet P, Gómez De Saravia S, Batistini P, Garcia M, Lavin P, et al. The quality of air at archives and the biodeterioration of photographs. Int Biodeterior Biodegradation. 2010;64(2):139–45. https://doi.org/10.1016/j.ibiod.2009.12.005.

Breitenbach M, Simon-Nobbe B. The allergens of *Cladosporium herbarum* and *Alternaria alternata*. Chem Immunol. 2002;81:48–72. https://doi.org/10.1159/000058862.

Cafarchia C, Camarda A, Iatta R, Danesi P, Favuzzi V, Di Paola G, et al. Environmental contamination by *aspergillus* spp. in laying hen farms and associated health risks for farm workers. J Med Microbiol. 2014;63(3):464–70.

Cai J, Li B, Yu W, Wang H, Du C, Zhang Y, et al. Household dampness-related exposures in relation to childhood asthma and rhinitis in China: a multicentre observational study. Environ Int. 2019;126:735–46. https://doi.org/10.1016/j.envint.2019.03.013.

Caillaud D, Leynaert B, Keirsbulck M, Nadif R. Indoor mould exposure, asthma and rhinitis: findings from systematic reviews and recent longitudinal studies. Eur Respir Rev. 2018;27(148):170137. https://doi.org/10.1183/16000617.0137-2017.

Campos V, Mendivil A. Calidad del aire interior de los centros de educación infantil del País Vasco. Revista Medico-Navarra. 2006;427:34–42.

Castelló MA, Almarales RLC, Marquez YL, González BL, Canosa JR, Flores TR, Rosado AL. Sensibilización a hongos anemófilos en trabajadores del Archivo y Biblioteca de la Universidad de La Habana. Archivos del Hospital Universitario" General Calixto García". 2020;8:2.

Cecchi L, D'Amato G, Ayres JG, Galan C, Forastiere F, Forsberg B. Projections of the effects of climate change on allergic asthma: the contribution of aerobiology. Allergy. 2010;65(9):1073–81. https://doi.org/10.1111/j.1398-9995.2010.02423.x.

Chacón ML. Estudio de los pacientes con diagnóstico de bronquiectasias, que siguen control en el servicio de neumología del Hospital Son Llátzer. Palma de Mallorca. 2012;

Chew GL, Saha S. Impacts of climate change on indoor allergens. In: Beggs PJ, editor. Impacts of climate change on allergens and allergic diseases. Cambridge: Cambridge University Press; 2016. p. 113–36.

Corden JM, Millington WM. The long- term trends and seasonal variation of the aeroallergen Alternaria in Derby, UK. Aerobiologia. 2001;17(2):127–36. https://doi.org/10.1023/A:1010876917512.

Corden JM, Millington WM, Mullins J. Long- term trends and regional variation in the aeroallergen *Alternaria* in Cardiff and Derby UK—are differences in climate and cereal production having an effect? Aerobiologia. 2003;19(3–4):191–9. https://doi.org/10.1023/B:AERO.0000006529.51252.2f.

Corry DB, Kheradmand F, Luong A, Pandit L. Immunological mechanisms of airway diseases and pathways to therapy. Clin Immunol. 2019:571–84. https://doi.org/10.1016/B978-0-7020-6896-6.00041-7.

Cottin V. Eosinophilic lung diseases. Clin Chest Med. 2016;37(3):535–56. https://doi.org/10.1016/j.ccm.2016.04.015.

Cruz R, Barthel M, Piontelli E, Fernandez G. Reportes Clínicos: Infección Rinosinusal probada por *Aspergillus flavus* y probable infección pulmonar por Emericella nidulans en pacientes inmunodeprimidos. Bol Micol. 2005;20:109–15.

D'Amato G, Cecchi L. Effects of climate change on environmental factors in respiratory allergic diseases. Clin Expl Allergy. 2008;38(8):1264–74. https://doi.org/10.4168/aair.2016.8.5.391.

D'Amato G, Chong-Neto HJ, Monge Ortega OP, Vitale C, Ansotegui I, Rosario N, Annesi-Maesano I. The effects of climate change on respiratory allergy and asthma induced by pollen and mold allergens. Allergy. 2020;75(9):2219–28. https://doi.org/10.1111/all.14476.

D'Amato G, Holgate ST, Pawankar R, Ledford DK, Cecchi L, Al-Ahmad M, Annesi-Maesano I. Meteorological conditions, climate change, new emerging factors, and asthma and related allergic disorders. A statement of the World Allergy Organization. World Allergy Organ J. 2015;8(1):1–52. https://doi.org/10.1186/s40413-015-0073-0.

D'Amato G, Ortega OPM, Annesi-Maesano I, D'Amato M, Dales RE, Cakmak S, Judek S. The role of fungal spores in thunderstorm asthma. Chest. 2003;123(3):745–50. https://doi.org/10.1378/chest.123.3.745.

Dales RE, Cakmak S, Judek S, Dann T, Coates F, Brook JR, et al. The role of fungal spores in thunderstorm asthma. Chest. 2003;123(3):745–50. https://doi.org/10.1378/chest.123.3.745.

Davies JM, Berman D, Beggs PJ, Ramón GD, Peter J, Katelaris CH, et al. Global climate change and pollen aeroallergens: a southern hemisphere perspective. Immunol Allergy Clin. 2021;41(1):1–16. https://doi.org/10.1016/j.iac.2020.09.002.

De Marco R, Accordini S, Antonicelli L. The gene-environment interactions in respiratory diseases (GEIRD) project. Int Arch Allergy Immunol. 2010;152(3):255–63. https://doi.org/10.1159/000283034.

Demain JG, Choi YJ, Oh JW. The impact of climate change on the pollen allergy and sporulation of allergic fungi. Curr Treat Opt Allergy. 2021:1–14. https://doi.org/10.1007/s40521-020-00277-5.

Denning DW, Pashley C, Hartl D, Wardlaw A, Godet C, Del Giacco S, et al. Fungal allergy in asthma–state of the art and research needs. Clin Transl Allergy. 2014;4(1):14. https://doi.org/10.1186/2045-7022-4-14.

Díaz A, Fabré D, Coutin G, González T. La sensibilización a hongos ambientales y su relación con enfermedades atópicas en escolares. Revista Cubana de Medicina General Integral. 2010a;26(4):20–9.

Díaz M, Gutiérrez J, Gutiérrez A, González MC, Vidal G, Zaragoza RM, et al. Caracterización aerobiológica de ambientes intramuros en presencia de cubiertas vegetales. Rev Int Contam Ambient. 2010b;26(4):279–89.

Diaz-Sanchez D, Proietti L, Polosa R. Diesel fumes and the rising prevalence of atopy: an urban legend? Curr Allergy Asthma Rep. 2003;3(2):146–52. https://doi.org/10.1007/s11882-003-0027-4.

Docampo S, Trigo MM, Recio M, Melgar M, Garcia-Sanchez J, Calderon-Ezquerro MC, et al. High incidence of aspergillus and Penicillium spores in the atmosphere of the cave of Nerja (Malaga, southern Spain). Aerobiologia. 2010;26(2):89–98. https://doi.org/10.1016/j.scitotenv.2010.10.048.

Donthi S, SivaSai KSR, Lakshmi VV, Latha GS, Murthy KJR. Prevalence of inhalant allergens in nasobronchial allergy in Hyderabad region: India. Int Sci Res J. 2011;3:192–9.

Eguiluz-Gracia I, Mathioudakis AG, Bartel S, et al. The need for clean air: the way air pollution and climate change affect allergic rhinitis and asthma. Allergy. 2020;75:2170–84. https://doi.org/10.1111/all.14177.

Fairs A, Agbetile J, Bourne M, Hargadon B, Monteiro WR, Morley JP, et al. Isolation of Aspergillus fumigatus from sputum is associated with elevated airborne levels in homes of patients with asthma. Indoor Air. 2013;23(4):275–84. https://doi.org/10.1111/ina.12020.

Fernández-Soto R, Navarrete-Rodríguez EM, Del-Rio-Navarro BE, Sienra-Monge JL, Meneses-Sánchez NA, Saucedo-Ramírez OJ. Fungal allergy: pattern of sensitization over the past 11 years. Allergol Immunopathol. 2018;46(6):557–64. https://doi.org/10.1016/j.aller.2018.01.005.

Fletcher MJ, Tsiligianni I, Kocks JW, Cave A, Chunhua C, de Sousa JC, van der Molen T. Improving primary care management of asthma: do we know what really works? NPJ Prim Care Respir Med. 2020;30(1):1–11. https://doi.org/10.1038/s41533-020-0184-0.

Forkel S, Beutner C, Schröder SS, Bader O, Gupta S, Fuchs T, et al. Sensitization against fungi in patients with airway allergies over 20 years in Germany. Int Arch Allergy Immunol. 2021;182(6):515–23. https://doi.org/10.1159/000512230.

Freye HB, King J, Litwin CM. Variations of pollen and mold concentrations in 1998 during the strong El Nino event of 1997–1998 and their impact on clinical exacerbations of allergic rhinitis, asthma, and sinusitis. Allergy Asthma Proc. 2001;22:239–47.

Fryen A, Mayser P, Glanz H, Fussle R, Breithaupt H, de Hoog GS. Allergic fungal sinusitis caused by *Bipolaris (Drechslera) hawaiiensis*. Eur Arch Otorhinolaryngol. 1999;256:330–4. https://doi.org/10.1007/s004050050157.

Gent JF, Triche EW, Holford TR, Belanger K, Bracken MB, Beckett WS, et al. Association of low-level ozone and fine particles with respiratory symptoms in children with asthma. JAMA. 2003;290(14):1859–67. https://doi.org/10.1001/jama.290.14.1859.

Glencross DA, Ho TR, Camina N, Hawrylowicz CM, Pfeffer PE. Air pollution and its effects on the immune system. Free Radic Biol Med. 2020:23–35. https://doi.org/10.1016/j.freeradbiomed.2020.01.179.

Gomes ML, Morais A, Rufo JC. The association between fungi exposure and hypersensitivity pneumonitis: a systematic review. Porto Biomed J. 2021;6(1):e17. https://doi.org/10.1097/j.pbj.0000000000000117.

Grammer LC. Hypersensitivity pneumonitis. Allergy Asthma Proc. 2019;40(6):425–8. https://doi.org/10.2500/aap.2019.40.4263.

Grayson JW, Cavada M, Harvey RJ. Clinically relevant phenotypes in chronic rhinosinusitis. Otolaryngol Head Neck Surg. 2019;48(1):1–10. https://doi.org/10.1186/s40463-019-0350-y.

Green BJ, Zinovia Mitakakis T, Tovey ER. Allergen detection from 11 fungal species before and after germination. J Allergy Clin Immunol. 2003;111:285–9. https://doi.org/10.1067/mai.2003.57.

Griffin DW. Atmospheric movement of microorganisms in clouds of desert dust and implications for human health. Clin Microbiol Rev. 2007;20:459–77. https://doi.org/10.1128/CMR.00039-06.

Grinn-Gofroń A, Çeter T, Pinar NM, Bosiacka B, Çeter S, Keçeli T, et al. Airborne fungal spore load and season timing in the central and eastern Black Sea region of Turkey explained by climate conditions and land use. Agric For Meteorol. 2020;295:108–91. https://doi.org/10.1016/j.agrformet.2020.108191.

Grinn-Gofroń A, Strzelczak A. Changes in concentration of Alternaria and Cladosporium spores during summer storms. Int. J. Biometeorol. 2013;57(5):759–768. https://doi.org/10.1007/s00484-012-0604-0.

Gutierrez-Rodriguez A, Postigo I, Guisantes JA, Sunen E, Martinez J. Identification of allergens homologous to alt a 1 from *Stemphylium botryosum* and *Ulocladium botrytis*. Med Mycol. 2011;49:892–6. https://doi.org/10.3109/13693786.2011.576350.

Harley KG, Macher JM, Lipsett M. Fungi and pollen exposure in the first months of life and risk of early childhood wheezing. Thorax. 2009;64:353–8. https://doi.org/10.1136/thx.2007.090241.

Hedayati MT, Pasqualotto AC, Warn PA, Bowyerand P, Denning DW. *Aspergillus flavus*: human pathogen, allergen and mycotoxin producer. Microbiology. 2007;153:1677–92. https://doi.org/10.1099/mic.0.2007/007641-0.

Hibbett DS, Binder M, Bischoff JF, Blackwell M, Cannon PF, Eriksson OE, et al. A higher-level phylogenetic classification of the fungi. Mycol Res. 2007;111(Pt 5):509–47. https://doi.org/10.1016/j.mycres.2007.03.004.

Hinson KFW, Moon AJ, Plummer NS. Bronchopulmonary aspergillosis: a review and report of eight new cases. Thorax. 1952;7:317.

Hulin M, Moularat S, Kirchner S, Robine E, Mandin C, Annesi-Maesano I. Positive associations between respiratory outcomes and fungal index in rural inhabitants of a representative sample of French dwellings. Int J Hyg Environ Health. 2013;216:155–62. https://doi.org/10.1016/j.ijheh.2012.02.011.

Ijadpanahsaravi M, Punt M, Wösten HA, Teertstra WR. Minimal nutrient requirements for induction of germination of *Aspergillus niger* conidia. Fungal Biol. 2021;125(3):231–8. https://doi.org/10.1016/j.funbio.2020.11.004.

Islam T, Gauderman WJ, Berhane K, McConnell R, Avol E, Peters JM, et al. Relationship between air pollution, lung function and asthma in adolescents. Thorax. 2007;62:957–63. https://doi.org/10.1136/thx.2007.078964.

Jaakkola MS, Aalto SA, Hyrkäs-Palmu H, Jaakkola JJ. Association between regular exercise and asthma control among adults: the population-based Northern Finnish Asthma Study. PLoS One. 2020;15(1):e0227983. https://doi.org/10.1371/journal.pone.0227983.

Jedryczka M, Strzelczak A, Grinn-Gofron A, Nowak M, Wolski T, Siwulski M, et al. Advanced statistical models commonly applied in aerobiology cannot accurately predict the exposure of people to *Ganoderma* spore-related allergies. Agric For Meteorol. 2015;201:209–17. https://doi.org/10.1016/j.agrformet.2014.11.015.

Jhan HT, Ballinger R, Jaleel A, Ting KH. Development and application of a socioeconomic vulnerability indicator framework (SVIF) for local climate change adaptation in Taiwan. Sustainability. 2020;12(4):1585.

Jubin V, Ranque S, Stremler Lebel N, Sarles J, Dubus JC. Risk factors for *aspergillus* colonization and allergic bronchopulmonary aspergillosis in children with cystic fibrosis. Pediatric Pulmonol. 2010;45(8):764–71. https://doi.org/10.1002/ppul.21240.

Karjalainen A, Kurppa K, Martikainen R, Klaukka T, Karjalainen J. Work is related to a substantial portion of adult-onset asthma incidence in the Finnish population. Am J Respir Crit Care Med. 2001;164:565–8. https://doi.org/10.1164/ajrccm.164.4.2012146.

Katelaris CH, Beggs PJ. Climate change: allergens and allergic diseases. J Intern Med. 2018;48(2):129–34. https://doi.org/10.1111/imj.13699.

Katial RK, Zhang Y, Jones RH, Dyer PD. Atmospheric mold spore counts in relation to meteorological parameters. Int J Biometeorol. 1997;41(1):17–22. https://doi.org/10.1007/s004840050048.

Kerbelker T. Moulds–indoor allergens and their role in atopic disease. Curr Opin Allergy Clin Immunol. 2019;32(1):16–21.

Kochar S, Ahlawat M, Dahiya P, Chaudhary D. Assessment of allergenicity to fungal allergens of Rohtak city, Haryana, India. Allergy Rhinol. 2014;5(2):52–6. https://doi.org/10.2500/ar.2014.5.0088.

Kolossa-Gehring M, Becker K, Conrad A, Ludecke A, Riedel S, Seiwert M, et al. German environmental survey for children (GerES IV)—first results. Int J Hyg Environ Health. 2007;210:535–40. https://doi.org/10.1016/j.ijheh.2007.07.018.

Lahoute C, Tonnel AB, Fournier E, Ramon P, Voisin C. Bronchopulmonary pathology with hypereosinophilia of fungal origin (excluding allergic bronchopulmonary aspergillosis). Poumon Coeur. 1983;39:87–93.

Lang-Yona N, Shuster-Meiseles T, Mazar Y, Yarden O, Rudich Y. Impact of urban air pollution on the allergenicity of *Aspergillus fumigatus* conidia: outdoor exposure study supported by laboratory experiments. Sci Total Environ. 2015;541:365–71. https://doi.org/10.1016/j.scitotenv.2015.09.058.

Levetin E, Horner WE, Scott JA, Workgroup EA. Taxonomy of allergenic fungi. J Allergy Clin Immunol. 2016;4(3):375–85. https://doi.org/10.1016/j.jaip.2015.10.012.

Lim SH, Chew FT, Sim SM, Huang YT, Goh DY, Tan HT. Allergens of *Bipolaris* species. Asian Pac J Allergy Immunol. 1995;13:101–5.

Maniwa K, Tanaka E, Taguchi Y, Oida K, Inoue T, Kato T, et al. A case of abrupt pulmonary infection by *Rhizopus microsporus* var. rhizopodiformis during treatment for bronchial asthma. Kansenshogaku Zasshi. 2002;76:396–9. https://doi.org/10.11150/kansenshogakuzasshi1970.76.396.

Mayaux P, Holmgren P, Achard F, Eva H, Stibig H, Branthomme A. Tropical forest cover change in the 1990s and options for future monitoring. Philos Trans R Soc B. 2005;360:373–84. https://doi.org/10.1098/rstb.2004.1590.

McAleer R, Kroenert DB, Elder JL, Froudist JH. Allergic bronchopulmonary disease caused by *Curvularia lunata* and *Drechslera hawaiiensis*. Thorax. 1981;36:338–44. https://doi.org/10.1136/thx.36.5.338.

McConnell R, Berhane K, Gilliland F. Asthma in exercising children exposed to ozone: a cohort study. Lancet. 2002;359:386–91. https://doi.org/10.1016/S0140-6736(02)07597-9.

Mendell MJ, Mirer AG, Cheung K, Tong M, Douwes J. Respiratory and allergic health effects of dampness, mold, and dampness-related agents: a review of the epidemiologic evidence. Environ Health Perspect. 2011;119(6):748–56. https://doi.org/10.1289/ehp.1002410.

Menezes EA, Gambale W, Macedo MS, Castro F, Paula CR, Croce J. Characterization of allergenic fractions from Drechslera monoceras. J Investig Allergol Clin Immunol. 1998;8:214–8.

Molina-Veloso A, Borrego-Alonso SF. Hongos alergénicos viables en un depósito documental del Archivo Nacional de Cuba. Rev Alerg Mex. 2017;64(1):40–51. https://doi.org/10.29262/ram.v64i1.234.

Montes-Angel IA, López-González B, Torres-Concepción J, Álvarez-Castelló M, Castro-Almarales RL, Rodríguez-Canosa JS. Characterization of allergic rhinitis in the "General Calixto Garcia" University Hospital. October 2016-October 2017. VacciMonitor. 2020;29(2):75–81.

Mota IA, Borrego LM. Allergic response to fungal exposure. In: Fernández SD, editor. Environmental mycology in public health. Academic Press; 2016. p. 35–43.

Muñoz X, Barreiro E, Bustamante V, Lopez-Campos JL, González-Barcala FJ, Cruz MJ. Diesel exhausts particles: their role in increasing the incidence of asthma. Reviewing the evidence of a causal link. Sci Total Environ. 2019;652:1129–38. https://doi.org/10.1016/j.scitotenv.2018.10.188.

Myszkowska D. Aerobiological studies – current state and future challenges. Alergoprofil. 2020;16(1):8–14. https://doi.org/10.24292/01.AP.161300320.

Navarro AM, Delgado J, Muñoz-Cano RM, Dordal MT, Valero A, Quirce S, ARD Study Group. Allergic respiratory disease (ARD), setting forth the basics: proposals of an expert consensus report. Clin Transl Allergy. 2017;7(1):16. https://doi.org/10.1186/s13601-017-0150-2.

Nogueira R, Melo N, Bastos HN, Martins N, Delgado L, Morais A, Mota PC. Hypersensitivity pneumonitis: antigen diversity and disease implications. Pulmonology. 2019;25(2):97–108. https://doi.org/10.1016/j.pulmoe.2018.07.003.

Ogawa H, Fujimura M, Satoh K, Makimura K. Re: integrated research on the association between climate change and Bjerkandera allergy. J Allergy Clin Immunol. 2013;1(5):543. https://doi.org/10.1016/j.jaip.2013.05.012.

Ogawa H, Fujimura M, Takeuchi Y, Makimura K. Possible roles of 2 basidiomycetous fungi in allergic fungal respiratory disease. J Allergy Clin Immunol. 2012;130:279–80. https://doi.org/10.1016/j.jaci.2012.03.051.

Oliveira M, Carmalho R. *Aspergillus fumigatus*: a mere airborne particle or a powerful biohazard? NACC (Bioloxía). 2014;21:57–64.

Olsson D, Forsberg B, Bråbäck L, Geels C, Brandt J, Christensen JH, et al. Early childhood exposure to ambient air pollution is associated with increased risk of paediatric asthma: an administrative cohort study from Stockholm. Sweden Environ Int. 2021;155:106667. https://doi.org/10.1016/j.envint.2021.106667.

Ortega-Rosas CI, Meza-Figueroa D, Vidal-Solano JR. Association of airborne particulate matter with pollen, fungal spores, and allergic symptoms in an arid urbanized area. Environ Geochem Health. 2021;43:1761–82. https://doi.org/10.1007/s10653-020-00752-7.

Osborne NJ, Thornton CR, Sharpe RA. Indoor fungal exposure and allergic respiratory disease. Curr Allergy Asthma Rep. 2015;15(12):1–9. https://doi.org/10.1007/s11882-015-0572-7.

Paavola J. Health impacts of climate change and health and social inequalities in the UK. Environ Health. 2017;16(1):61–8. https://doi.org/10.1186/s12940-017-0328-z.

Park JH, Cox-Ganser JM, Kreiss K, White SK, Rao CY. Hydrophilic fungi and ergosterol associated with respiratory illness in a water-damaged building. Environ Health Perspect. 2008;116:45–50. https://doi.org/10.1289/ehp.10355.

Park SK, Park KW, Mo JH, Baek BJ, Shim WS, Jung HJ, et al. Clinicopathological and radiological features of chronic rhinosinusitis with eosinophilic mucin in Chungcheong province of Korea. Mycopathologia. 2019;184(3):423–31. https://doi.org/10.1007/s11046-019-00340-z.

Patella V, Florio G, Magliacane D, Giuliano A, Crivellaro MA, Di Bartolomeo D, et al. Urban air pollution and climate change: "The Decalogue: Allergy Safe Tree" for allergic and respiratory diseases care. Clin Mol Allergy. 2018;16(1):1–11. https://doi.org/10.1186/s12948-018-0098-3.
Pawankar R, Mori S, Ozu C, Kimura S. Overview on the pathomechanisms of allergic rhinitis. Asia Pac Allergy. 2011;1(3):157. https://doi.org/10.5415/apallergy.2011.1.3.157.
Peden D, Reed CE. Environmental and occupational allergies. J Allergy Clin Immunol. 2010;125(2):150–60. https://doi.org/10.1016/j.jaci.2009.10.073.
Pendrey CGA, Beaton L, Kneebone JA. General practice in the era of planetary health: responding to the climate health emergency. Aust J Gen Pract. 2020;49(8):520. https://doi.org/10.31128/AJGP-01-20-5207.
Peter Mshelia L, Selamat J, Iskandar Putra Samsudin N, Rafii MY, Abdul Mutalib NA, Nordin N, Berthiller F. Effect of temperature, water activity and carbon dioxide on fungal growth and mycotoxin production of acclimatised isolates of *fusarium verticillioides and F. graminearum*. Toxins. 2020;12(8):478. https://doi.org/10.3390/toxins12080478.
Pfavayi LT, Sibanda EN, Mutapi F. The pathogenesis of fungal-related diseases and allergies in the African population: the state of the evidence and knowledge gaps. Int Arch Allergy Immunol. 2020;181(4):257–69. https://doi.org/10.1159/000506009.
Poole JA, Barnes CS, Demain JG, Bernstein JA, Padukudru MA, Sheehan WJ, et al. Impact of weather and climate change with indoor and outdoor air quality in asthma: A Work Group Report of the AAAAI Environmental Exposure and Respiratory Health Committee. J Allergy Clin Immunol. 2019;143(5):1702–10. https://doi.org/10.1016/j.jaci.2019.02.018.
Prescott SL. Early-life environmental determinants of allergic diseases and the wider pandemic of inflammatory noncommunicable diseases. J Allergy Clin Immunol. 2013;131(1):23–30. https://doi.org/10.1016/j.jaci.2012.11.019.
Pulimood TB, Corden JM, Bryden C, Sharples L, Nasser SM. Epidemic asthma and the role of the fungal mold Alternaria alternata. J Allergy Clin Immunol. 2007;120(3):610–7. https://doi.org/10.1016/j.jaci.2007.04.045.
Pyrri I, Kapsanaki-Gotsi E. A comparative study on the airborne fungi in Athens, Greece, by viable and non-viable sampling methods. Aerobiologia. 2007;23:3–15. https://doi.org/10.1007/s10453-006-9039-6.
Quintero E, Rivera-Mariani F, Bolaños-Rosero B. Analysis of environmental factors and their effects on fungal spores in the atmosphere of a tropical urban area (San Juan, Puerto Rico). Aerobiologia. 2010;26(2):113–24. https://doi.org/10.1007/s10453-009-9148-0.
Raphoz M, Goldberg MS, Garneau M, Heguy L, Valois MF, Guay F. Associations between atmospheric concentrations of spores and emergency department visits for asthma among children living in Montreal. Arch Environ Occup Health. 2010;65:201–10. https://doi.org/10.1080/19338241003730937.
Reid CE, Gamble JL. Aeroallergens, allergic disease, and climate change: impacts and adaptation. EcoHealth. 2009;6(3):458–70. https://doi.org/10.1007/s10393-009-0261-x.
Reinmuth-Selzle K, Kampf CJ, Lucas K, Lang-Yona N, Fröhlich-Nowoisky J, Shiraiwa M, Pöschl U. Air pollution and climate change effects on allergies in the anthropocene: abundance, interaction, and modification of allergens and adjuvants. Environ Sci Technol. 2017;51(8):4119–41. https://doi.org/10.1021/acs.est.6b04908.
Reponen T, Lockey J, Bernstein DI, Vesper SJ, Levin L, Khurana Hershey GK, et al. Infant origins of childhood asthma associated with specific molds. J Allergy Clin Immunol. 2012;130(3):639–44. https://doi.org/10.1016/j.jaci.2012.05.030.
Reponen T, Vesper S, Levin L, Johansson E, Ryan P, Burkle J. High environmental relative moldiness index during infancy as a predictor of asthma at 7 years of age. Ann Allergy Asthma Immunol. 2011;107:120–6. https://doi.org/10.1016/j.anai.2011.04.018.
Rivera-Mariani FE, Nazario-Jiménez S, López-Malpica F, Bolaños-Rosero B. Sensitization to airborne ascospores, basidiospores, and fungal fragments in allergic rhinitis and asthmatic subjects in San Juan. Puerto Rico Int Arch Allergy Immunol. 2011;155(4):322–34. https://doi.org/10.1159/000321610.

Rivera-Mariani FE, Bolaños-Rosero B. Allergenicity of airborne basidiospores and ascospores: need for further studies. Aerobiologia. 2012;28(2):83–97. https://doi.org/10.1007/s10453-011-9234-y.

Rocha A, Alvarado MA, Gutiérrez R, Salcedo SM, Moreno S. Variación temporal de esporas de *Alternaria, Cladosporium, Coprinus, Curvularia* y *Venturia* en el aire del área metropolitana de Monterrey, Nuevo León. México Rev Int Contam Ambient. 2013;29(2):155–65.

Ryan TJ, Whitehead LW, Connor TH, Burau KD. Survey of the asp f 1 allergen in office environments. Appl Occup Environ Hyg. 2001;16:679–84. https://doi.org/10.1080/10473220119089.

Sabino R, Veríssimo C, Viegas C, Viegas S, Brandão J, Alves-Correia M, et al. The role of occupational aspergillus exposure in the development of diseases. Med Mycol. 2019;57(2):196–205. https://doi.org/10.1093/mmy/myy090.

Sadys M, Skjøth C, Kennedy R. Back-trajectories show export of airborne fungal spores (*Ganoderma* sp.) from forests to agricultural and urban areas in England. Atmos Environ. 2014;84:88–99. https://doi.org/10.1016/j.atmosenv.2013.11.015.

Salim R, Runco R. Aspergilosis sinusal no invasiva por Aspergillus parasiticus en niño inmunocomprometido. Bol Micol. 2008;23:1–7. https://doi.org/10.22370/bolmicol.2008.23.0.115.

Santos RK, da Fonseca Loureiro GDJ, Barros LB. Allergic bronchopulmonary aspergillosis in Brazil. Int Arch Med. 2021;14 https://doi.org/10.3823/2631.

Sarofim MC, Saha S, Hawkins MD, Mills DM, Hess J, Horton R, et al. Chapter 2: temperature-related death and illness. In: The impacts of climate change on human health in the United States: a scientific assessment. Washington, DC: US Global Change Research Program; 2016. p. 43–68.

Sautour M, Sixt N, Dalle F, L'Ollivier C, Fourquenet V, Calinon C, et al. Profiles and seasonal distribution of airborne fungi in indoor and outdoor environments at a French hospital. Sci Total Environ. 2009;407:3766–71. https://doi.org/10.1016/j.scitotenv.2009.02.024.

Scott J, Untereiner WA, Wong B, Straus NA, Malloch D. Genotypic variation in *Penicillium chrysogenum* from indoor environments. Mycologia. 2004;96:1095–105.

Segura-Medina P, Vargas MH, Aguilar-Romero JM, Arreola-Ramírez JL, Miguel-Reyes JL, Salas-Hernández J. Mold burden in house dust and its relationship with asthma control. Respir Med. 2019;150:74–80. https://doi.org/10.1016/j.rmed.2019.02.014.

Sharpe RA, Bearman N, Thornton CR, Husk K, Osborne NJ. Indoor fungal diversity and asthma: a meta-analysis and systematic review of risk factors. J Allergy Clin Immunol. 2015;135:110–22. https://doi.org/10.1016/j.jaci.2014.07.002.

Shea KM, Truckner RT, Weber RW, Peden DB. Climate change and allergic disease. J Allergy Clin Immunol. 2008;122(3):443–53. https://doi.org/10.1016/j.jaci.2008.06.032.

Sheffield PE, Weinberger KR, Kinney PL. Climate change, aeroallergens, and pediatric allergic disease. Mt Sinai J Med. 2011;78(1):78–84. https://doi.org/10.1002/msj.20232.

Shelton BG, Kirkland KH, Flanders WD, Morris GK. Profiles of airborne fungi in buildings and outdoor environments in the United States. Appl Environ Microbiol. 2002;68:1743–53. https://doi.org/10.1128/AEM.68.4.1743-1753.2002.

Shen HD, Tam MF, Chou H. *Penicillium* allergy. In: Shen HD, editor. Mold allergy, biology and pathogenesis. Elsevier; 2005. p. 69–75.

Shi C, Miller JD. Characterization of the 41kDa allergen Aspv 13, a subtilisin-like serine protease from *aspergillus versicolor*. Mol Immunol. 2011;48:1827–34. https://doi.org/10.1016/j.molimm.2011.05.010.

Sidel FB, Bouziane H, Trigo M, Haskouri E, Bardei F, Redouane A, et al. Airborne fungal spores of *Alternaria*, meteorological parameters and predicting variables. Int J Biometeorol. 2015;59(3):339–46. https://doi.org/10.1007/s00484-014-0845-1.

Sindt C, Besancenot JP, Thibaudon M. Airborne Cladosporium fungal spores and climate change in France. Aerobiologia. 2016;32:53–68. https://doi.org/10.1007/s10453-016-9422-x.

Singh N, Pursell KJ. Combination therapeutic approaches for the management of invasive aspergillosis in organ transplant recipients. Mycoses. 2008;51:99–108. https://doi.org/10.1111/j.1439-0507.2007.01479.x.

Sircar G, Chakrabarti HS, Saha B, Gupta-Bhattacharya S. Identification of aeroallergens from *Rhizopus oryzae*: an immunoproteomic approach. J Proteome. 2012;77:455–68. https://doi.org/10.1016/j.jprot.2012.09.023.

Smith KR. National burden of disease in India from indoor air pollution. Proc Natl Acad Sci U S A. 2000;97:13286–93. https://doi.org/10.1073/pnas.97.24.13286.

Soeria-Atmadja D, Onell A, Borga A. IgE sensitization to fungi mirrors fungal phylogenetic systematics. J Allergy Clin Immunol. 2010a;125:1379–86. https://doi.org/10.1016/j.jaci.2010.02.028.

Soeria-Atmadja D, Önell A, Kober A, Matsson P, Gustafsson MG, Hammerling U. Multivariate statistical analysis of large-scale IgE antibody measurements reveals allergen extract relationships in sensitized individuals. J Allergy Clin Immunol. 2007;120(6):1433–40. https://doi.org/10.1016/j.jaci.2007.07.021.

Sofiev M, Prank M. Impacts of climate change on aeroallergen dispersion, transport, and deposition. In: Beggs PJ, editor. Impacts of climate change on allergens and allergic diseases. Cambridge: Cambridge University Press; 2016. p. 50–73.

Stamper CE, Hoisington AJ, Gomez OM, Halweg-Edwards AL, Smith DG, Bates KL, et al. The microbiome of the built environment and human behavior: implications for emotional health and Well-being in postmodern western societies. Int Rev Neurobiol. 2016;131:289–323. https://doi.org/10.1016/bs.irn.2016.07.006.

Tham R, Vicendese D, Dharmage SC, Hyndman RJ, Newbigin E, Lewis E, et al. Associations between outdoor fungal spores and childhood and adolescent asthma hospitalizations. J Alergol Inmunol Clin. 2017;139(4):1140–7. https://doi.org/10.1016/j.jaci.2016.06.046.

Troutt C, Levetin E. Correlation of spring spore concentrations and meteorological conditions in Tulsa, Oklahoma. Int. J. Biometeorol. 2001;45(2):64–74. https://doi.org/10.1007/s004840100087.

Tu Q, Yuan M, He Z, Deng Y, Xue K, Wu L, et al. Fungal communities respond to long-term CO2 elevation by community reassembly. Appl Environ Microbiol. 2015;81(7):2445–54. https://doi.org/10.1128/aem.04040-14.

Tummon F, Arboledas LA, Bonini M, Guinot B, Hicke M, Jacob C, et al. The need for pan-European automatic pollen and fungal spore monitoring: a stakeholder workshop position paper. Clin Transl Allergy. 2021;11(3):e12015. https://doi.org/10.1002/clt2.12015.

Tuuminen T, Lohi J. Revising the criteria for occupational mold-related disease: arguments, misconceptions and facts. EMJ Allergy Immunol. 2018;3(1):128–35.

Twaroch TE, Curin M, Valenta R, Swoboda I. Mold allergens in respiratory allergy: from structure to therapy. Allergy Asthma Immunol Res. 2015;7(3):205–20. https://doi.org/10.4168/aair.2015.7.3.205.

Vasakova M, Selman M, Morell F, Sterclova M, Molina-Molina M, Raghu G. Hypersensitivity pneumonitis: current concepts of pathogenesis and potential targets for treatment. Am J Respir Crit Care Med. 2019;200(3):301–8. https://doi.org/10.1164/rccm.201903-0541PP.

Villacís M. Chapter 3. Instruments, Policies and practices to develop climate resilience. In: Villamarín G, Pacha MJ, Vásquez A, Villacís M, Wilkinson E. Synthesis report: contributions of the climate-resilient cities in Latin America initiative. Ecuador. Fundación Futuro Latinoamericano. 2019. p. 36–44.

Watts MM, Grammer LC. Hypersensitivity pneumonitis. Allergy Asthma Proc. 2019;40(6):23–8. https://doi.org/10.2500/aap.2019.40.4263.

Weber RW. On the cover: *Epicoccum purpurascens*. Ann Allergy Asthma Immunol. 2006;96:A6. https://doi.org/10.1016/s1081-1206(10)61029-0.

Weinberger K, Lovasi G, Perzanowski M, Sheehan D, Acosta L, Kheirbek I, et al. Association of early life exposure to tree pollen with the development of allergic sensitization, asthma, wheeze, and rhinitis in the New York City Neighborhood Allergy and Asthma Study. In ISEE Conference Abstracts. 2016.

Williams PB, Barnes CS, Portnoy JM, Environmental Allergens Workgroup. Innate and adaptive immune response to fungal products and allergens. J Allergy Clin Immunol Pract. 2016;4(3):386–95. https://doi.org/10.1016/j.jaip.2015.11.016.

Wolf J, O'Neill NR, Rogers CA, Muilenberg ML, Ziska LH. Elevated atmospheric carbon dioxide concentrations amplify Alternaria alternata sporulation and total antigen production. Environ Health Perspect. 2010;118(9):1223–8. https://doi.org/10.1289/ehp.0901867.

Xu YY, Xue T, Li HR, Guan K. Retrospective analysis of epidemic thunderstorm asthma in children in Yulin, Northwest China. Pediatr Res. 2021;89(4):958–61. https://doi.org/10.1038/s41390-020-0980-9.

Zhang Z, Reponen T, Hershey GKK. Fungal exposure and asthma: IgE and non-IgE-mediated mechanisms. Curr Allergy Asthma Rep. 2016;16(12):1–12. https://doi.org/10.1007/s11882-016-0667-9.

Ziska LH. Impacts of climate change on allergen seasonality. In: Beggs PJ, editor. Impacts of climate change on allergens and allergic diseases. Cambridge: Cambridge University Press; 2016. p. 92–112.

Ziska LH. An overview of rising CO_2 and climatic change on aeroallergens and allergic diseases. Allergy Asthma Immunol Res. 2020;12(5):771–82. https://doi.org/10.4168/aair.2020.12.5.771.

Ziska LH, Makra L, Harry SK, Bruffaerts N, Hendrickx M, Coates F, et al. Temperature-related changes in airborne allergenic pollen abundance and seasonality across the northern hemisphere: a retrospective data analysis. Lancet Planet Health. 2019;3(3):e124–31. https://doi.org/10.1016/S2542-5196(19)30015-4.

Zubeldia JM, Senent CJ, Jáuregui I, Baeza ML. Libro de las enfermedades alérgicas. Fundación BBVA; 2012. https://www.fbbva.es/microsite/alergiasfbbva/index.html

Chapter 14
Mycotoxins and Climate Change

Karen Saez Gomez, Elsa Castañeda Roldán, Raúl Ávila Sosa, and Ricardo Munguía-Pérez

Contents

Abbreviations

AFB1, AFB2, AFG1, AFG2	Aflatoxins
DON	Deoxynivalenol

K. Saez Gomez · E. Castañeda Roldán · R. Munguía-Pérez (✉)
Centro de Investigaciones en Ciencias Microbiológicas, Laboratorio de Micología, Posgrado en Ciencias Ambientales, Instituto de Ciencias-Benemérita Universidad Autónoma de Puebla, Pue., Puebla, Mexico
e-mail: karen.saezgomez@viep.com.mx; ricardo.munguia@correo.buap.mx

R. Ávila Sosa
Departamento de Bioquímica-Alimentos, Facultad de Ciencias Químicas, Benemérita Universidad Autónoma de Puebla, Pue., Puebla, Mexico
e-mail: raul.avila@correo.buap.mx

M. G. Frías-De-León et al. (eds.), *The Impact of Climate Change on Fungal Diseases*, Fungal Biology, https://doi.org/10.1007/978-3-030-89664-5_14

EA	Ergot alkaloids
EC	European Commission
FAO	Food and Agriculture Organization
FB1, FB2, FB3	Fumonisin
FDA	Food and Drug Administration
IARC	International Agency for Research on Cancer
NOM	Official Mexican Standard
OTA	Ochratoxin A
PAT	Patulin
WMO	World Meteorological Organization
ZEN	Zearalenone

14.1 Introduction

Climate change has been predicted to significantly impact on food safety, which is determined by three key components: sufficient food availability, access to these food supplies, and quality and use in terms of nutritional and cultural perspectives (Medina et al. 2017). During the last few decades, anthropogenic activities have caused environmental transformations, generating disturbances that modified microbial community structures and affected the function of diverse ecosystems (Dikhoba et al. 2019; Kaisermann et al. 2015; Perrone et al. 2020). Climate change has especially influenced mycotoxin production and has caused a spatial-temporal risk. This phenomenon varies according to every region or country (Lee et al. 2018). Moreover, crops are an ideal substrate for fungal growth and mycotoxin production that are at the beginning of the food supply chain (Warnatzsch et al. 2020).

The geographic expansion of microbes and vectors into new territories has been made possible by the increased human travel and sustained warmer temperatures at higher latitudes. It exposes all planetary life to diseases previously unknown in particular habitats (Coates and Norton 2021). In an agroecosystem, weather is a key factor. Regions are vulnerable from temperature variability and extreme events, like heat waves, droughts, floods, cyclones, and wildfires (IPCC 2014; Lee et al. 2018). There are several direct and indirect effects on crop health after interactions with global change drivers (Chakraborty et al. 2021). For example, environmental conditions have a significant effect on field crops and stored grains that can range from pathogens' sporulation to virulence gene expression in stressed situations. Thus, climate change and environmental strain cause significant consequences for mycotoxin production (Medina et al. 2017; Perrone et al. 2020). Unfortunately, mycotoxins represent a risk to human and animal health due to their mutagenic, teratogenic, and carcinogenic effects (Iamanaka et al. 2007; Murphy et al. 2006). Therefore, it is important to understand how and why mycotoxins occur.

14.2 The Discovery of Mycotoxins

The term *mycotoxin* comes from Greek "mýkēs" and from the Latin expression "toxicum," that is, "toxic fungus." Mycotoxins are toxic, secondary metabolites produced naturally by certain filamentous fungal species or even fungi strains within the same genus. These metabolites are produced during the stationary stage of the growth phase, which is associated with differentiation and sporulation (González-Osnaya et al. 2006; Heussner and Bingle 2015; Janik et al. 2020). In addition, they possess diverse structures, are chemically stable and heat-resistant, and can persist during food processing (Sacco et al. 2020; WHO 2020). Exposure to mycotoxins can cause a syndrome called mycotoxicosis, which can affect various systems depending on where the contact took place (Peraica 2015). One of the first known cases of mycotoxicosis was caused by food poisoning in Europe in the Middle Ages. It was called "Saint Anthony's Fire" due to the burning sensation that people felt in addition to seizures, hallucinations, nausea, vomiting, psychosis, and delusions (Ramírez Quintero 2018). Today, that form of mycotoxicosis refers to ergotism, caused by the consumption of contaminated rye with *Claviceps purpurea* alkaloids (Serrano-Coll and Cardona-Castro 2015).

There are about 400 mycotoxins, produced by approximately 100 fungi, which have been identified. The most infamous of these metabolites, in terms of toxicity and occurrence, are aflatoxins (AFB1, AFB2, AFG1, AFG2), ochratoxin A (OTA), fumonisin (FB1, FB2, FB3) deoxynivalenol (DON), zearalenone (ZEN), patulin (PAT), and ergot alkaloids (EA) (Alshannaq and Yu 2017; Berthiller et al. 2007; Deng et al. 2021; Paterson and Lima 2010).

14.3 Principal Toxigenic Fungi and their Metabolites

There are two classifications of toxigenic fungi: field or plant, which contaminates crops and generate metabolites before harvest; and storage or saprophyte, which increases the production of metabolites if there are no adequate control measures (Table 14.1). Despite this classification, the primary source of contamination originates in the field (Deng et al. 2021; Sadhasivam et al. 2017).

Aflatoxin production occurs within a wide temperature range of 12 to 42 °C (with an optimum temperature of 28–30 °C). There are around 20 types of aflatoxins; but only six are studied regularly due to their ability to affect a broad range of agricultural crops and for their negative effects on both human and animal health (Martínez et al. 2013). These six aflatoxins are AFB1, AFB2, AFG1, AFG2, AFM1, and AFM2. The letters "B" and "G" refer to the blue and green fluorescence generated when ultraviolet light is present, and the letter "M" signifies that the toxin has been identified in milk. These metabolites can produce mutagenic, teratogenic, and

Table 14.1 Fungal species and their mycotoxins

Fungal species	Mycotoxins	References
Aspergillus parasiticus	B_1, B_2, G_1, and G_2 aflatoxins	Campos et al. (2017); García-Cela et al. (2015); Hernández-Martínez and Navarro-Blasco (2015); Liu et al. (2021); Peromingo et al. (2019)
Aspergillus flavus	B_1 and B_2 aflatoxins	Campos et al. (2017); Ghali et al. (2009); Hernández-Martínez and Navarro-Blasco (2015); Liu et al. (2021)
Fusarium sporotrichioides	T-2 toxin	Ostry et al. (2020); Yli-Mattila et al. (2011)
Fusarium graminearum	Deoxynivalenol, zearalenone	Shao et al. (2021); Wang et al. (2020); Williams et al. (1994)
F. verticillioides (Fusarium moniliforme)	Fumonisin B_1	Chen et al. (2021); Hu et al. (2021)
Penicillium verrucosum	Ochratoxin A	Meerpoel et al. (2021); Park et al. (2019)
Aspergillus ochraceus	Ochratoxin A	Park et al. (2019); Paterson and Lima (2010)
Penicillium expansum	Patulin	Hermosillo et al. (2015); Xing et al. (2021)
Claviceps purpurea	Ergot alkaloids	Králová et al. (2021); van der Linde et al. (2016)

carcinogenic effects and have also been linked to chronic and acute liver diseases (Iamanaka et al. 2007; Murphy et al. 2006). The most researched aflatoxin is AFB1M; the International Agency for Research on Cancer (IARC) classifies it as a Group 1 human carcinogenic substance (IARC 1993).

Aspergillus are habitually present in the human environment, but it wasn't until the middle of the nineteenth century that the fungi were discovered and revealed to be involved in decomposition processes. These cosmopolitan fungi can proliferate in almost any environment of the world, ranging from arid to humid to temperate to tropical, with soil as the primary reservoir (Mamo et al. 2020). While the fungi do cause human and animal diseases, they also produce fermenting agents capable of creating important secondary metabolites (Hedayati et al. 2007). After *A. fumigatus*, *A. flavus* is responsible for causing the most cases of invasive aspergillosis in humans (Piontelli 2014). *A. flavus* is also the highest aflatoxin-producing species; therefore, its precise identification is important: Colonies are yellowish green with a powdery texture and radial furrows (Guevara-Robles et al. 2007; Luo et al. 2018). Microscopically, they have irradiated and uniseriate conidial heads, rough walls, spherical vesicle, globose conidia, and septate mycelium (Abarca 2000).

Another mycotoxin producer is *Fusarium*, which infects field crops and causes diseases and yield reduction. It is one of the more widely studied genera, so its genetic and phenotypic characterization is clearly defined (Desai et al. 2020). It presents in its macromorphology a white mycelium at its start which then forms pigments that are colored gray to violet. Under the microscope, there are abundant oval microconidia with flat bases, grouped in chains, showing an apex and a basal cell (foot shape) (de la Torre-Hernández et al. 2014). There are around 15 types of fumonisins produced mainly by *F. verticillioides*, which is also the most researched species because it produces one of the most toxic mycotoxins in the genus, FB1 (other frequently studied mycotoxins are FB2 and FB3). According to IARC, it is classified within Group 2B as a possibly carcinogenic substance in humans, associated with liver and esophageal cancer (Luo et al. 2018; Serrano-Coll and Cardona-Castro 2015).

Ochratoxins are metabolites produced by species like *A. ochraceus* and *P. verrucosum* (Clemente et al. 2017). Five types of ochratoxins have been described: A, B, C, α, and β. The most toxic is ochratoxin A, which is considered a possibly carcinogenic substance, classified within Group 2B according to the IARC (González-Osnaya et al. 2006). It is associated with nephrotoxic, mutagenic, and carcinogenic effects (Omotayo et al. 2019). *A. ochraceus* has been frequently isolated from peanuts, green coffee, grapes, and soybeans (Clemente et al. 2017). Through a microscope, finely rough globose conidia are observed as well as rough hyphae 3.5 μm conidiophore, and from the globose vesicle, chains of conidia radiate with a phialides of 5 μm (Borges et al. 2011; Guarniz-Benites and Valdez-Arana 2019). *P. verrucosum* is found in salty food. Many isolates of this species have been reclassified due to their notable similarity. A polyphasic approach is necessary to classify and identify the species of this genus (Cabañes et al. 2010; Coton et al. 2019). Fungal colonies are opaque to dark green and grow from 0 to 30 °C (Pitt 2014). In its micromorphology, *P. verrucosum* presents septate hyphae with finely rough walls, terverticylated conidiophores, and spherical spores with smooth walls in unbranched chains (Guarniz-Benites and Valdez-Arana 2019).

14.4 Food Mycotoxins and its Toxicological Effect

The Food and Agriculture Organization (FAO) suggested that mycotoxin contamination affected 25% of crops worldwide (FAO 2015). However, due to the improvement and sensitivity of analytical technology and methods, we now know that contamination levels are actually 60–80% due to fungal growth and mycotoxin production influenced by climatic conditions, relative humidity, the presence of water, temperature, and inadequate storage (Eskola et al. 2020; Juan et al. 2020). In addition, insects and other pests are compromising food security worldwide, causing large losses of stored grains. Some of the products contaminated by mycotoxins are presented in Table 14.2.

Drought conditions, cultivation systems, and some agronomic practices can generate stress in plants, increasing their susceptibility to fungal contamination. These conditions determine the severity of mycotoxin infection (Pinotti et al. 2016). The presence of toxigenic fungi and mycotoxins is not the same between crops, since the species and strains maintain various capacities to infect a specific host plant (Perrone et al. 2020). Given the importance of these toxic metabolites to global health, tolerance levels have been regulated in several countries. Table 14.3 reports the maximum content safe for human consumption of the most essential mycotoxins in cereals and products based on regulated tolerance levels.

Table 14.2 Main mycotoxins found in food

Mycotoxin	Food product detected	References
Aflatoxin	Peanut, corn, wheat, nuts, milk, eggs, white cheese, spices, figs, sorghum, pistachios, coffee	Adegbeye et al. (2020); Ghali et al. (2009); Liu et al. (2021); Móricz et al. (2007); Paterson and Lima (2010)
Ochratoxin A	Wheat, dried fruit, barley, oats, rye, barley, corn, rice, dried beans, cheese, pork, coffee, raisins, grapes, nuts, wine, cocoa, fresh sweet peppers, herbal medicines	Meerpoel et al. (2021); Park et al. (2019); Paterson and Lima (2010); Zhai et al. (2021)
Patulin	Apples, pears, and their by-products	Huang and Peng (2021); Paterson and Lima (2010); Xing et al. (2021)
Trichothecenes	Corn, wheat, barley, oats	Aniołowska et al. (2021); Paterson and Lima (2010)
Zearalenone	Corn, wheat, granulated commercial feed	Wang et al. (2020); Paterson and Lima (2010)

Table 14.3 Maximum level of mycotoxins in cereals and their by-products for human consumption

Mycotoxin	Cereals and by-products	Regulator1	Maximum level	References
Aflatoxin B1	Cereals	EC	2 μg/kg	Pinotti et al. (2016)
Total aflatoxins		NOM	20 μg/kg	Reyes Velázquez et al. (2009)
	Cereals	EC	4 μg/kg	Pinotti et al. (2016)
Deoxinivalenol	Raw corn, except for raw corn intended to be processed by wet milling	EC	1750 μg/kg	Pinotti et al. (2016)
Zearalenone	Raw corn, except for raw corn intended to be processed by wet milling	EC	350 μg/kg	Pinotti et al. (2016)
Ochratoxin A	Cereals	EC	5 μg/kg	Ravelo Abreu et al. (2011)
	Derivatives made from cereals	EC	3 μg/kg	Ravelo Abreu et al. (2011)
Fumonisins B1 + B2 + B3	Corn by-products	FDA	2 mg/kg	Chen et al. (2021)
Fumonisins B1 + B2	Unprocessed corn	EC	4 mg/kg	Chen et al. (2021)

EC The European Commission, *FDA* Food and Drug Administration, *NOM* Official Mexican Standard

Mycotoxin concentrations among crop fields, food storage, and products for sale are uneven and are dependent on fungal growth characteristics (Zhou et al. 2020). These concentrations are now considered a global challenge due to the strategies used for their management in crops. New and emerging strains in food and feed have generated worldwide concern. There is evidence of new fungal genotypes with higher levels of aggressiveness and altered production of mycotoxins due to the ability of fungi to adapt to changing conditions (Moretti et al. 2019; Perrone et al. 2020).

Mycotoxin's primary effects on human and animal health are immunotoxic, teratogenic, nephrotoxic, and carcinogenic (Fig. 14.1). Due to their adverse health effects, the most studied mycotoxins are aflatoxins, ochratoxin A, fumonisins, trichothecenes, and zearalenone (Kalagatur et al., 2020; Omotayo et al. 2019). For example, the aflatoxigenic fungi found in peanuts and corn crops could cause up to 20% of liver cancers worldwide (Laursen 2014).

Metabolites easily penetrate the cell membrane and cause genome-producing mutations in the nucleotide sequence (Dey et al. 2021). Thus, chronic toxicity begins; it is commonly associated with mycotoxin exposure. As a consequence of low-dose exposure for a prolonged period, most cases develop cancer in various organs. Aflatoxin residues are retained in animal tissues creating risks to human health due to consumption of contaminated food (Dey et al. 2021; Selim et al. 2014). Table 14.4 shows some of the toxic effects caused by the most studied mycotoxins.

Because mycotoxins possess lipophilic properties and are able to bind with plasma proteins, they can persist in the body when there is chronic or repeated exposure (Sirot et al. 2013). In many countries, mycotoxins are strictly regulated based on their carcinogenicity and toxicity; they exhibit different toxicities in varying doses (Zhou et al. 2020). In recent studies, it has been shown that the most affected organs are the liver and intestines because they are the first defense in the body

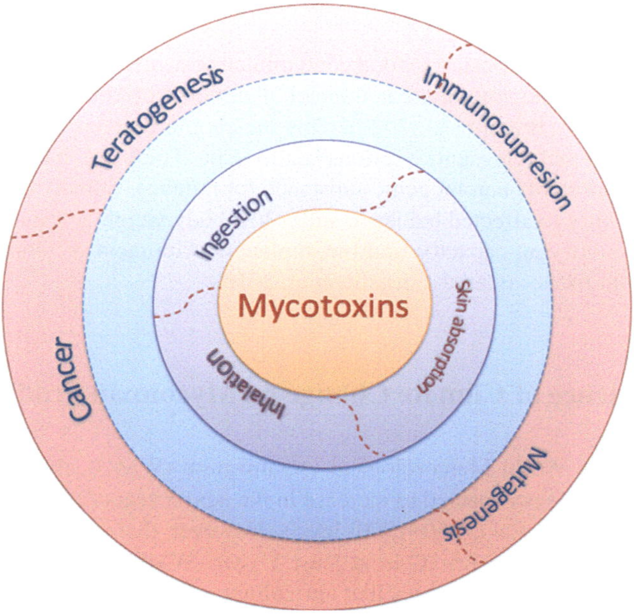

Fig. 14.1 Pathways of exposure to mycotoxins in humans and animals, and main health effects

Table 14.4 Mycotoxins' main health effects

Mycotoxin	Health effect	Reference
Aflatoxin	Hepatotoxic, immunotoxic, mutagenic, teratogenic, and carcinogenic	Ghali et al. (2009); Hernández-Martínez and Navarro-Blasco (2015); Liu et al. (2021); Móricz et al. (2007); Omotayo et al. (2019); von Hertwig et al. (2020)
Ochratoxin A	Carcinogenic, genotoxic, immunosuppressive, teratogenic, nephrotoxic, neurotoxic, and inducer of lower urinary tract disease	Meerpoel et al. (2021); Omotayo et al. (2019); Park et al. (2019); Sakin et al. (2018)
Fumonisin	Carcinogenic, hepatotoxic, nephrotoxic, neurotoxic, reproductive toxicity, and immunosuppressive	Chen et al. (2021); Hu et al. (2021); Omotayo et al. (2019)
Deoxinivalenol	Nausea, vomiting, diarrhea, inhibits protein synthesis, reproductive effects, and toxicosis	Omotayo et al. (2019); Shao et al. (2021)
Trichothecenes	Hepatotoxic, genotoxic, nephrotoxic, cytotoxic, and immunosuppressive	Lemos et al. (2021); Omotayo et al. (2019)
Zearalenone	Carcinogenic, hormonal imbalance, and reproductive effects	Omotayo et al. (2019)
Patulin	Neurotoxic, hepatotoxic, nephrotoxic, mutagenic, teratogenic, and immunosuppressive	Huang and Peng (2021); Omotayo et al. (2019); Xing et al. (2021)

against toxins. Therefore, they carry out an extremely vital role in the metabolism of metabolites (Dey et al. 2021).

During childhood, exposure to food contaminants can affect health over the coming years. Breastfeeding mothers can transfer aflatoxin M1 (AFLM1) to their babies through their milk (Braun et al. 2020). This metabolite remains the hydroxylated product of AFB1 and represents a serious health issue. The IARC has classified it in Group 2B as a possibly carcinogenic substance for humans. Equally dangerous is a high susceptibility to affected babies to suffer low body weight, elevated metabolic rate, lower detoxifying capacity, and incomplete development of vital organs and nervous system tissues (Cantú-Cornelio et al. 2016).

14.5 Influence of Climate Change on Mycotoxin Production

According to the World Meteorological Organization (WMO), the annual global temperature average is expected to increase in the next 5 years (2020–2024) at least 1 °C above pre-industrial (1850–1900) levels. However, there is a 20% probability that this rise will exceed 1.5 °C in at least 1 year (WMO 2020). This change is caused by anthropogenic activities like agriculture intensification, fossil fuel combustion, and excess use of chemical fertilizers, all actions that increase the emissions of nitrous oxide, ammonia, and CO_2 (Chourasiya et al. 2021). In addition, more frequent extreme weather events including heat waves, droughts, and floods establish circumstances that cause existing infectious microorganisms to flourish and new infections to emerge. Furthermore, in many regions, the myriad consequences of climate change drive mass intra- and international human migration, which disturbs the regional healthcare infrastructure and the habitats of microbes, vectors, and animal reservoirs (Coates and Norton 2021).

Expected global climate change includes an increase in temperature and modifications in rainfall patterns which shift the range of species, even in presently colder regions. For instance, CO_2 level has increased from 280 to 400 ppm, a rise which has contributed to global warming effects (Chourasiya et al. 2021; Nesic et al. 2015). Local farming communities fortunately possess enough experience to produce groups of productive crops in the new, warmer environment. If the climate shifts further from now, more adjustments to current practices will be needed to maintain productivity (Chakraborty et al. 2021).

Ultimately, the shifting of farming and other agricultural practices under global climate change will modify the outcome of plant-soil microbe interactions. Environmental disturbances can cause changes in microbial community structures, affecting the ecosystem's functioning (Kaisermann et al. 2015). In tropical and arid regions where atmospheric temperatures are close to crop physiological maximums, elevated temperatures are more immediately damaging by increasing heat stress and water loss by evaporation (Chakraborty et al. 2021). Moreover, climate represents the principal driver in agroecosystems, with temperature and relative humidity being key environmental factors (Fig. 14.2) in the colonization of mycotoxigenic fungi and mycotoxin production (Peter Mshelia et al. 2020; Perrone et al. 2020; Valencia-Quintana et al. 2020).

14.5.1 Temperature

Related to fungal growth, some species were found to be affected solely by temperature, while others were affected by a joint impact of both temperature and rainfall. Climate change was also found to influence morphological alterations in fungal spores. For example, at the beginning of autumn, fungi are driven toward greater water accumulation, resulting in size enlargement. On the other hand, spores produced toward the end of the season are characterized by a reduction in size. Fungal spore seasons were shorter when fungi encounter propitious weather conditions that facilitated their growth, spore production, release, and dispersal (Kauserud et al. 2011; Sadyś et al. 2016). Indeed, Sadyś et al. (2016) reported elevated levels of highly allergenic *Alternaria* and *Cladosporium* species, as well as other fungi, in the following next two decades (2016–2036).

Abiotic factors can stimulate or inhibit mycotoxin production; climate change affects the host-pathogen interaction that favored it (Desai et al. 2020; Janik et al. 2020). For example, a toxigenic fungi could disappear from the environment but appear in another region previously not at risk (Perrone et al. 2020). It has been seen

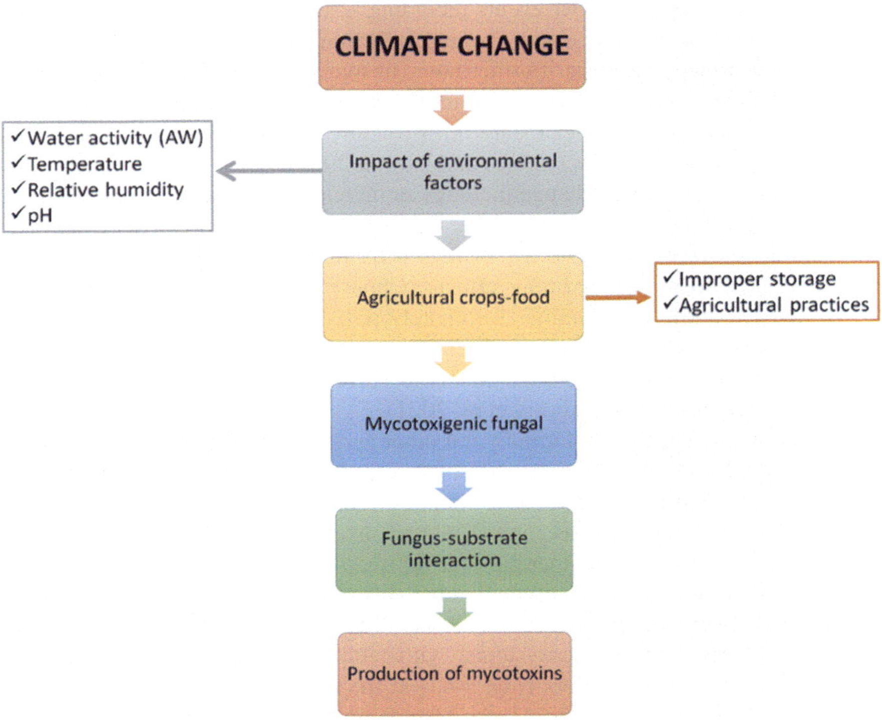

Fig. 14.2 Environmental factors that influence the proliferation of mycotoxigenic fungi and mycotoxin production

that the interaction of these two factors – mycotoxigenic fungi and climate – can stimulate or suppress the biosynthetic genes involved in mycotoxins' production (Abdelmohsen et al. 2021; Medina et al. 2017).

Temperature increases metabolic processes and biomass of plants and microorganisms within their respective biological limits (Beauchamp et al. 1992), and consequently may modify interactions between them. Temperature represents a key factor influencing composition and functioning of soil microbiota (Landa et al. 2013). Bailly et al. (2018) reported that hot and dry climatic conditions favored the production of aflatoxins in France. They observed that 6% of fields and 15% of corn silos sampled were contaminated. Furthermore, Abdelmohsen et al. (2021) mentioned that *Penicillium verrucosum*'s growth and survival required three abiotic factors: temperature, CO_2, and matrix/solute stress (which are related with the climate they interact). *Fusarium* exists in soil and crop residues for a longer time than other fungi; plant stress due to drought is one of the foremost factors that increases its severity. *Aspergillus* ear rot is another plant disease in the drought-prone, nonirrigated cornfields. This condition is primarily caused by *Aspergillus flavus* and *A. parasiticus* (Chakraborty et al. 2021). The extremely elevated temperatures in summer, along with the prolonged heat wave in tropical and even higher latitudes, have contributed to widespread trouble in the world cereal markets (Chakraborty et al. 2021).

14.5.2 Water Availability

Floods preceded by droughts represent a threat to agriculture and food safety since they affect crop production and the supply and quality of food and feed, as well as negatively impacting plants' physiology (Lee et al. 2018; Nesic et al. 2015; Yazid et al. 2020). Moisture conditions boost the spread of weeds and their biomass, which constitute a menace for crop production. The soilborne microbial diseases and pests, like certain mycotoxins, increase as moisture content does. This situation directly causes worsening of crop annual production. When atmospheric water content increases, fungal species regenerate from spores present in agricultural soils, affecting more significantly the crops growing in it, for example, foliar fungal diseases in wheat (Chakraborty et al. 2021).

There is much convincing evidence of the substantial effect that water availability has on microbial survival in soil and plant growth. In fact, water availability, irrigation regimes, and drought stress may affect several soil properties and directly or indirectly affect fungal activities that are of ecological significance such as percolation, adsorption/desorption, growth rate, sedimentation, chemotaxis, motility, and transport through different soil types (Landa et al. 2013). Moreover, seawater inundation in the coastal regions remains another problem in field irrigation. This may be managed by producing salt-resistant plants like some paddy varieties. Finally, high-speed winds caused by monsoons or hurricanes can spread, over thousands of kilometers, fungal spores for diseases, like stem rust in wheat caused by the pathogen *Puccinia graminis* (Chakraborty et al. 2021).

14.5.3 Soil Composition

A key aspect to remember is that change in soil composition is positively correlated to air temperature. Moreover, soil composition is a function of its water content. The higher the soil water content, the lower the temperature. Hence, soil temperature represents a function of crop water uptake and, consequently, of crop cycle duration; the latter shortens when the temperature is optimal for plants. Further, air temperature may frequently reach values above the optimum for an organism's growth. Soil acts as a buffer, that is, compared to the air temperature, there will be a markedly lower temperature increase in the top layer and below. In other words, when considering short- to medium-term time horizons, the range of fluctuation of soil temperature tends to be still either below or close to the optimum. Consequently, in most of the conditions evaluated, the growth rates of organisms in the soil are rarely projected to decrease. Unfortunately, the opposite is true for the aboveground organisms where decline is frequent at the most elevated temperatures (Manici et al. 2014).

There is an evidence that some fungal species responses to diverse environmental conditions could be very heterogeneous, potentially increasing their harmfulness (Broders et al. 2009). This leads to considering soilborne fungal pathogens as components in the complexity of crop-climate-environment interactions, which makes projecting the net outcome of climate change in agriculture difficult. Soilborne fungal pathogens remain the causal agents of root rot in herbaceous and fruit tree crops and represent the critical biotic components of yield decline in intensively cultivated areas. Their specific impact on yield losses is not easy to evaluate, given the difficulties in distinguishing their role from, for example, unfertile soil or abiotic stresses. Moreover, soilborne pathogens can reduce crop quality in both direct and indirect ways (Manici et al. 2014). Therefore, if global warming continues, fungal species will display a larger response with an earlier start of mycelium growth but with typically lower and belated to spore production. Such phenological changes may have important implications on fungal communities from the biodiversity perspective (Damialis et al. 2015).

14.6 Conclusion

Climate change is crucial to the proliferation of mycotoxigenic fungi and the production of mycotoxins. It can generate disturbances that modify the microbial ecosystem, making the regions previously considered suppressive for the production of mycotoxins now propitious for these metabolites. Hence, climate change has a vast impact on toxigenic fungi and their metabolites.

In addition, mycotoxins compromise food safety, causing serious public health problems up to and including death. In most cases, chemicals harmful to the environment are used to eliminate the fungi. Currently, however, the aim of many farmers is to guarantee food free from agrochemicals and toxins. Therefore, there is a great need for the application of innovative control measures to reduce the level of pesticides and mycotoxins in food products. Discovering new control methods that have low or no toxicity and little persistence in the environment is of utmost relevance and must be a priority worldwide.

References

Abarca ML. Taxonomía e identificación de especies implicadas en la aspergilosis nosocomial [Taxonomy and identification of the species involved in nosocomial aspergillosis]. Rev Iberoam Micol. 2000;17(3):S79–84.

Abdelmohsen S, Verheecke-Vaessen C, Garcia-Cela E, Medina A, Magan N. Dynamics of solute/matric stress interactions with climate change abiotic factors on growth, gene expression and ochratoxin a production by Penicillium verrucosum on a wheat-based matrix. Fungal Biol. 2021;125(1):62–8. https://doi.org/10.1016/j.funbio.2020.10.005.

Adegbeye MJ, Reddy PRK, Chilaka CA, Balogun OB, Elghandour MMMY, Rivas-Caceres RR, Salem AZM. Mycotoxin toxicity and residue in animal products: prevalence, consumer exposure and reduction strategies—a review. Toxicon. 2020;177:96–108. https://doi.org/10.1016/j.toxicon.2020.01.007.

Alshannaq A, Yu JH. Occurrence, toxicity, and analysis of major mycotoxins in food. Int J Environ Res Public Health. 2017;14(6):632. https://doi.org/10.3390/ijerph14060632.

Aniołowska M, Steininger M, Campos WEO, Rosas LB, Neto AAP, Mello RA, et al. Overpressured layer chromatographic determination of aflatoxin B1, B2, G1 and G2 in red paprika. Food Chem Toxicol. 2021;147(1):111280. https://doi.org/10.1016/j.fm.2019.01.024.

Bailly S, Mahgubi AE, Carvajal-Campos A, Lorber S, Puel O, Oswald IP, et al. Occurrence and identification of aspergillus section Flavi in the context of the emergence of aflatoxins in French maize. Toxins (Basel). 2018;10(12):525. https://doi.org/10.3390/toxins10120525.

Beauchamp RO Jr, St Clair MB, Fennell TR, Clarke DO, Morgan KT, Kari FW. A critical review of the toxicology of glutaraldehyde. Crit Rev Toxicol. 1992;22(3–4):143–74. https://doi.org/10.3109/10408449209145322.

Berthiller F, Sulyok M, Krska R, Schuhmacher R. Chromatographic methods for the simultaneous determination of mycotoxins and their conjugates in cereals. Int J Food Microbiol. 2007;119(1–2):33–7. https://doi.org/10.1016/j.ijfoodmicro.2007.07.022.

Borges VB, Maia MCA, Couto MAPG, Vital HC, Souza MCL. Morphological changes of Aspergillus ochraceus irradiated on peanut grains. INAC 2011: International Nuclear Atlantic Conference Nuclear Energy: New Jobs for a Better Life. 2011. http://inis.iaea.org/search/search.aspx?orig_q=RN:43046322

Braun D, Ezekiel CN, Marko D, Warth B. Exposure to mycotoxin-mixtures via breast milk: an ultra-sensitive LC-MS/MS biomonitoring approach. Front Chem. 2020;8:423. https://doi.org/10.3389/fchem.2020.00423.

Broders KD, Wallhead MW, Austin GD, Lipps PE, Paul PA, Mullen RW, et al. Association of soil chemical and physical properties with Pythium species diversity, community composition, and disease incidence. Phytopathology. 2009;99(8):957–67. https://doi.org/10.1094/PHYTO-99-8-0957.

Cabañes FJ, Bragulat MR, Castellá G. Ochratoxin a producing species in the genus Penicillium. Toxins (Basel). 2010;2(5):1111–20. https://doi.org/10.3390/toxins2051111.

Campos WEO, Rosas LB, Neto AP, Mello RA, Vasconcelos AA. Extended validation of a sensitive and robust method for simultaneous quantification of aflatoxins B1, B2, G1 and G2 in Brazil nuts by HPLC-FLD. J Food Comp Anal. 2017;60:90–6. https://doi.org/10.1016/j.jfca.2017.03.014.

Cantú-Cornelio F, Aguilar-Toalá JE, de León-Rodríguez CI, Esparza-Romero J, Vallejo-Cordoba B, González-Córdova AF, et al. Occurrence and factors associated with the presence of aflatoxin M1 in breast milk samples of nursing mothers in Central Mexico. Food Control. 2016;62:16–22. https://doi.org/10.1016/j.foodcont.2015.10.004.

Chakraborty A, Singh J, Upadhyay H. Fungal diversity and ecology: major challenges for crop production in present climate scenario. In: Yadav AN, editor. Recent trends in mycological research. Springer Nature; 2021. p. 89–107. https://doi.org/10.1007/978-3-030-60659-6_4.

Chen J, Wei Z, Wang Y, Long M, Wu W, Kuca K. Fumonisin B1: mechanisms of toxicity and biological detoxification progress in animals. Food Chem Toxicol. 2021;149:111977. https://doi.org/10.1016/j.fct.2021.111977.

Chourasiya D, Gupta MM, Sahni S, Oehl F, Agnihotri R, Buade R, et al. Unraveling the AM fungal community for understanding its ecosystem resilience to changed climate in agroecosystems. Symbiosis. 2021; https://doi.org/10.1007/s13199-021-00761-9.

Clemente I, Aznar M, Nerín C. Effect of an active label based on benzyl isothiocyanate on the morphology and ochratoxins production of Aspergillus ochraceus. Food Res Int. 2017;101:61–72. https://doi.org/10.1016/j.foodres.2017.08.060.

Coates SJ, Norton SA. The effects of climate change on infectious diseases with cutaneous manifestations. Int J Womens Dermatol. 2021;7(1):8–16. https://doi.org/10.1016/j.ijwd.2020.07.005.

Coton M, Auffret A, Poirier E, Debaets S, Coton E, Dantigny P. Production and migration of ochratoxin A and citrinin in Comté cheese by an isolate of Penicillium verrucosum selected among Penicillium spp. mycotoxin producers in YES medium. Food Microbiol. 2019;82:551–9. https://doi.org/10.1016/j.fm.2019.03.026.

Damialis A, Mohammad AB, Halley JM, Gange AC. Fungi in a changing world: growth rates will be elevated, but spore production may decrease in future climates. Int J Biometeorol. 2015;59(9):1157–67. https://doi.org/10.1007/s00484-014-0927-0.

de la Torre-Hernández ME, Sánchez-Rangel D, Galeana-Sánchez E, Plasencia-de la Parra J. Fumonisinas –Síntesis y función en la interacción Fusarium verticillioides-maíz. Rev Esp Cienc Quim Biol. 2014;17(1):77–91. https://doi.org/10.1016/s1405-888x(14)70321-3.

Deng LZ, Sutar PP, Mujumdar AS, Tao Y, Pan Z, Liu YH, et al. Thermal decontamination technologies for microorganisms and mycotoxins in low-moisture foods. Annu Rev Food Sci Technol. 2021;12:287–305. https://doi.org/10.1146/annurev-food-062220-112934.

Desai S, Dubey SC, Prasad RD. Impacts of climate change on Fusarium species Vis-à-Vis adaptation strategies. Indian Phytopathol. 2020;73(4):593–603. https://doi.org/10.1007/s42360-020-00258-3.

Dey DK, Chang SN, Kang SC. The inflammation response and risk associated with aflatoxin B1 contamination was minimized by insect peptide CopA3 treatment and act towards the beneficial health outcomes. Environ Pollut. 2021;268(Pt B):115713. https://doi.org/10.1016/j.envpol.2020.115713.

Dikhoba PM, Mongalo NI, Elgorashi EE, Makhafola TJ. Antifungal and anti-mycotoxigenic activity of selected South African medicinal plants species. Heliyon. 2019;5(10):e02668. https://doi.org/10.1016/j.heliyon.2019.e02668.

Eskola M, Kos G, Elliott CT, Hajšlová J, Mayar S, Krska R. Worldwide contamination of food-crops with mycotoxins: validity of the widely cited 'FAO estimate' of 25. Crit Rev Food Sci Nutr. 2020;60(16):2773–89. https://doi.org/10.1080/10408398.2019.1658570.

FAO. The state of agricultural commodity markets 2015–2016. Trade and food security: Achieving a better balance between national priorities and the collective good 2015.

García-Cela E, Marin S, Sanchis V, Crespo-Sempere A, Ramos AJ. Effect of ultraviolet radiation A and B on growth and mycotoxin production by Aspergillus carbonarius and Aspergillus parasiticus in grape and pistachio media. Fungal Biol. 2015;119(1):67–78. https://doi.org/10.1016/j.funbio.2014.11.004.

Ghali R, Belouaer I, Hdiri S, Ghorbel H, Maaroufi K, Hedilli A. Simultaneous HPLC determination of aflatoxins B1, B2, G1 and G2 in Tunisian sorghum and pistachios. J Food Compos Anal. 2009;22(7–8):751–5. https://doi.org/10.1016/j.jfca.2009.04.009.

González-Osnaya LR, Catalá-Gregori AI, Soriano Del Castillo JM, Moltó-Cortés JC, Mañes-Vinuesa J. Nomenclatural amendments in the mycotoxicology field. Rev Iberoam Micol. 2006;23(2):64–6. https://doi.org/10.1016/s1130-1406(06)70015-8.

Guarniz-Benites J, Valdez-Arana JDC. Morphological identification of mycotoxigenic fungi in accessions of quinoa (Chenopodium quinoa Wild.) of the peruvian coast and sierra. Scientia Agropecuaria. 2019;10(3):327–36. https://doi.org/10.17268/sci.agropecu.2019.03.02.

Guevara-Robles M, Urcia-Ausejo F, Casquero-Cavero J. Manual de procedimientos y técnicas de laboratorio para la identificación de los principales hongos causantes de micosis humanas. Ministerio de Salud; 2007.

Hedayati MT, Pasqualotto AC, Warn PA, Bowyer P, Denning DW. Aspergillus flavus: human pathogen, allergen and mycotoxin producer. Microbiology (Reading). 2007;153(Pt 6):1677–92. https://doi.org/10.1099/mic.0.2007/007641-0.

Hermosillo G, Quezada M, Moreno J, Bustamante P, Trejo M. Detección de patulina en manzana "Golden Delicious" y en productos derivados elaborados industrial y artesanalmente comercializados en México. Rev Iberoam Tecnol Postcosecha. 2015;16(2):281–6.

Hernández-Martínez R, Navarro-Blasco I. Surveillance of aflatoxin content in dairy cow feedstuff from Navarra (Spain). Anim Feed Sci Technol. 2015;200(1):35–46. https://doi.org/10.1016/j.anifeedsci.2014.12.002.

Heussner A, Bingle L. Comparative Ochratoxin toxicity: a review of the available data. Toxins. 2015;7(10):4253–82. https://doi.org/10.3390/toxins7104253.

Hu Z, Yuan K, Zhou Q, Lu C, Du L, Liu F. Mechanism of antifungal activity of Perilla frutescens essential oil against Aspergillus flavus by transcriptomic analysis. Food Control. 2021;123:107703. https://doi.org/10.1016/j.foodcont.2020.107703.

Huang C, Peng B. Photocatalytic degradation of patulin in apple juice based on nitrogen-doped chitosan-TiO2 nanocomposite prepared by a new approach. LWT. 2021;140:110726. https://doi.org/10.1016/j.lwt.2020.110726.

Iamanaka B, de Menezes H, Vicente E, Leite R, Taniwaki M. Aflatoxigenic fungi and aflatoxins occurrence in sultanas and dried figs commercialized in Brazil. Food Control. 2007;18(5):454–7. https://doi.org/10.1016/j.foodcont.2005.12.002.

IARC. Evaluation of carcinogenic risks to humans. 1993. https://publications.iarc.fr/Book-And-Report-Series/Iarc-Monographs-On-The-Identification-Of-Carcinogenic-Hazards-To-Humans. Accessed 15 Nov 2020.

IPCC. Cambio climático Informe de síntesis. Contribución de los Grupos de trabajo I, II y III al Quinto Informe de Evaluación del Grupo Intergubernamental de Expertos sobre el Cambio Climático. 2014. https://www.ipcc.ch/site/assets/uploads/2018/02/SYR_AR5_FINAL_full_es.pdf.

Janik E, Niemcewicz M, Ceremuga M, Stela M, Saluk-Bijak J, Siadkowski A, et al. Molecular aspects of mycotoxins-a serious problem for human health. Int J Mol Sci. 2020;21(21):8187. https://doi.org/10.3390/ijms21218187.

Juan C, Mannai A, Salem H, Oueslati S, Berrada H, Juan A, et al. Mycotoxins presence in pre- and post-fermented silage from Tunisia. Arab J Chem. 2020;13(8):6753–61. https://doi.org/10.1016/j.arabjc.2020.06.029.

Kaisermann A, Maron PA, Beaumelle L, Lata JC. Fungal communities are more sensitive indicators to non-extreme soil moisture variations than bacterial communities. Appl Soil Ecol. 2015;86:158–64. https://doi.org/10.1016/j.apsoil.2014.10.009.

Kalagatur NK, Gurunathan S, Kamasani JR, Gunti L, Kadirvelu K, Mohan CD, et al. Inhibitory effect of C. zeylanicum, C. longa, O. basilicum, Z. officinale, and C. martini essential oils on growth and ochratoxin A content of A. ochraceus and P. verrucosum in maize grains. Biotechnol Rep (Amst). 2020;27:e00490. https://doi.org/10.1016/j.btre.2020.e00490.

Kauserud H, Heegaard E, Halvorsen R, Boddy L, Høiland K, Stenseth NC. Mushroom's spore size and time of fruiting are strongly related: is moisture important? Biol Lett. 2011;7(2):273–6. https://doi.org/10.1098/rsbl.2010.0820.

Králová M, Frébortová J, Pěnčík A, Frébort I. Overexpression of Trp-related genes in Claviceps purpurea leading to increased ergot alkaloid production. New Biotechnol. 2021;61:69–79. https://doi.org/10.1016/j.nbt.2020.11.003.

Landa BB, Montes-Borrego M, Navas-Cortés JA. Use of PGPR for controlling soilborne fungal pathogens: assessing the factors influencing its efficacy. In: Maheshwari DK, editor. Bacteria in agrobiology: disease management. Berlin Heidelberg: Springer; 2013. p. 1–495. https://doi.org/10.1007/978-3-642-33639-3.

Laursen L. A preventable cancer. Nature. 2014;516:S2–3. https://doi.org/10.1038/516S2a.

Lee HS, Kwon N, Kim Y, Lee H. Prediction of mycotoxin risks due to climate change in Korea. Appl Biol Chem. 2018;61(4):389–96. https://doi.org/10.1007/s13765-018-0370-8.

Lemos AC, Borba VS, Badiale-Furlong E. The impact of wheat-based food processing on the level of trichothecenes and their modified forms. Trends Food Sci Technol. 2021;111:89–99. https://doi.org/10.1016/j.tifs.2021.02.056.

Liu Q, Li X, Wu R, Xiao X, Xing F. Development of an on-spot and rapid recombinase polymerase amplification assay for Aspergillus flavus detection in grains. Food Control. 2021;125 https://doi.org/10.1016/j.foodcont.2021.107957.

Luo Y, Liu X, Li J. Updating techniques on controlling mycotoxins—a review. Food Control. 2018;89:123–32. https://doi.org/10.1016/j.foodcont.2018.01.016.

Mamo FT, Abate BA, Tesfaye K, Nie C, Wang G, Liu Y. Mycotoxins in Ethiopia: a review on prevalence, economic and health impacts. Toxins. 2020;12(10):1–21. https://doi.org/10.3390/toxins12100648.

Manici LM, Bregaglio S, Fumagalli D, Donatelli M. Modelling soil borne fungal pathogens of arable crops under climate change. Int J Biometeorol. 2014;58(10):2071–83. https://doi.org/10.1007/s00484-014-0808-6.

Martínez H, Hernández S, Reyes C, Vázquez G. El género Aspergillus y sus micotoxinas en maíz en México problemática y perspectivas. Rev Mex Fitopatol. 2013;31(2):126–46.

Medina A, Akbar A, Baazeem A, Rodriguez A, Magan N. Climate change, food security and mycotoxins: do we know enough? Fungal Biol Rev. 2017;31(3):143–54.

Meerpoel C, Vidal A, Andjelkovic M, De Boevre M, Tangni EK, Huybrechts B, et al. Dietary exposure assessment and risk characterization of citrinin and ochratoxin a in Belgium. Food Chem Toxicol. 2021;147:111914. https://doi.org/10.1016/j.fct.2020.111914.

Moretti A, Pascale M, Logrieco A. Mycotoxin risks under a climate change scenario in Europe. Trends Food Sci Technol. 2019;84:38–40. https://doi.org/10.1016/j.tifs.2018.03.008.

Móricz ÁM, Fatér Z, Otta KH, Tyihák E, Mincsovics E. Overpressured layer chromatographic determination of aflatoxin B1, B2, G1 and G2 in red paprika. Microchem J. 2007;85:140–4. https://doi.org/10.1016/j.microc.2006.03.007.

Murphy P, Hendrich S, Landgren C, Bryant C. Food mycotoxins: an update. J Food Sci. 2006;71(5):R51–65. https://doi.org/10.1111/j.1750-3841.2006.00052.x.

Nesic K, Milicevic D, Nesic V, Ivanovic S. Mycotoxins as one of the foodborne risks most susceptible to climatic change. Procedia Food Sci. 2015;5:207–10. https://doi.org/10.1016/j.profoo.2015.09.058.

Omotayo OP, Omotayo AO, Mwanza M, Babalola OO. Prevalence of mycotoxins and their consequences on human health. Toxicol Res. 2019;35(1):1–7. https://doi.org/10.5487/TR.2019.35.1.001.

Ostry V, Dofkova M, Blahova J, Malir F, Kavrik R, Rehurkova I, et al. Dietary exposure assessment of sum deoxynivalenol forms, sum T-2/HT-2 toxins and zearalenone from cereal-based foods and beer. Food Chem Toxicol. 2020;139:111280. https://doi.org/10.1016/j.fct.2020.111280.

Park S, Lim W, You S, Song G. Ochratoxin A exerts neurotoxicity in human astrocytes through mitochondria-dependent apoptosis and intracellular calcium overload. Toxicol Lett. 2019;313:42–9. https://doi.org/10.1016/j.toxlet.2019.05.021.

Paterson RRM, Lima N. How will climate change affect mycotoxins in food? Food Res Int. 2010;43(7):1902–14. https://doi.org/10.1016/j.foodres.2009.07.010.

Peraica M. Mycotoxicoses. In: Environmental mycology in public health: fungi and mycotoxins risk assessment and management. Elsevier; 2015. p. 458. https://doi.org/10.1016/B978-0-12-411471-5.00005-3.

Peromingo B, Andrade MJ, Delgado J, Sánchez-Montero L, Núñez F. Biocontrol of aflatoxigenic aspergillus parasiticus by native Debaryomyces hansenii in dry-cured meat products. Food Microbiol. 2019;82:269–76. https://doi.org/10.1016/j.fm.2019.01.024.

Perrone G, Ferrara M, Medina A, Pascale M, Magan N. Toxigenic fungi and mycotoxins in a climate change scenario: ecology, genomics, distribution, prediction and prevention of the risk. Microorganisms. 2020;8(10):1496. https://doi.org/10.3390/microorganisms8101496.

Peter Mshelia L, Selamat J, Iskandar Putra Samsudin N, Rafii MY, Abdul Mutalib NA, Nordin N, et al. Effect of temperature, water activity and carbon dioxide on fungal growth and mycotoxin production of acclimatised isolates of fusarium verticillioides and F. graminearum. Toxins. 2020;12(8):478. https://doi.org/10.3390/toxins12080478.

Pinotti L, Ottoboni M, Giromini C, Dell'Orto V, Cheli F. Mycotoxin contamination in the EU feed supply chain: a focus on cereal byproducts. Toxins. 2016;8(2):45. https://doi.org/10.3390/toxins8020045.

Piontelli E. Agentes comunes en las aspergilosis humanas: conceptos primarios en la diferenciación de sus complejos de especies. Boletín Micológico. 2014;29(2):63–100. https://doi.org/10.22370/bolmicol.2014.29.2.868.

Pitt JI. Mycotoxins: Ochratoxin a. in Encyclopedia of food safety. Academic Press. 2014:304–9. https://doi.org/10.1016/B978-0-12-378612-8.00191-8.

Ramírez Quintero JD. Sobre el mal de los ardientes o del Fuego de San Antonio. Acta Médica Colombiana. 2018;43(3):156–60. https://doi.org/10.36104/amc.2018.975.

Ravelo Abreu A, Rubio Armendáriz C, Gutiérrez Fernández AJ, Hardisson de la Torre A. La ocratoxina a en alimentos de consumo humano. Revisión Nutricion Hosp. 2011;26(6):1215–26. https://doi.org/10.3305/nh.2011.26.6.5381.

Reyes Velázquez W, Patricio Martínez S, Isaías Espinosa VH, Nathal Vera MA, Lucas Palacios E, Rojo F. Aflatoxinas totales en raciones de bovinos y AFM1 en leche cruda obtenida en establos del estado de Jalisco. México Tecnica Pecuaria en Mexico. 2009;47(2):223–30.

Sacco C, Donato R, Zanella B, Pini G, Pettini L, Marino MF, et al. Mycotoxins and flours: effect of type of crop, organic production, packaging type on the recovery of fungal genus and mycotoxins. Int J Food Microbiol. 2020;334:108808. https://doi.org/10.1016/j.ijfoodmicro.2020.108808.

Sadhasivam S, Britzi M, Zakin V, Kostyukovsky M, Trostanetsky A, Quinn E, et al. Rapid detection and identification of mycotoxigenic fungi and mycotoxins in stored wheat grain. Toxins. 2017;9(10):302. https://doi.org/10.3390/toxins9100302.

Sadyś M, Kennedy R, West JS. Potential impact of climate change on fungal distributions: analysis of 2 years of contrasting weather in the UK. Aerobiologia. 2016;32(1):127–37. https://doi.org/10.1007/s10453-015-9402-6.

Sakin F, Tekeli İO, Yipel M, Kürekci C. Occurrence and health risk assessment of aflatoxins and ochratoxin a in Sürk, a Turkish dairy food, as studied by HPLC. Food Control. 2018;90:317–23. https://doi.org/10.1016/j.foodcont.2018.03.012.

Selim KM, El-hofy H, Khalil RH. The efficacy of three mycotoxin adsorbents to alleviate aflatoxin B1-induced toxicity in Oreochromis niloticus. Aquacult Int. 2014;22(2):523–40. https://doi.org/10.1007/s10499-013-9661-6.

Serrano-Coll HA, Cardona-Castro N. Micotoxicosis y micotoxinas: generalidades y aspectos básicos. Rev CES Med. 2015;29(1):143–52.

Shao J, Pei Z, Jing H, Wang L, Jiang C, Du X, et al. Physiological and molecular plant pathology antifungal activity of myriocin against fusarium graminearum and its inhibitory effect on deoxynivalenol production in wheat grains. Physiol Mol Plant Pathol. 2021;114:101635. https://doi.org/10.1016/j.pmpp.2021.101635.

Sirot V, Fremy JM, Leblanc JC. Dietary exposure to mycotoxins and health risk assessment in the second French total diet study. Food Chem Toxicol. 2013;52:1–11. https://doi.org/10.1016/j.fct.2012.10.036.

Valencia-Quintana R, Milić M, Jakšić D, Klarić MŠ, Tenorio-Arvide MG, Pérez-Flores GA, et al. Environment changes, aflatoxins, and health issues, a review. Int J Environ Res Pub Health. 2020;17(21):1–10. https://doi.org/10.3390/ijerph17217850.

van der Linde EJ, Pešicová K, Pažoutová S, Stodůlková E, Flieger M, Kolařík M. Ergot species of the Claviceps purpurea group from South Africa. Fungal Biol. 2016;120(8):917–30. https://doi.org/10.1016/j.funbio.2016.05.006.

von Hertwig AM, Iamanaka BT, Amorim Neto DP, Rezende JB, Martins LM, Taniwaki MH, et al. Interaction of Aspergillus flavus and A. parasiticus with Salmonella spp. isolated from peanuts. Int J Food Microbiol. 2020;328:108666. https://doi.org/10.1016/j.ijfoodmicro.2020.108666.

Wang C, Wang Y, Wang Y, Wang Z, Zhang L, Liang Y, et al. The ADP-ribosylation factor-like small GTPase FgArl1 participates in growth, pathogenicity and DON production in Fusarium graminearum. Fungal Biol. 2020;124(11):969–80. https://doi.org/10.1016/j.funbio.2020.08.004.

Warnatzsch EA, Reay DS, Camardo Leggieri M, Battilani P. Climate change impact on aflatoxin contamination risk in Malawi's maize crops. Front Sustainable Food Syst. 2020;4:1–13. https://doi.org/10.3389/fsufs.2020.591792.

WHO. Micotoxinas. Micotoxinas. https://www.who.int/es/news-room/fact-sheets/detail/mycotoxins 2020.
Williams KC, Blaney BJ, Peters RT. Pigs fed Fusarium-infected maize containing zearalenone and nivalenol with sweeteners and bentonite. Livestock Prod Sci. 1994;39(3):275–81. https://doi.org/10.1016/0301-6226(94)90207-0.
WMO. New climate predictions. https://public.wmo.int/en/media/press-release/new-climate-predictions-assess-global-temperatures-coming-five-years#:~:text=Geneva%2C 9 July 2020 – The predictions issued by the World 2020.
Xing M, Chen Y, Li B, Tian S. Characterization of a short-chain dehydrogenase/reductase and its function in patulin biodegradation in apple juice. Food Chem. 2021;348 https://doi.org/10.1016/j.foodchem.2021.129046.
Yazid SNE, Jinap S, Ismail SI, Magan N, Samsudin NIP. Phytopathogenic organisms and mycotoxigenic fungi: why do we control one and neglect the other? A biological control perspective in Malaysia. Compr Rev Food Sci Food. 2020;19(2):643–69. https://doi.org/10.1111/1541-4337.12541.
Yli-Mattila T, Ward TJ, O'Donnell K, Proctor RH, Burkin AA, Kononenko GP, Gavrilova OP, et al. Fusarium sibiricum sp. nov, a novel type A trichothecene-producing Fusarium from northern Asia closely related to F. sporotrichioides and F. langsethiae. Internat. J Food Microbiol. 2011;147(1):58–68. https://doi.org/10.1016/j.ijfoodmicro.2011.03.007.
Zhai S, Zhu Y, Feng P, Li M, Wang W, Yang L, et al. Ochratoxin A: its impact on poultry gut health and microbiota, an overview1. In: Poultry science. Poultry Science Association Inc.; 2021. https://doi.org/10.1016/j.psj.2021.101037.
Zhou S, Xu L, Kuang H, Xiao J, Xu C. Immunoassays for rapid mycotoxin detection: state of the art. Analyst. 2020;145(22):7088–102. https://doi.org/10.1039/d0an01408g.

Chapter 15
New Tools in Laboratory Diagnosis of Invasive Fungal Infections

Rosa Areli Martínez-Gamboa

Contents

Abbreviations

BDG	(1,3)-β-D-Glucan
CE	Conformité Européenne
CSF	Cerebrospinal fluid
EAPCRI	European *Aspergillus* PCR Initiative
EIA	Enzyme immunoassay

R. A. Martínez-Gamboa (✉)
Laboratorio de Microbiología Clínica, Instituto Nacional de Ciencias Médicas y Nutrición "Salvador Zubirán", CDMX, Mexico City, Mexico

M. G. Frías-De-León et al. (eds.), *The Impact of Climate Change on Fungal Diseases*, Fungal Biology, https://doi.org/10.1007/978-3-030-89664-5_15

EORTC/MSGERC	The European Organisation for Research and Treatment of Cancer and the Mycoses Study Group Education and Research Consortium
FDA	Food and Drug Administration
GM	Galactomannan
HIV	Human immunodeficiency virus
HRM	High resolution melting analysis
IA	Invasive aspergillosis
IC	Invasive candidiasis
IFIs	Invasive fungal infections
IHAG	International Histoplasmosis Advocacy Group
ITS	Internal transcribed spacer
MALDI-TOF MS	Matrix-assisted laser desorption/ionization time-of-flight mass spectrometry
NGS	Next-generation sequencing
PCP	*Pneumocystis jirovecii* pneumonia
PNA-FISH	In situ hybridization using fluorescently labeled nucleic acid probes
qPCR	Quantitative PCR amplification
SNPs	Single nucleotide polymorphisms
VOCs	Volatile organic compounds
WGS	Whole genome sequencing

15.1 Introduction

Fungi emerged about 1.6 million years ago, and it is considered that, on planet earth, there are approximately two million species, of which only a small percentage have been formally described. Only around 600 species have been described as responsible for any disease, and less than 30 of them are responsible for 99% of infections in humans (Leading International Fungal Education 2021).

Most fungi grow well at temperatures between 12 and 30 °C; however, there is a wide range of temperature tolerance among different species. Fungal infections comprise a wide variety of clinical presentations ranging from superficial to severe deep infections, which are characterized by the ability of the fungus to grow at human body temperature (thermotolerance at 37 °C). Most environmental fungi are not able to grow at this temperature, so it has been argued that human body temperature, as well as the immune system, provides resistance or protection against many ubiquitous fungi. Increased environmental temperature could mean an advantage in the selection of thermotolerant fungi existing in the environment and, therefore, lead to increased exposure of humans (Rickerts 2019).

Invasive fungal infections have become increasingly important with the emergence of immunosuppressive diseases and treatments. Among invasive fungal infections (IFIs), invasive candidiasis (IC) accounts for 70% of all IFIs, followed by cryptococcosis (20%) and aspergillosis (10%). However, we cannot ignore *P. jirovecii*, responsible for a considerable percentage of infections in patients with human immunodeficiency virus (HIV) infection, as well as endemic mycoses caused by *Histoplasma*, *Coccidioides*, *Paracoccidioides*, and *Blastomyces*, which also play an important role. Also, the recent increase in IFIs caused by *Zygomycetes*, *Scedosporium*, and *Fusarium* is a cause for concern (Ibáñez-Martínez et al. 2017).

Candida and *Pneumocystis* are commensal fungi that live on human skin and mucous membranes, and behave as opportunistic pathogens in immunosuppressed patients, while *Aspergillus*, *Cryptococcus*, *Histoplasma*, *Coccidioides*, *Paracoccidioides*, and *Blastomyces* are fungi that are mainly found in the environment; therefore, they are exposed to environmental and climatic factors to which they must adapt if these undergo changes. Significant changes in the epidemiology of systemic mycoses, possibly related to climate change, have been described in the last 20 years, for example, the simultaneous and independent appearance of *C. auris* on three continents; the appearance of *Cryptococcus gattii* as responsible for cryptococcosis in regions with warm climates where it had not previously been reported; the extension of endemic areas of *Histoplasma capsulatum*; the increased incidence of blastomycosis in humans and dogs due to selection of thermotolerant fungi in the environment; increased numbers of human cases of coccidioidomycosis following extreme weather conditions such as sandstorms resulting in increased release of infectious particles from the soil; and climatic variability correlating with the frequency of cases of paracoccidioidomycosis (Rickerts 2019).

Traditional diagnostic methods, such as culture and histopathology, are considered the reference standard for the diagnosis of IFI; however, they have low sensitivity and require prolonged time, even weeks, to observe development. On the other hand, tests such as galactomannan detection for the diagnosis of aspergillosis, (1,3)-β-D-glucan (BDG) for the detection of *Candida* spp. and other fungi, as well as antigen detection tests for *Cryptococcus* spp., *H. capsulatum*, and other dimorphic fungi have established themselves as important diagnostic approaches (Ibáñez-Martínez et al. 2017) and are increasingly being implemented in routine clinical practice given their high performance and the rapidity with which results are obtained. All these tests have enabled the diagnosis of specific diseases, for which there must be a high clinical suspicion of the organism causing the infection. However, with the emergence of new fungal species in the clinical setting, there is an urgent need for tools that allow the detection of a broad spectrum of unknown fungal species.

Next-generation sequencing (NGS) is a culture-free molecular tool that allows amplification of specific targets or whole genome sequencing (WGS). This tool has been widely used in the research field. The great advantage of WGS is that it can be applied directly to the patient's sample; allows the detection and identification not only of a particular or specific organism, but of many organisms, which may or may not be known; and allows the detection of virulence or resistance factors, giving all these the opportunity to perform microbiome analysis, taxonomic analysis, outbreak analysis, characterization, and surveillance of pathogens, all from the results of a single test (Deurenberg et al. 2017).

The following are the methods available for diagnostic use and some of those that are in more advanced stages of validation for the diagnosis of IFIs caused by the fungi currently known to be the main culprits of IFIs.

15.2 Laboratory Diagnosis of Invasive Fungal *Candida* Infection

Bloodstream infections caused by *Candida* spp. are serious and prevalent nosocomial infections worldwide, and despite the availability of antifungals, mortality reaches 30–70% in some settings, partly due to delayed diagnosis (Garey et al. 2006).

There are different tests available for the diagnosis of invasive fungal infections by *Candida* spp. (Table 15.1). Blood culture remains the reference or gold standard method for the diagnosis of candidemia despite its low sensitivity (50–60%) (Ibáñez-Martínez et al. 2017). Currently, there is a wide variety of commercial systems, manual and automated, for the identification of *Candida* spp. each with their own advantages and disadvantages (Neppelenbroek et al. 2014). However, among the most widely distributed are the automated VITEK 2 system (bioMérieux, Marcy d'Etoile, France), and the recently introduced matrix-assisted laser desorption/ionization time-of-flight mass spectrometry (MALDI-TOF MS). The VITEK 2 system allows the correct identification of the main *Candida* spp. responsible for IFIs (Ochiuzzia et al. 2014), as well as the determination of antifungal susceptibility

Table 15.1 Diagnostic tests for invasive fungal infection by *Candida* spp.

	Method	Samples	Results	Sensitivity	Specificity
Detection and identification directly from positive blood culture					
PNA-FISH	In situ hybridization using fluorescently labeled nucleic acid probes	Positive blood culture	90 min	92.3–100% depending on the species	94.8–100% depending on the species
FilmArray sepsis	Multiplex real-time PCR assay	Positive blood culture	70 min	100%	100%
MALDI-TOF MS	Mass *spectrometry*	Positive blood culture	30 min	96%	97.5–98%
Molecular tests directly from blood sample					
T2*Candida* Panel	T2 magnetic resonance	Blood	3–5 h	91.1%	99.4%
SeptiFast	Multiplex real-time PCR assay	Blood	3.5–5 h	43–95%	60–100%
MagicPlex Sepsis	Multiplex real-time PCR assay	Blood	3.5–5 h	37–65%	77–92%
SepsiTest	PCR and Sanger sequencing	Blood	8–12 h	21–85%	58–95%

(Borghi et al. 2010). On the other hand, MALDI-TOF MS is being widely used in the identification of microorganisms by protein analysis through the creation of a genus- and species-specific mass spectrum and has come to revolutionize identification, since results are obtained in just a few minutes, with high specificity. It has high percentages of correct identification (97.5%) when compared to nucleotide sequencing (Sendid et al. 2013). The identification directly from the positive blood culture represents a great advantage since it provides results 24–72 hours earlier than conventional subculture-based methods (Spanu et al. 2012). Unfortunately, the number of protein spectra contained in the MALDI-TOF databases is limited, so in many cases identification is not possible (Sendid et al. 2013).

Molecular methods for identification from positive blood culture include in situ hybridization using fluorescently labeled nucleic acid probes (PNA-FISH), which has allowed direct identification of the five most frequent *Candida* spp. in approximately 90 min. The assay has high sensitivity and specificity and is approved by the Food and Drug Administration (FDA) and the Conformité Européenne (CE) for diagnostic use (Stone et al. 2013). On the other hand, the FilmArray sepsis (bioMérieux, Marcy l'Etoile, France) is a system that also allows direct identification from yeast-positive blood culture by a multiplex polymerase chain reaction (PCR) that detects 24 etiological agents of sepsis. It requires 5 min of handling, and approximately 60 minutes to obtain results. This assay has a sensitivity and specificity of 100% for the *Candida* spp. included in the panel (Simor et al. 2018).

Regardless of the laboratory's identification system, the user must ensure that the laboratory can identify *C. auris*, which is a species capable of causing serious infections and hospital outbreaks, colonizing patients for many months, and persisting in the environment and is multidrug-resistant (Center for Disease Control and Prevention 2020).

The use of molecular techniques for DNA amplification directly from blood samples is the best option to achieve substantial reduction in turnaround time. Currently, there are many multiplex PCRs, several of which are CE labeled; however, only one—T2*Candida* Panel—has FDA approval. The T2*Candida* Panel is the first and only FDA-cleared assay to detect and differentiate the five most common *Candida* species directly from whole blood samples, providing results in as little as 3–5 h. The T2*Candida* is a fully automated multiplex PCR-based system and has an overall sensitivity greater than 90% (Pfaller et al. 2016). The LightCycler SeptiFast (Roche Molecular System, Germany), another automated multiplex PCR amplification system, can identify, directly from the blood sample, 25 microorganisms (including *Candida* spp. and *Aspergillus fumigatus*) responsible for approximately 90% of all bloodstream infections. Sensitivity and specificity for this assay are variable, probably because the patient groups studied are quite different (Opota et al. 2015). The Magicplex Sepsis real-time test (Seegene, Korea) detects more than 90 pathogens, including six fungi (*C. albicans*, *C. tropicalis*, *C. parapsilosis*, *C. glabrata*, *C. krusei*, and *A. fumigatus*). The sensitivity of this PCR is extremely low (37–65%), and the specificity is moderately acceptable (77–92%) (Opota et al. 2015). The SepsiTest (Molzym, Bremen, Germany) is a semiautomated broad-range PCR-based system using universal oligonucleotides targeting bacterial and fungal rRNA. The system requires several nonautomated processes, which delay the result and increase the risk of sample contamination, and its sensitivity and specificity are highly variable (Opota et al. 2015). The latter molecular methods require extensive validations in the clinical setting to determine the performance of each of them.

15.3 Laboratory Diagnosis of Invasive Pulmonary Aspergillosis

Aspergillosis constitutes a spectrum of diseases ranging from noninvasive forms to invasive pulmonary aspergillosis, with possible dissemination in immunocompromised hosts. Despite advances in the diagnosis and treatment of these infections, mortality rates remain high, especially in immunocompromised hosts (Denning 1996). Diagnosis of this infection remains a challenge and requires a high index of clinical suspicion.

There are different diagnostic tests for invasive pulmonary aspergillosis (Fig. 15.1). The gold standard in the laboratory diagnosis of invasive aspergillosis (IA) remains detection by culture or observation by histopathology; however, one test that has been critical is the detection of galactomannan (GM) (Donnelly et al. 2020). The GM antigen is a component of the cell wall of *Aspergillus*, so its detection has been quite useful for the diagnosis of invasive aspergillosis, although it is also a component found in the cell walls of *H. capsulatum* and *Fusarium* spp. (Tortorano et al. 2012).

The Platelia GM-EIA assay (Platelia *Aspergillus* EIA Bio-Rad) is an immunoassay that allows the detection of GM in both serum and bronchoalveolar lavage (BAL). The sensitivity and specificity of GM-EIA vary greatly depending on the

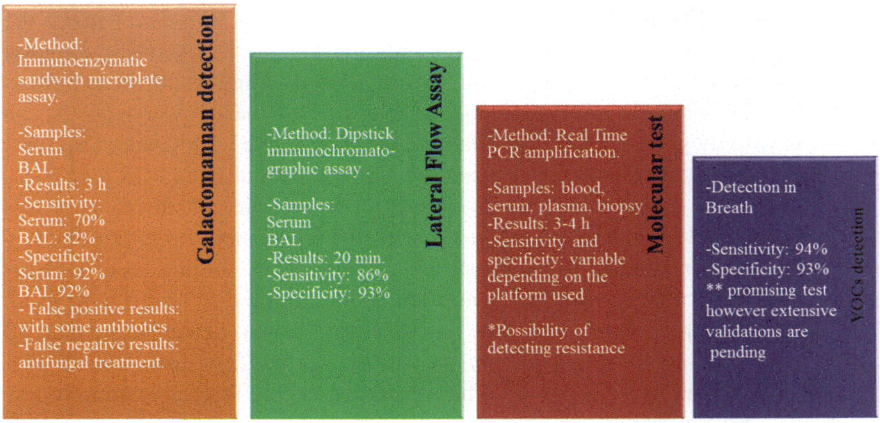

Fig. 15.1 Laboratory diagnosis of invasive pulmonary aspergillosis. *VOCs* volatile organic compounds, *BAL* bronchoalveolar lavage

population studied, being higher in patients with hematological malignancies. The determination in serum has a sensitivity of 70% and a specificity of 92%, while in BAL is 82% and 92%, respectively. However, when this test is performed in patients without neutropenia, the sensitivity can range from 44% to 90% (Cadena et al. 2016).

On the other hand, one of the recently introduced methods is the detection of *Aspergillus* by lateral flow tests, which have come to revolutionize the diagnosis of fungal infections, since the detection is performed directly on the clinical sample in just a few minutes, and in many cases, without the need for special laboratory equipment. This test consists of a sandwich immunochromatographic assay that uses monoclonal antibodies to detect the extracellular glycoprotein antigen or GM secreted by actively growing *Aspergillus*. It is designed in an individual format, which allows the sample to be analyzed at the time it arrives at the laboratory. This test has a sensitivity and specificity for proven/probable aspergillosis in solid organ transplant recipients of 86% (95% CI, 76–93%) and 93% (95% CI, 89–96%), respectively (Heldt and Hoenigl 2017).

PCR detection of *Aspergillus* is advised if the recommendations of the European *Aspergillus* PCR Initiative (EAPCRI) regarding the method of DNA extraction from blood, serum, and plasma samples are followed (White et al. 2010). There are multiple commercial PCR platforms for the detection of *Aspergillus* directly in clinical samples: MycoGENIE (Ademtech), AsperGenius (PathoNostics), Fungiplex (Renishaw), and SeptiFast (Roche), which are validated for blood, BAL, and even biopsy samples (MycoGENIE). Most of them have a standardized amplification system which, when combined with EAPCRI recommendations, provides a fully standardized approach (Ibáñez-Martínez et al. 2017). The AsperGenius real-time multiplex PCR can detect all clinically relevant *Aspergillus* species directly from BAL, serum, and plasma samples. In addition, it can detect, using fluorescent probes and melting curve analysis, the most prevalent mutations (TR_{34}/L98H and TR_{46}/Y121F/T289A) in the cyp51A gene that confer azole resistance. Its overall sensitivity, specificity, PPV, and NPV are 84.2%, 91.4%, 76.2%, and 94.6%, respectively (Chong et al. 2015). When GM and PCR tests are performed in patients at high risk for invasive aspergillosis (IA), and both are negative, the need for antifungal agents can be obviated with a negative predictive value of 100%, while the presence of at least two positive results is highly suggestive of active infection with a positive predictive value of 88% (Arvanitis et al. 2015).

Recently, interest has arisen in the measurement of volatile organic compounds (VOCs) in the breath of patients with suspected IA. This test has the great advantage of being noninvasive unlike tests that require samples such as BAL or biopsies.

A. fumigatus produces a wide variety of VOCs, among which, four (α-trans-bergamotene, β-trans-bergamotene, β-vatirenene, and trans-geranylacetone) have been able to differentiate patients with and without IA with 94% sensitivity and 93% specificity (Koo et al. 2014). Unfortunately, the number of studies related to VOCs is very scarce; therefore, extensive validations are still required to better understand the true performance of the test in different populations, to determine if there is any relationship between the production of these compounds and the size of lung lesions, to understand the interaction of these VOCs with antifungal agents or other drugs, as well as to identify the profile of these compounds to differentiate between colonized and infected patients.

15.4 Laboratory Diagnosis of *Cryptococcus* Infection

Cryptococcosis is an infection caused by *Cryptococcus neoformans* and *Cryptococcus gattii*. Cryptococcosis is responsible for 15% of deaths in HIV patients (Rajasingham et al. 2017). Early diagnosis of cryptococcal infection is critical to improve clinical outcomes. Culture, although the gold standard in diagnosis, has low sensitivity (Abassi et al. 2015). The recent introduction of tests such as antigen detection and DNA detection by PCR has greatly improved the diagnosis of this infection. The detection of *C. neoformans* and *C. gattii* capsular antigen in serum, plasma, blood, and cerebrospinal fluid (CSF) by lateral flow-based testing has greatly improved the diagnosis of cryptococcosis. It is an easy test, is quick to perform, and has demonstrated an overall sensitivity greater than 99% (Rajasingham et al. 2019).

Molecular detection of *Cryptococcus* directly from a CSF sample has also come to play an important role in the diagnosis of cryptococcosis. The FilmArray meningitis/encephalitis panel (bioMérieux, Marcy l'Etoile, France) consists of a multiplex PCR that detects not only *Cryptococcus* but also the 14 main pathogens (bacteria, viruses, and fungi) causing central nervous system infections. It has high sensitivity and specificity and is FDA and CE approved (Rhein et al. 2016).

15.5 Laboratory Diagnosis of *Pneumocystis Jirovecii* Infection

P. jirovecii pneumonia (PCP) is a life-threatening fungal infection in immunocompromised patients (Cordonnier et al. 2016). Among the most recently implemented laboratory tests for the diagnosis of PCP are BDG detection and DNA amplification by PCR.

(1,3)-β-D-Glucan is an important component of the fungal cell wall and is found in the serum of patients with different fungal infections, including IC, IA, invasive fusariosis, and *P. jirovecii* infection, so its detection is not specific to a particular infection (Yoshida et al. 1997). BDG detection is adequate to diagnose probable IFI, if it is performed considering the patient's risk factors. A cutoff point >80 pg/mL is recommended to determine positivity, provided it is performed with the Fungitell test (Associates of Cape Cod, Falmouth, MA). Confidence in obtaining true positive results increases with repeat positive tests and with values well above the threshold for positivity. However, test results should always be interpreted in the context of the clinical characteristics of the individual patient (Donnelly et al. 2020).

P. jirovecii can colonize the respiratory tract, so it can be detected by DNA amplification methods such as PCR (Maskell et al. 2003). To differentiate between colonization and pneumonia, quantitative PCR amplification (qPCR) is preferred, since the higher the fungal load, the higher the probability that the diagnosis is pneumonia. A CT value greater than 35 excludes PCP, while a CT value less than 31 excludes colonization with a sensitivity of 80% and a specificity of 80% in HIV-negative patients (Fauchier et al. 2016). However, further studies are required to validate cutoff points to differentiate colonization from disease.

Combined strategies, using qPCR and BDG detection, seem promising in the diagnosis of this disease, since these tests have been observed to have a high negative predictive value, resulting in the exclusion of PCP when both determinations are negative (Alanio and Bretagne 2014).

15.6 Laboratory Diagnosis of Disseminated Histoplasmosis

H. capsulatum infection mainly affects patients with human immunodeficiency virus (HIV) infection with low CD4+ cell counts (<150 cells/μL) in whom it commonly presents as disseminated histoplasmosis. These patients may die within a few weeks if not diagnosed promptly and treatment is not initiated early (Adenis et al. 2014).

Clinical diagnosis of histoplasmosis is challenging, especially in non-endemic areas, as the signs and symptoms are often nonspecific and may be indistinguishable from those of disseminated tuberculosis (Lofgren et al. 2012).

There are several recently introduced laboratory tests for the diagnosis of histoplasmosis (Fig. 15.2). The gold standard for diagnosis is isolation of the fungus by culture or observation of characteristic intracellular yeasts by histopathology. Unfortunately, the sensitivity of these is low, and culture requires several weeks of incubation and level 3 laboratory safety (Guimarães et al. 2006). Urine antigen detection tests have high sensitivity, specificity, and a remarkable negative predictive value, and results are available within hours.

Histoplasma antigen detection in urine using the clarus *Histoplasma* GM Enzyme Immunoassay kit (Immuno-Mycologics [IMMY], Norman, OK, USA), which uses monoclonal antibodies, has high sensitivity, specificity, and negative predictive value (98%, 97%, 100%, respectively) in patients with HIV and histoplasmosis. This kit has FDA and CE approval for in vitro diagnostic use. This test requires only approximately 3 h to perform (Cáceres et al. 2018; Martínez-Gamboa et al. 2021). On the other hand, MiraVista (MiraVista Diagnostic Laboratories) recently designed a lateral flow-based urine *Histoplasma* antigen detection test. This test has the great advantage of being presented in a single format, requiring less than 1 min to perform, and 40 min to read the result, requiring no special laboratory equipment or supplies, allowing it to truly function as a point-of-care test. The sensitivity and specificity of this test are greater than 90% (Cáceres et al. 2020; Martínez-Gamboa et al. 2021). In 2019, the *Histoplasma* antigen detection test was included in the

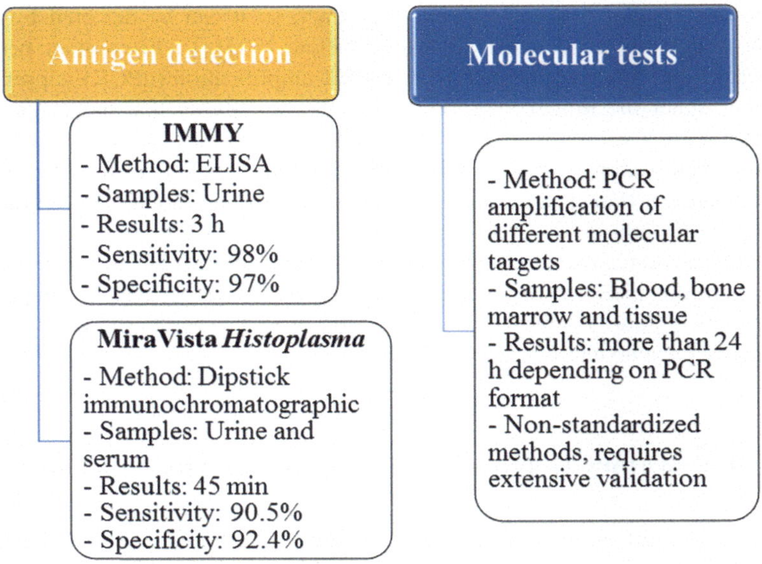

Fig. 15.2 Recently introduced tests for the diagnosis of histoplasmosis

second edition of the WHO list of essential diagnostics (WHO 2019). Likewise, the International Histoplasmosis Advocacy Group (IHAG) has as a goal for the year 2025 that at least one laboratory in each of the Latin American countries has available rapid tests (antigen detection or molecular test) for the diagnosis of histoplasmosis (Cáceres et al. 2019). Therefore, it is important that laboratories manage the inclusion of these tests in their diagnostic test portfolio.

For the molecular detection of *Histoplasma*, there are several PCRs which have used various molecular targets such as Hcp100, ITS, 18S, and 1281-1283$_{220}$SCAR; however, none of them is commercially available, because they are not standardized or approved, so extensive validation is still required to determine which is the best molecular target to amplify, which is the best method of DNA extraction to use, which is the sample with better yields, as well as which is the best PCR platform to implement (Martínez-Gamboa et al. 2021).

15.7 Laboratory Diagnosis of Blastomycosis

Culture remains the standard reference method for diagnosis of blastomycosis; however, the organism may take several days or weeks to grow and test sensitivity is low (Hage et al. 2019). Other tests currently used are antibody and antigen detection. Antibody detection has the disadvantage of false negative results in acutely infected and immunosuppressed patients. On the other hand, antigen detection has been the most widely used test. The sensitivity reported for this, in urine samples, ranges from 76% to 93% and in serum from 55.6% to 82%, which also depends on the clinical manifestation under study (Frost and Novicki 2015). Additionally, publications on new diagnostic tests for blastomycosis are scarce. The existing molecular tests are not commercially available and have not been validated (Sidamonidze et al. 2012).

15.8 Laboratory Diagnosis of Coccidioidomycosis

Coccidioidomycosis is endemic to the desert regions of the southwestern United States, particularly Arizona and California. Between 1998 and 2012, there was a significant increase in human cases of the disease in both states. The incidence can be as high as 200 cases per one million population (Herrmann et al. 2011). Diagnosis of the disease is often delayed due to the lack of rapid diagnostic tests (Grys et al. 2018). Microscopy and culture lack sensitivity. Serology is probably the most requested diagnostic test. Several methodologies are available, including the complement fixation test, the immunodiffusion (ID) test, and, more recently, the enzyme immunoassay (EIA). However, none of these are sufficiently sensitive and specific, so the results must be interpreted in the clinical context (Laniado-Laborín et al. 2019). The two tests currently available for antibody detection that have FDA approval for diagnostic use are Meridian-EIA (Meridian Premier-EIA) and IMMY-EIA (IMMY) (Grys et al. 2018). The diagnosis of this disease has also been innovated recently with the introduction of "point-of-care" tests (Donovan et al. 2020); however, further validation is required to know their actual performance.

15.9 Pan-Fungal PCR

Eukaryotic rRNA gene complex is composed of conserved and non-conserved regions that are phylogenetically informative. The 18S, 5.8S, and 28S ribosomal genes (conserved regions) are separated by internal transcribed spacer (ITS). The ITS and the D1/D2 region of the 28S subunit (non-conserved regions) are the most variable and phylogenetically informative. These sequences are considered the gold standard for fungal identification (Kidd et al. 2020; Perlin and Wiederhold 2017) and are the most widely used nowadays.

Amplification of these regions is pan-fungal, which allows the detection of most, if not all, fungi (including rare or new species) (Freeman Weiss et al. 2021). The amplification product is then subjected to sequencing, and the resulting sequence is compared with existing sequences in various databases (Prakash et al. 2017) to find homologies and arrive at identification. One of the largest and most widely used databases is GenBank which contains over 500,000 ITS and 28S sequence records and contains sequences of at least 5000 fungal species (Wickes and Wiederhold 2018).

Pan-fungal amplification can be performed conventionally by end-point PCR amplification with subsequent Sanger sequencing or by using NGS. On the other hand, amplification can be performed from the DNA of the clinical isolate, which allows the identification of the organism, or it can be performed directly from the clinical sample, which allows both the detection and identification of the fungus present in it. This strategy has been widely used on isolates and has allowed the classification and reclassification of several species, as well as the description of new ones (Li et al. 2020). However, pan-fungal PCR directly from clinical samples for diagnostic purposes is challenging because, if performed from a non-sterile sample, the amplification may represent the presence of commensal fungi of uncertain clinical significance (Zeller et al. 2017) or of environmental fungi that may have contaminated the sample. On the other hand, direct amplification from the clinical specimen is low, generally successful only in cases where the fungus is observed by histopathology. The European Organisation for Research and Treatment of Cancer and the Mycoses Study Group Education and Research Consortium (EORTC/MSGERC) recommends that to establish proven invasive fungal disease, amplification and sequencing should be performed only when the fungus is observed by histopathology (Donnelly et al. 2020). Pan-fungal amplification and WGS of specimens where no fungal structures are observed can help establish a positive diagnosis as long as clinical information supports the diagnosis.

The use of pan-fungal real-time PCR in combination with high resolution melting (HRM) analysis for subsequent species identification in 104 BAL samples from immunosuppressed patients (including 18 patients with proven or probable invasive fungal diseases) allowed the amplification and identification of *Candida* spp., *Aspergillus* spp., and *Fusarium* spp. However, it was not able to detect mucoromycetes. The sensitivity, specificity, positive predictive value, and negative predictive value of the method were 67%, 100%, 100%, and 94%, respectively. The positive detection of yeasts in three patients was probably due to BAL contamination of the oral cavity since none of these patients had evidence of a *Candida* spp. infection (Bezdicek et al. 2016).

It should be clarified that some genera require the amplification and sequencing of secondary or additional genes to achieve the correct discrimination between their species (Kidd et al. 2020). On the other hand, it should be kept in mind that although the databases have a large number of sequences, these only represent a minority of the millions of existing species on the planet, so it is not surprising that we can find isolates whose sequences have low levels of similarity with the existing ones, in which case, the suspicion of being in front of a new species is high.

15.10 Next-Generation Sequencing

NGS, also called "high-throughput sequencing" or second- and third-generation sequencing, is a set of genetic sequencing techniques that improve on the original Sanger sequencing process, and encompasses all technologies aimed at carrying out massive sequencing on a large scale.

There are several companies (Illumina, Thermo Fisher, Oxford Nanopore, Pacific Biosciences) that have multiple and varied NGS sequencing platforms, each with their own particularities. Whole genome sequencing is not performed in a single read, so before sequencing, the DNA (human and fungal) extracted directly from the clinical sample undergoes fragmentation. The common or second-generation platforms have a read capacity ranging between 100 and 1000 bases, whereas the recently introduced so-called third-generation platforms (MinION, Oxford Nanopore and Sequel, Pacific Biosciences) can generate fragment reads as large as 200 kb (Deurenberg et al. 2017). Read size is particularly important when sequencing, because the longer the sequence length, the larger also the genome fragment covered, and therefore there are fewer reads to overlap, which will facilitate assembly in subsequent bioinformatics analysis. The main weakness of nanopore sequencing is the error rate, which makes this technology less accurate than common WGS platforms (Lu et al. 2016).

NGS is slowly moving from the research field into clinical diagnostics. Although the costs of these technologies have decreased dramatically in the last decade, they are only available at a few reference centers. The University Medical Center Groningen in the Netherlands is one center that introduced NGS several years ago for routine diagnostics. Most applications are outbreak investigation and genotyping of highly resistant microorganisms. NGS is requested, through a general workflow by clinical microbiologists or infectious disease specialists in collaboration with molecular microbiologists and infection control professionals (Deurenberg et al. 2017).

In the field of mycology, NGS is playing an important role in the characterization of new pathogens, outbreak investigation, mycobiome analysis, and detection of new resistance genes.

15.10.1 Characterization of New Pathogens

Candida auris is a yeast first described in 2009 when it was recovered from the outer ear of a patient in Japan. This *Candida* is multidrug-resistant, can cause large hospital outbreaks, and thrives at high temperatures (40 °C). *C. auris* is an emerging fungus representing a global health problem as it is now distributed on all continents (Center for Disease Control and Prevention 2021). *C. auris* is one of the examples where WGS has been of great help in characterizing and understanding the emergence of a new fungal species that currently represents a global health problem. Chapter 10 presents detailed information regarding this microorganism.

15.10.2 Outbreak Characterization

WGS has also played an important role in identifying outbreaks, including cases of fungal infections caused by species without a previously established typing system. WGS was useful in detecting an outbreak of *Exserohilum rostratum*. *Exserohilum rostratum* is an ubiquitous, filamentous, thermotolerant fungus common in areas with abundant grasses, vegetables, and cereals. It can produce phaeohyphomycotic disease in both animals and humans, producing keratitis, arthritis, meningitis, and pulmonary or disseminated infection, especially in immunocompromised patients (Cruz-Choappa and Piontelli 2020). WGS was used to investigate an outbreak of fungal meningitis caused by this fungus following the use of a batch of contaminated methylprednisolone acetate injections. Analysis of 35 isolates (22 recovered from patients with meningitis or spinal abscesses, 6 recovered from contaminated methylprednisolone vials, and 7 isolates unrelated to the outbreak) revealed high clonality [only 8 single nucleotide polymorphisms (SNPs)] among the 28 isolates associated with the outbreak, while evidencing the existence of thousands of SNPs among the unrelated isolates, thus confirming the source of infection. This study confirmed the usefulness of WGS for the epidemiological investigation of fungal infections caused by phytopathogenic species that are rarely involved in cases of human infection (Litvintseva et al. 2014).

15.10.3 Resistance Mechanism Analysis

WGS analysis of five sequential *Aspergillus* isolates recovered from a patient with invasive pulmonary aspergillosis showed one mutation in 3/5 isolates, while 22 mutations were detected in 3/3 sequential isolates from a patient with aspergilloma, in one of which mutations were found in the *cyp51A*, *afyap1*, and *aldA* genes. WGS also revealed dynamic alterations (a large deletion, in a region containing 11 genes), in one isolate from the aspergilloma patient occurring in the *A. fumigatus* genome within its host during infection and treatment (Hagiwara et al. 2014).

WGS has been used to characterize strains with resistance to amphotericin B, a last-line antifungal for the treatment of severe IFIs. However, although the prevalence of resistance is relatively low, high rates of resistance have been described in some countries in recent years. The mechanism of action of this antifungal has not been fully elucidated. WGS analysis of 196 *A. fumigatus* isolates from 11 countries detected more than 400,000 SNPs and generated clustering of these isolates into three groups. Susceptibility data from 71/196 isolates showed that cluster 1 consisted of susceptible isolates only; cluster 2 consisted of susceptible isolates and all resistant isolates (except one), while cluster 3 consisted of susceptible isolates and one resistant isolate. Targeted analysis of 22 genes, previously recognized as involved in amphotericin B resistance in other fungi, showed 60 SNPs significantly associated with resistance to this antifungal. These SNPs represent promising candidates from which to investigate putative molecular mechanisms of amphotericin B resistance and may have potential use in the development of rapid diagnostic markers for the clinical detection of amphotericin B resistance in *A. fumigatus* (Fan et al. 2020).

15.10.4 Mycobiome

The study of the microbiota by conventional culture-based methods is limited because many fungal species cannot be cultured. NGS has been widely used to understand the relationship of the microbiome to health or disease. Fungi comprise only 0.1% of the total microbiome, yet they are believed to play a particularly important role in maintaining microbial communities and physiological processes in humans (Zoll et al. 2016).

Analysis by NGS of the respiratory tract microbiome of people with HIV, lung transplant recipients, and healthy individuals showed an increased representation of medically relevant organisms such as *Candida*, *Cryptococcus*, and *Aspergillus* in subjects with increasingly severe pulmonary and immune problems. NGS enabled the characterization of fungal communities in the human respiratory tract and allowed exploration of their interactions with bacterial communities in health and disease situations (Bittinger et al. 2014).

15.10.5 Diagnosis of Fungal Infections

Metagenomics is being introduced to the clinical laboratory for the diagnosis of fungal diseases. The use of MinION from Oxford Nanopore Technologies directly on BAL samples from patients with PCP allowed the detection of *P. jirovecii* and the exploration of the mycobiome present in the sample. Among the taxa detected were *Aspergillus* spp., *Candida* spp., *Lobosporangium* spp., and *Malassezia* spp., which are part of the microbiome of the respiratory tract and skin of healthy individuals, so their finding is probably due to contamination of the sample during collection (Irinyi et al. 2020).

The clinical impact of metagenomic next-generation sequencing (mNGS) of plasma cell-free DNA for the diagnosis of infectious diseases was evaluated in 82 samples from immunosuppressed patients. The positivity rate was 50/82 (61.0%), with 25 (50.0%) showing two or more organisms. Test results generated positive impact in 6 (7.3%), negative impact in 3 (3.7%), and no impact in 71 (86.6%) samples and were indeterminate in 2 (2.4%). In several samples, several organisms were identified that did not necessarily play a pathogenic role. The mean turnaround time from sample collection to result was 3.1 days (SD: 1.2), and the time from sample receipt to result was 1.5 days (SD: 0.9) for a subset of 20 samples. Cases with a positive result and clinical impact involved bacteria and/or fungi. In ten patients who underwent 16 additional repeat tests, only one was associated with clinical impact. The actual impact of this test as currently being used in clinical practice is limited. Further studies are needed to determine the patient populations in which this technology will have the greatest impact, as well as to define the complementary role of mNGS to conventional microbiological methods and how best to integrate mNGS into current testing algorithms (Hogan et al. 2021).

WGS appears to be a potentially useful diagnostic technique for cases where clinical suspicion of IFI is high. Currently, these tools are mainly available in reference laboratories, which may delay diagnosis. It is expected that, in the next few years, NGS will be available in most laboratories, and the turnaround time for the diagnosis of IFIs caused by already known species, and even those we do not yet know, will be drastically reduced. To achieve this, high-quality reference sequence bases, advances in instrumentation and bioinformatics, cost reduction, and the existence of external quality controls to validate the results are required.

15.11 Conclusion

IFIs have become more important in recent years. There is a change in the epidemiology of endemic invasive fungal infections, and new species have emerged as important pathogens worldwide, which has been argued to be a consequence of climate change. Currently, there are diagnostic tests that have been widely validated and approved for the clinical diagnosis of IFIs, which are undoubtedly playing a particularly important role in the clinic. Mass sequencing is a strategy that has been widely used in the field of microbiology for research purposes and has proven to have great advantages over traditional diagnostic methods. It is urgent that these tools be made available in clinical laboratories to perform a faster diagnosis and have the complete characterization of the microorganism from the results of a single test, in addition to having the possibility of detecting and characterizing new pathogens.

References

Abassi M, Boulware DR, Rhein J. Cryptococcal meningitis: diagnosis and management update. Curr Trop Med Rep. 2015;2(2):90–9. https://doi.org/10.1007/s40475-015-0046-y.

Adenis A, Nacher M, Hanf M, Vantilcke V, Boukhari R, Blachet D, et al. HIV-associated histoplasmosis early mortality and incidence trends: from neglect to priority. PLoS Negl Trop Dis. 2014;8(8):e3100. https://doi.org/10.1371/journal.pntd.0003100.

Alanio A, Bretagne S. Diagnosis of pneumocystis jirovecii pneumonia: role of β-D-glucan detection and PCR. Curr Fungal Infect Rep. 2014;8:322–30. https://doi.org/10.1007/s12281-014-0198-6.

Arvanitis M, Anagnostou T, Mylonakis E. Galactomannan and polymerase chain reaction-based screening for invasive aspergillosis among high-risk hematology patients: a diagnostic meta-analysis. Clin Infect Dis. 2015;61(8):1263–72. https://doi.org/10.1093/cid/civ555.

Bezdicek M, Lengerova M, Ricna D, Weinbergerova B, Kocmanova I, Volfova P, et al. Rapid detection of fungal pathogens in bronchoalveolar lavage samples using panfungal PCR combined with high resolution melting analysis. Med Mycol. 2016;54(7):714–24. https://doi.org/10.1093/mmy/myw032.

Bittinger K, Charlson ES, Loy E, Shirley DJ, Haas AR, Laughlin A, et al. Improved characterization of medically relevant fungi in the human respiratory tract using next-generation sequencing. Genome Biol. 2014;15(10):487. https://doi.org/10.1186/s13059-014-0487-y.

Borghi E, Iatta R, Sciota R, Biassoni C, Cuna T, Montagna MT, et al. Comparative evaluation of the Vitek 2 yeast susceptibility test and CLSI broth microdilution reference method for testing antifungal susceptibility of invasive fungal isolates in Italy: the GISIA3 study. J Clin Microbiol. 2010;48(9):3153–7. https://doi.org/10.1128/JCM.00952-10.

Cáceres DH, Adenis A, de Souza JVB, Gomez BL, Cruz KS, Pasqualotto AC, The International Histoplasmosis Advocacy Group (iHAG), et al. The Manaus declaration: current situation of histoplasmosis in the Americas, Report of the II Regional Meeting of the International Histoplasmosis Advocacy Group. Curr Fungal Infect Rep. 2019;13:244–9. https://doi.org/10.1007/s12281-019-00365-3.

Cáceres DH, Gómez BL, Tobón AM, Chiller TM, Lindsley MD. Evaluation of a Histoplasma antigen lateral flow assay for the rapid diagnosis of progressive disseminated histoplasmosis in Colombian patients with AIDS. Mycoses. 2020;63(2):139–44. https://doi.org/10.1111/myc.13023.

Cáceres DH, Samayoa BE, Medina NG, Tobón AM, Guzmán BJ, Mercado D, et al. Multicenter validation of commercial antigenuria reagents to diagnose progressive disseminated histoplasmosis in people living with HIV/AIDS in two Latin American countries. J Clin Microbiol. 2018;56(6):e01959–17. https://doi.org/10.1128/JCM.01959-17.

Cadena J, Thompson GR 3rd, Patterson TF. Invasive Aspergillosis: current strategies for diagnosis and management. Infect Dis Clin N Am. 2016;30(1):125–42. https://doi.org/10.1016/j.idc.2015.10.015.

Center for Disease Control and Prevention. https://www.cdc.gov/drugresistance/solutions-initiative/stories/cdc-response-to-global-threat.html. (2020). Accessed on Mar 24, 2021.

Center for Disease Control and Prevention. https://www.cdc.gov/fungal/candida-auris/tracking-c-auris.html#world. (2021). Accessed on Mar 24, 2021.

Chong GL, van de Sande WW, Dingemans GJ, Gaajetaan GR, Vonk AG, Hayette MP, et al. Validation of a new Aspergillus real-time PCR assay for direct detection of Aspergillus and azole resistance of Aspergillus fumigatus on bronchoalveolar lavage fluid. J Clin Microbiol. 2015;53(3):868–74. https://doi.org/10.1128/JCM.03216-14.

Cordonnier C, Cesaro S, Maschmeyer G, Einsele H, Donnelly JP, Alanio A, et al. Fifth European Conference on Infections in Leukemia (ECIL-5), a joint venture of the European Group for Blood and Marrow Transplantation (EBMT), the European Organization for Research and Treatment of Cancer (EORTC), the Immunocompromised Host Society (ICHS) and the European LeukemiaNet (ELN). Pneumocystis jirovecii pneumonia: still a concern in patients with haematological malignancies and stem cell transplant recipients. J Antimicrob Chemother. 2016;71(9):2379–85. https://doi.org/10.1093/jac/dkw155.

Cruz-Choappa R, Piontelli E. Exserohilum rostratum (Drechsler) Leonard y Suggs. Rev Chil Infectol. 2020;37(1):45–6. https://doi.org/10.4067/S0716-10182020000100045.

Denning DW. Therapeutic outcome in invasive aspergillosis. Clin Infect Dis. 1996;23:608–15. https://doi.org/10.1093/clinids/23.3.608.

Deurenberg RH, Bathoorn E, Chlebowicz MA, Couto N, Ferdous M, García-Cobos S, et al. Application of next generation sequencing in clinical microbiology and infection prevention. J Biotechnol. 2017;243:16–24. https://doi.org/10.1016/j.jbiotec.2016.12.022.

Donnelly JP, Chen SC, Kauffman CA, Steinbach WJ, Baddley JW, Verweij PE, et al. Revision and update of the consensus definitions of invasive fungal disease from the European Organization for Research and Treatment of Cancer and the Mycoses Study Group Education and Research Consortium. Clin Infect Dis. 2020;71(6):1367–76. https://doi.org/10.1093/cid/ciz1008.

Donovan FM, Ramadan FA, Khan SA, Bhaskara A, Lainhart WD, Narang AT, et al. Comparison of a novel rapid lateral flow assay to enzyme immunoassay results for early diagnosis of coccidioidomycosis. Clin Infect Dis. 2020:ciaa1205. https://doi.org/10.1093/cid/ciaa1205.

Fan Y, Wang Y, Xu J. Comparative genome sequence analyses of geographic samples of aspergillus fumigatus—relevance for amphotericin B resistance. Microorganisms. 2020;8(11):1673. https://doi.org/10.3390/microorganisms8111673.

Fauchier T, Hasseine L, Gari-Toussaint M, Casanova V, Marty PM, Pomares C. Detection of Pneumocystis jirovecii by quantitative PCR to Differentiate Colonization and Pneumonia in immunocompromised HIV-positive and HIV-negative patients. J Clin Microbiol. 2016;54(6):1487–95. https://doi.org/10.1128/JCM.03174-15.

Freeman Weiss Z, Leon A, Koo S. The evolving landscape of fungal diagnostics, current and emerging microbiological approaches. J Fungi (Basel). 2021;7(2):127. https://doi.org/10.3390/jof7020127.

Frost HM, Novicki TJ. Blastomyces antigen detection for diagnosis and management of blastomycosis. J Clin Microbiol. 2015;53(11):3660–2. https://doi.org/10.1128/JCM.02352-15.

Garey KW, Rege M, Pai MP, Mingo DE, Suda KJ, Turpin RS, et al. Time to initiation of fluconazole therapy impacts mortality in patients with candidemia: a multi-institutional study. Clin Infect Dis. 2006;43(1):25–31. https://doi.org/10.1086/504810.

Grys TE, Brighton A, Chang YH, Liesman R, Bolster LaSalle C, Blair JE. Comparison of two FDA-cleared EIA assays for the detection of Coccidioides antibodies against a composite clinical standard. Med Mycol. 2018; https://doi.org/10.1093/mmy/myy094.

Guimarães AJ, Nosanchuk JD, Zancopé-Oliveira RM. Diagnosis of histoplasmosis. Braz. J Microbiol. 2006;37(1):1–13. https://doi.org/10.1590/S1517-83822006000100001.

Hage CA, Carmona EM, Epelbaum O, Evans SE, Gabe LM, Haydour Q, et al. Microbiological laboratory testing in the diagnosis of fungal infections in pulmonary and critical care practice. An Official American Thoracic Society Clinical Practice Guideline. Am J Respir Crit Care Med. 2019;200(5):535–50. https://doi.org/10.1164/rccm.201906-1185ST. Erratum in: Am J Respir Crit Care Med 2019;200(10):1326

Hagiwara D, Takahashi H, Watanabe A, Takahashi-Nakaguchi A, Kawamoto S, Kamei K, et al. Whole-genome comparison of Aspergillus fumigatus strains serially isolated from patients with aspergillosis. J Clin Microbiol. 2014;52(12):4202–9. https://doi.org/10.1128/JCM.01105-14.

Heldt S, Hoenigl M. Lateral flow assays for the diagnosis of invasive aspergillosis: current status. Curr Fungal Infect Rep. 2017;11(2):45–51. https://doi.org/10.1007/s12281-017-0275-8.

Herrmann JA, Kostiuk SL, Dworkin MS, Johnson YJ. Temporal and spatial distribution of blastomycosis cases among humans and dogs in Illinois (2001–2007). J Am Vet Med Assoc. 2011;239:335–43. https://doi.org/10.2460/javma.239.3.335.

Hogan CA, Yang S, Garner OB, Green DA, Gomez CA, Dien Bard J, et al. Clinical impact of metagenomic next-generation sequencing of plasma cell-free DNA for the diagnosis of infectious diseases: a multicenter retrospective cohort study. Clin Infect Dis. 2021;72(2):239–45. https://doi.org/10.1093/cid/ciaa035.

Ibáñez-Martínez E, Ruiz-Gaitán A, Pemán-García J. Update on the diagnosis of invasive fungal infection. Rev Esp Quimioter. 2017;30(Suppl 1):16–21.

Irinyi L, Hu Y, Hoang MTV, Pasic L, Halliday C, Jayawardena M, et al. Long-read sequencing based clinical metagenomics for the detection and confirmation of Pneumocystis jirovecii directly from clinical specimens: a paradigm shift in mycological diagnostics. Med Mycol. 2020;58(5):650–60. https://doi.org/10.1093/mmy/myz109.

Kidd SE, Chen SC, Meyer W, Halliday CL. A new age in molecular diagnostics for invasive fungal disease: are we ready? Front Microbiol. 2020;10:2903. https://doi.org/10.3389/fmicb.2019.02903.

Koo S, Thomas HR, Daniels SD, Lynch RC, Fortier SM, Shea MM, et al. A breath fungal secondary metabolite signature to diagnose invasive aspergillosis. Clin Infect Dis. 2014;59(12):1733–40. https://doi.org/10.1093/cid/ciu725.

Laniado-Laborín R, Arathoon EG, Canteros C, Muñiz-Salazar R, Rendon A. Coccidioidomycosis in Latin America. Med Mycol. 2019;57(Supplement_1):S46–55. https://doi.org/10.1093/mmy/myy037.

Leading International Fungal Education (LIFE). http://life-worldwide.org/fungal-diseases. Accessed on Mar 24, 2021.

Li AH, Yuan FX, Groenewald M, Bensch K, Yurkov AM, Li K, et al. Diversity and phylogeny of basidiomycetous yeasts from plant leaves and soil: proposal of two new orders, three new families, eight new genera and one hundred and seven new species. Stud Mycol. 2020;96:17–140. https://doi.org/10.1016/j.simyco.2020.01.002.

Litvintseva AP, Hurst S, Gade L, Frace MA, Hilsabeck R, Schupp JM, et al. Whole-genome analysis of Exserohilum rostratum from an outbreak of fungal meningitis and other infections. J Clin Microbiol. 2014;52(9):3216–22. https://doi.org/10.1128/JCM.00936-14.

Lofgren SM, Kirsch EJ, Maro VP, Morrissey AB, Msuya LJ, Kinabo GD, et al. Histoplasmosis among hospitalized febrile patients in northern Tanzania. Trans R Soc Trop Med Hyg. 2012;106(8):504–7. https://doi.org/10.1016/j.trstmh.2012.05.009.

Lu H, Giordano F, Ning Z. Oxford nanopore MinION sequencing and genome assembly. Genomics Proteomics Bioinformatics. 2016;14(5):265–79. https://doi.org/10.1016/j.gpb.2016.05.004.

Martínez-Gamboa A, Niembro-Ortega MD, Torres-González P, Santiago-Cruz J, Velázquez-Zavala NG, Rangel-Cordero A, et al. Diagnostic accuracy of antigen detection in urine and molecular assays testing in different clinical samples for the diagnosis of progressive disseminated histoplasmosis in patients living with HIV/AIDS: a prospective multicenter study in Mexico. PLoS Negl Trop Dis. 2021;15(3):e0009215. https://doi.org/10.1371/journal.pntd.0009215.

Maskell NA, Waine DJ, Lindley A, Pepperell JC, Wakefield AE, Miller RF, et al. Asymptomatic carriage of pneumocystis jirovecii in subjects undergoing bronchoscopy: a prospective study. Thorax. 2003;58(7):594–7. https://doi.org/10.1136/thorax.58.7.594.

Neppelenbroek KH, Seó RS, Urban VM, Silva S, Dovigo LN, Jorge JH, et al. Identification of Candida species in the clinical laboratory: a review of conventional, commercial, and molecular techniques. Oral Dis. 2014;20(4):329–44. https://doi.org/10.1111/odi.12123.

Ochiuzzia ME, Cataldib S, Guelfandb L, Maldonadob I, Arechavalab A. Evaluación del sistema Vitek 2 para la identificación de las principales especies de levaduras del género Candida. Rev Arg Microbiol. 2014;46(2):107–10. https://doi.org/10.1016/S0325-7541(14)70057-4.

Opota O, Jaton K, Greub G. Microbial diagnosis of bloodstream infection: towards molecular diagnosis directly from blood. Clin Microbiol Infect. 2015;21(4):323–31. https://doi.org/10.1016/j.cmi.2015.02.005.

Perlin DS, Wiederhold NP. Culture-independent molecular methods for detection of antifungal resistance mechanisms and fungal identification. J Infect Dis. 2017;216(suppl_3):S458–65. https://doi.org/10.1093/infdis/jix121.

Pfaller MA, Wolk DM, Lowery TJ. T2MR and T2Candida: novel technology for the rapid diagnosis of candidemia and invasive candidiasis. Future Microbiol. 2016;11(1):103–17. https://doi.org/10.2217/fmb.15.111.

Prakash PY, Irinyi L, Halliday C, Chen S, Robert V, Meyer W. Online databases for taxonomy and identification of pathogenic fungi and proposal for a cloud-based dynamic data network platform. J Clin Microbiol. 2017;55(4):1011–24. https://doi.org/10.1128/JCM.02084-16.

Rajasingham R, Smith RM, Park BJ, Jarvis JN, Govender NP, Chiller TM, et al. Global burden of disease of HIV-associated cryptococcal meningitis: an updated analysis. Lancet Infect Dis. 2017;17(8):873–81. https://doi.org/10.1016/S1473-3099(17)30243-8.

Rajasingham R, Wake RM, Beyene T, Katende A, Letang E, Boulware DR. Cryptococcal meningitis diagnostics and screening in the era of point-of-care laboratory testing. J Clin Microbiol. 2019;57(1):e01238-18. https://doi.org/10.1128/JCM.01238-18.

Rhein J, Bahr NC, Hemmert AC, Cloud JL, Bellamkonda S, Oswald C, et al. ASTRO-CM team. Diagnostic performance of a multiplex PCR assay for meningitis in an HIV-infected population in Uganda. Diagn Microbiol Infect Dis. 2016;84(3):268–73. https://doi.org/10.1016/j.diagmicrobio.2015.11.017.

Rickerts V. Klimawandel und epidemiologie systemischer pilzinfektionen [climate change and systemic fungal infections]. Bundesgesundheitsblatt Gesundheitsforschung Gesundheitsschutz. 2019;62(5):646–51. https://doi.org/10.1007/s00103-019-02931-z.

Sendid B, Ducoroy P, François N, Lucchi G, Spinali S, Vagner O, et al. Evaluation of MALDI-TOF mass spectrometry for the identification of medically-important yeasts in the clinical laboratories of Dijon and Lille hospitals. Med Mycol. 2013;51(1):25–32. https://doi.org/10.3109/13693786.2012.693631.

Sidamonidze K, Peck MK, Perez M, Baumgardner D, Smith G, Chaturvedi V, et al. Real-time PCR assay for identification of Blastomyces dermatitidis in culture and in tissue. J Clin Microbiol. 2012;50(5):1783–6. https://doi.org/10.1128/JCM.00310-12.

Simor AE, Porter V, Mubareka S, Chouinard M, Katz K, Vermeiren C, Fattouh R, et al. Rapid identification of Candida species from positive blood cultures by use of the film array blood culture identification panel. J Clin Microbiol. 2018;56(12):e01387–18. https://doi.org/10.1128/JCM.01387-18.

Spanu T, Posteraro B, Fiori B, D'Inzeo T, Campoli S, Ruggeri A, et al. Direct MALDI-TOF mass spectrometry assay of blood culture broths for rapid identification of Candida species causing bloodstream infections: an observational study in two large microbiology laboratories. J Clin Microbiol. 2012;50(1):176–9. https://doi.org/10.1128/JCM.05742-11.

Stone NR, Gorton RL, Barker K, Ramnarain P, Kibbler CC. Evaluation of PNA-FISH yeast traffic light for rapid identification of yeast directly from positive blood cultures and assessment of clinical impact. J Clin Microbiol. 2013;51(4):1301–2. https://doi.org/10.1128/JCM.00028-13.

Tortorano AM, Esposto MC, Prigitano A, Grancini A, Ossi C, Cavanna C, et al. Cross-reactivity of Fusarium spp. in the Aspergillus Galactomannan enzyme-linked immunosorbent assay. J Clin Microbiol. 2012;50(3):1051–3. https://doi.org/10.1128/JCM.05946-11.

White PL, Bretagne S, Klingspor L, Melchers WJ, McCulloch E, Schulz B, et al. European Aspergillus PCR Initiative. Aspergillus PCR: one step closer to standardization. J Clin Microbiol. 2010;48(4):1231–40. https://doi.org/10.1128/JCM.01767-09.

Second WHO model list of essential in vitro diagnostics. World Health Organization. 2019. https://www.who.int/medical_devices/publications/Standalone_document_v8.pdf Accessed on 24, 2021.

Wickes BL, Wiederhold NP. Molecular diagnostics in medical mycology. Nat Commun. 2018;9(1):5135. https://doi.org/10.1038/s41467-018-07556-5.

Yoshida M, Obayashi T, Iwama A, Ito M, Tsunoda S, Suzuki T, et al. Detection of plasma (1 --> 3)-beta-D-glucan in patients with fusarium, trichosporon, Saccharomyces and Acremonium fungaemias. J Med Vet Mycol. 1997;35(5):371–4.

Zeller I, Schabereiter-Gurtner C, Mihalits V, Selitsch B, Barousch W, Hirschl AM, et al. Detection of fungal pathogens by a new broad range real-time PCR assay targeting the fungal ITS2 region. J Med Microbiol. 2017;66(10):1383–92. https://doi.org/10.1099/jmm.0.000575.

Zoll J, Snelders E, Verweij PE, Melchers WJ. Next-generation sequencing in the mycology lab. Curr Fungal Infect Rep. 2016;10:37–42. https://doi.org/10.1007/s12281-016-0253-6.

Chapter 16
Natural Antifungal Products: Another Option for Antifungal Resistance

Rigoberto Hernández-Castro and Juan Xicohtencatl-Cortes

Contents

16.1 Introduction

An important element associated with climate change is the redistribution and colonization of microorganisms, specifically pathogenic fungi in geographic areas where they had not been previously reported. Climate change has generated an increase in temperature, as well as an increase in rainfall and relative humidity in various geographical areas; however, in other countries, it has caused long periods of droughts, which allows the more efficient development of fungi in new habitats. Likewise, the increase in resistance is a feature that further complicates this picture globally. An increased frequency of fungal infections, associated with the

R. Hernández-Castro (✉)
Departamento de Ecología de Agentes Patógenos, Hospital General Dr. Manuel Gea González, CDMX, Mexico City, Mexico
e-mail: rhc@unam.mx

J. Xicohtencatl-Cortes
Laboratorio de Bacteriología Intestinal, Hospital Infantil de México Dr. Federico Gómez, CDMX, Mexico City, Mexico

M. G. Frías-De-León et al. (eds.), *The Impact of Climate Change on Fungal Diseases*, Fungal Biology, https://doi.org/10.1007/978-3-030-89664-5_16

widespread use of antifungals, has resulted in the development of antifungal resistance. Levels of resistance have increased significantly in the last decade, which has important implications for morbidity, mortality, as well as medical care in the community.

Mammals and fungi share some of the same molecular processes; this similarity is the main reason for the high toxicity associated with antifungal therapy (Ostrosky-Zeichner et al. 2010; Pierce et al. 2013). So it's necessary to look for new, safer, and more powerful alternative therapies to combat fungal infections. There are numerous reports on the antifungal activity of various plant extracts, particularly aromatic components. Most of it has been made using crude extracts and active components that are not clearly identified or well characterized. However, in some cases, purified components responsible for the activity have been identified and studied (Lopes et al. 2017).

The therapies commonly called alternative, or complementary, have been used for years, where species of plants with medicinal properties have been investigated to evaluate the viability, sustainability, and affordability of the use of natural medicines. Plants of terrestrial or marine origin represent an inexhaustible source of bioactive molecules, as well as their volatile and nonvolatile extracts, which are also a valuable source of therapeutic compounds that are candidates for antifungal therapies (Orafidiya et al. 2001; Rojas et al. 2006; Schmidt et al. 2008).

The importance of natural products in the search and development of new therapeutic tools is evident. In this sense, medicinal plants and their derivatives are an important piece for pharmacological research and drug development. These natural products can be used directly as therapeutic agents, as well as a source of raw materials for synthesis, or they can serve as prototypes for new pharmacologically active models. In this context, some progress has been made in the development of pharmaceutical formulations capable of transporting and releasing bioactive compounds in infected tissues (Scorzoni et al. 2016; Soares et al. 2014; Wilson and Danishefsky 2006).

Phenolic compounds and alkaloids represent the main classes of bioactive antifungal compounds isolated from nonvolatile natural extracts, while monoterpenes and sesquiterpenes are the main compounds responsible for the antifungal activity of natural volatile extracts (essential oils) (Negri et al. 2014).

16.2 Nonvolatile Natural Extracts

16.2.1 Phenolic Compounds

Phenolic compounds are a group of secondary metabolites with a central characteristic, the presence of at least one hydroxyl group associated with an aromatic ring in its structure. These secondary metabolites are widely distributed in nature and most abundantly in plants, where they are normally found as esters or glycosides. Phenolic compounds are extremely diverse, both in terms of chemical structure and biological functions. Those with the most promising anti-dermatophytic activity come from natural sources and include lignans and neolignans, coumarins, flavonoids, quinones, and tannins (Butts et al. 2017; Lopes et al. 2017; Vermerris and Nicholson 2008).

Among the phenolic compounds, thymol is the main component of thyme extracts, and it has been shown to have the ability to weaken the integrity of the cell membrane, as well as to decrease the ergosterol content and leave it sensitive to a wide variety of fungi, pathogens. In addition to this, thymol can generate synergy with amphotericin B, fluconazole, and itraconazole in *C. albicans* and *C. neoformans* (Faria et al. 2011; Guo et al. 2009).

Carvacrol, also found in thyme extracts, shares the same chemosensitization antimycotic mechanism (Pinto et al. 2006). Eugenol and methyleugenol, which are present in high levels in basil and clove extracts, are also associated with the interruption of ergosterol biosynthesis, as well as generating synergy with fluconazole against fluconazole resistant and sensitive *C. albicans* strains. In addition to the mechanism of action against the cell membrane, thymol and eugenol have been shown to inhibit plasma membrane ATPase resulting in acidification and adverse effect against fungi (Ahmad et al. 2010a, 2010b).

Other phenolic compounds found in plant extracts include berberine, curcumin, and dihydrobenzaldehydes. All three molecules enhance the activity of fluconazole against *C. albicans* and *C. neoformans* by inhibiting the response to oxidative stress. Interestingly, another phenolic compound such as cinnamaldehyde, which is structurally similar, acts through a different mechanism; in vitro studies have shown that it inhibits *Saccharomyces cerevisiae* b-(1,3)-glucan synthase and chitin synthase (Bang et al. 2000; Faria et al. 2011; Iwazaki et al. 2010; Sharma et al. 2010).

Allicin, a component of garlic extract, can synergize with amphotericin B and fluconazole against *C. albicans*, where the main mechanism of action appears to be disruption of vacuolar function, but only when ergosterol is present in the cell membrane (Borjihan et al. 2009; Guo et al. 2010). Epigallocatechin-O-gallate, found in black tea, is a dihydrofolate reductase inhibitor that in turn inhibits ergosterol biosynthesis. In combination with amphotericin B, it improved antifungal activity and has been shown to improve outcome in a murine model of systemic candidiasis (Han 2007).

Tannins and salicylic acid are polyphenol molecules extracted from *Gaultheria procumbens*, *Rhamnus purshiana*, and *Anacardium pulsatilla*, which showed high antifungal activity. Another phenolic compound such as the geranilated biphenyl 3-hydroxy-4-geranyl-5-methoxybiphenyl derived from the green fruits of purple mangosteen (*Garcinia mangostana*) presented a strong antifungal action and a series of biological activities against fungi (Dharmaratne et al. 2005). Another example is the polyisoprenylated benzophenone from an ethanol extract of Cuban propolis that showed significant antimicrobial and antifungal activity against a wide variety of bacteria and yeasts (Rubio et al. 1999). Likewise, another phenolic compound is 4-hydroxyphenyl-6-O-[(3R)-3,4-dihydroxy-2-methylenebutanoyl]-D-glucopyranoside isolated from the foliage of *Toronia toru*, with high antimicrobial activity (Perry and Brennan 1997). Some of the prenylindoles isolated from *Monodora angolensis* and *Isolona cauliflora* exhibited antifungal and antimalarial activities (Nkunya et al. 2004). Finally, eriosemaones A–D also showed good antifungal activity (Ma et al. 1995) (Table 16.1).

Table 16.1 Plant extracts used as antifungal treatment

Extracts	Compound class	Main element	Antifungal mechanism	Fungus type	Reference
Nonvolatile	Phenolic	Lignans, neolignans, coumarins, flavonoids, quinones, and tannins	Interruption of ergosterol biosynthesis and plasma membrane ATPase Inhibition of response to oxidative stress and b-(1,3)-glucan synthase and chitin synthase	*C. albicans*, *C. neoformans*, and *Saccharomyces cerevisiae*	Guo et al. (2009), Faria et al. (2011), Ahmad et al. (2010a, 2010b), Sharma et al. (2010), Bang et al. (2000)
	Alkaloids	Cycloanine and piperidine	Interruption of ergosterol biosynthesis	*Microsporum canis*, *T. longifusum*, *T. mentagrophytes*, *A. flavus*, *F. solani*, and *Candida* spp.	Lohombo-Ekomba et al. (2004), Ferheen et al. (2005), Xiao et al. (2014)
Volatile	Monoterpenes	Thymol, carvacrol, gerianol, and citronellol	Interruption of ergosterol biosynthesis Inhibition of synthesis of DNA, RNA, proteins, and polysaccharides	*C. albicans*, *C. glabrata*, *C. tropicalis*, *C. neoformans*, *Fusarium graminearum*, *A. fumigatus*, *M. canis*, *T. rubrum*, and *T. mentagrophytes*	Bajpai et al. (2009), Danielli et al. (2013), Miron et al. (2014)
	Sesquiterpenes	Farnesol	Interruption of ergosterol biosynthesis Inhibition action on membrane ATPases	*T. mentagrophytes*, *C. albicans*, and *C. neoformans*	Roemer and Krysan (2014), Mahboubi and Kazempour (2015), Moraes et al. (2015), Pippi et al. (2015)

16.2.2 Alkaloids

Alkaloids are organic nitrogenous compounds derived from amino acids and have unique pharmacological properties at low doses. Several reports on the antifungal activity of alkaloids against a wide variety of fungal species have been reported, most in relation to pathogenic yeast strains (Cordell 2009; Emile et al. 2007; Meng et al. 2009; Singh et al. 2000). *Albertisia villosa*, a subtropical medicinal plant widely used in African medicine, contains three alkaloids of the bisbenzyl isoquinoline type: cycloanine, cocsolin, and N-desmethylcycloanine, cycloanine being the most abundant with 85% and the one that showed the best activity of the crude extract against *Microsporum canis*. Likewise, it was used as a hair tonic with good results, controlling dandruff in 90% and preventing alopecia in 80% (Lohombo-Ekomba et al. 2004).

Two new piperidine alkaloids, haloxyline A and B, isolated from the chloroform-soluble fraction of *Haloxylon salicornicum*, showed antifungal activity against *Trichophyton longifusum*, *Aspergillus flavus*, *M. canis*, *Fusarium solani*, and strains of *Candida* spp. Haloxyline B was slightly more potent, which can be attributed to the presence of an additional alcoholic moiety on the C-3 carbon (Ferheen et al. 2005). On the other hand, the ethanolic extract of the bark of *Phellodendron amurense* presented antifungal activity in vitro against *T. mentagrophytes* similar to that of clotrimazole, as well as a partial efficacy of the extract in rabbits experimentally infected with *T. mentagrophytes*, which suggests that it may be an alternative for the treatment of dermatophytosis (Xiao et al. 2014).

16.3 Natural Volatile Extracts

16.3.1 Essential Oils

Essential oils are mixtures of volatile elements contained in various parts of the plant: flowers, buds, stems, leaves, seeds, twigs, roots, fruits, bark, and wood. Essential oils are stored in cavities, secretory cells, epidermic cells, canals, or glandular trichomes. The biological activity and the fragrance of essential oils are mainly due to two molecules, terpenes and phenylpropanoids, major elements of essential oils (Bakkali et al. 2008; Tariq et al. 2019).

In yeast, essential oils establish a membrane potential across the cell wall and disrupt ATP assembly, causing cell wall damage. Essential oils have the power to penetrate and break down the fungal cell wall and cell membrane through a permeabilization process, which ends in the disintegration of mitochondrial membranes (Aleksic and Knezevic 2014).

In most cases, these oils are obtained by steam distillation or by pressing the pericarp of the fruits (Silva-Santos et al. 2006). Likewise, the biological activity of an extract often cannot be experimentally reproducible because the chemical substances of the components may be different. In addition, the type of solvent used to obtain the extract and the extraction method can result in the loss of some compounds. On the other hand, some studies have shown changes in the chemical composition during the different stages of plant growth, and others have shown diversity based on the geographic area of origin (Binns et al. 2002; Fuentefria et al. 2018; Naghdi et al. 2008; Negri et al. 2014; Zhao et al. 2013). In addition, its chemical composition depends on numerous factors, mainly the geographical location of the plants, ecotype or variety, biological and physicochemical properties of the soil, climate, nutrition, stress during growth, fertilizers used, and seasonal changes (Copetta et al. 2011; Raut and Karuppayil 2014).

The mechanism of action of essential oils on microorganisms has not been fully explained. However, it is known that the components of essential oils, due to their lipophilic properties, destroy the cell wall and cytoplasmic membrane of bacteria and fungi, which results in the leakage of cytoplasm and its subsequent coagulation (Basak and Guha 2018; Burt 2004; Hyldgaard et al. 2012; Nazzaro et al. 2013). Likewise, they inhibit the synthesis of DNA, RNA, proteins, and polysaccharides of a great variety of fungi and bacteria (Karpinski 2020) (Table 16.2).

16.3.2 Terpenoids

The most important terpenes present are monoterpenes (C10) (monoterpene hydrocarbons and oxygenated monoterpenes) and sesquiterpenes (C15) (sesquiterpene hydrocarbons and oxygenated sesquiterpenes). Monoterpenes, formed from the coupling of two isoprene units, are the most representative molecules of many essential oils (Bakkali et al. 2008; Lopes et al. 2017). Terpenoids are substances present in essential oils, and the terpene compounds of vegetable oils are frequently monoterpenes (approximately 90% of the volatile oil) and sesquiterpenes. Other terpenoids, such as diterpenes, are found only in volatile oils and can be extracted using solvents (Mello et al. 2003).

In several studies, it has been reported that terpenoids may show antioxidant and antimicrobial properties against pathogenic fungi, including *Candida* and dermatophytes (Barchiesi et al. 2008; Hammer et al. 2000; Marcos-Arias et al. 2011; Palmeira-de-Oliveira et al. 2009; Pauli 2006). However, there are few terpene compounds with well-established antifungal activity and well-described pharmacological properties. Among these are nine diterpenoids of *Clerodendrum eriophyllum* that showed good antifungal activity against *Candida* spp., *C. neoformans*, and *A. fumigatus*; also reported is the isolation and determination of a new diterpenoid abiethane (2-hydroxy-8,12-abiethadiene-3,11,14-trione). On the other hand, it has been observed that the terpenes of *Ocimum sanctum*, including methyl chavicol and linalool, *affected the synthesis of ergosterol and generate damage to the cell membrane* in *Candida* species, making it a good candidate as an antifungal agent (Amber et al. 2010; Machumi et al. 2010).

There is a wide variety of extracts derived from *Curcuma zedoaria*, *Psidium guajava*, *Aristolochia cymbifera*, *Plectranthus barbatus*, *Lippia alba*, *Hydrocotyle bonariensis*, *Herreria salsaparrilha*, *Mentha X piperita*, *Eleutherine bulbosa*, *Baccharis trimera*, *Calamintha adscendens*, *Albizia inundata*, *Bauhinia forficata*, *Cymbopogon citratus*, *Plectranthus grandis*, and *Euphorbia hirta* L., among others which have shown antifungal activity against *Candida* spp. (Pessini et al. 2003; Rajeh et al. 2010; Shinobu-Mesquita et al. 2011; Tempone et al. 2008).

Specifically, the essential oils obtained from *Achillea millefolium* and *Curcuma longa* demonstrated substantial inhibition of *C. albicans*, *C. glabrata*, and *C. tropicalis* (Ribeiro et al. 2010). *Nandina domestica* Thunb essential oil, containing 1-indolizinecarbazole, 2-pentanone, monophenol, aziridine, methylcarbinol, ethanone, furfural, 1-hydroxy-4-methylbenzene, 2(5H)-furanone, and 3,5-dimethylpyrazole as components, showed high levels of antifungal activity in vitro, and severely affected spore germination of *M. canis*, *T. rubrum*, and *T. mentagrophytes* (Bajpai et al. 2009).

On the other hand, some examples of compounds of natural volatile extracts (essential oils) are the lipophilic extracts of *Hypericum carinatum* with in vitro antifungal activity against strains of *Candida* spp., *Cryptococcus neoformans*, and *Rhodotorula mucilaginosa* (Barros et al. 2013) and the volatile oil of *Stenachaenium megapotamicum* with antifungal action against *Epidermophyton floccosum* and *T. rubrum* (Danielli et al. 2013); the mechanism of action of these compounds has not yet been completely defined (Table 16.1).

Within this group, monoterpenes (volatile oils with antifungal action against *Candida* spp. and dermatophytes; Miron et al. (2014)) also stand out due to their antifungal potential; the aqueous extract of *Euphorbia tirucalli* L. (De Oliveira et al. 2014); parnafungins (Roemer and Krysan 2014); anthraquinones from *Morinda tomentosa* with good activity against *C. albicans* (Favre-Godal et al. 2014); Brazilian red propolis against non-*Candida albicans* strains n-hexane extract (Pippi et al. 2015); the water-insoluble fraction of *Uncaria tomentosa* Willd (Moraes et al. 2015); and the extracts of *Artemisia sieberi* Besser with action against dermatophytes (Mahboubi and Kazempour 2015). In most cases, the mechanism of action was related to the destabilization of the fungal cell wall and membrane (Mahboubi and Kazempour 2015; Miron et al. 2014; Moraes et al. 2015; Pippi et al. 2015; Roemer and Krysan 2014).

Recently, antifungal activity has also been reported for *Acca sellowiana* with specific activity against *Candida glabrata* (Machado et al. 2016); *Nectandra* spp., with defined action against dermatophytes (Danielli et al. 2017); and *Artemisia sieberi* (Mahboubi and Kazempour 2015) and essential oils from *Zataria multiflora* (Mahboubi et al. 2017). The mechanism was also affecting the cell wall and membrane of fungi (Danielli et al. 2017; Machado et al. 2016), as well as inhibiting the production of fungal elastase (Mahboubi and Kazempour 2015).

16.4 Conclusion

Despite the activity of natural products, with some advantages over commercial drugs, there is still a long way to go to overcome their drawbacks and their use in therapeutics. Some examples are its efficacy from in vitro studies which has been confirmed using human and animal infection models.

Acknowledgments RH-C and JX-C thank the Consejo Nacional de Ciencia y Tecnología (CONACyT) for the support to the realization of this chapter.

References

Ahmad A, Khan A, Khan LA, Manzoor N. In vitro synergy of eugenol and methyleugenol with fluconazole against clinical Candida isolates. J Med Microbiol. 2010a;59:1178–84. https://doi.org/10.1099/jmm.0.020693-0.

Ahmad A, Khan A, Yousuf S, Khan LA, Manzoor N. Proton translocating ATPase mediated fungicidal activity of eugenol and thymol. Fitoterapia. 2010b;81:1157–62. https://doi.org/10.1016/j.fitote.2010.07.020.

Aleksic V, Knezevic P. Antimicrobial and antioxidative activity of extracts and essential oils of Myrtus communis L. Microbiol Res. 2014;169:240–54. https://doi.org/10.1016/j.micres.2013.10.003.

Amber K, Aijaz A, Immaculata X, Luqman KA, Nikhat M. Anticandidal effect of Ocimum sanctum essential oil and its synergy with fluconazole and ketoconazole. Phytomedicine. 2010;17:921–5. https://doi.org/10.1016/j.phymed.2010.02.012.

Bajpai VK, Yoon JI, Kang SC. Antifungal potential of essential oil and various organic extracts of Nandina domestica thunb. Against skin infectious fungal pathogens. Appl Microbiol Biotechnol. 2009;83:1127–33. https://doi.org/10.1007/s00253-009-2017-5.

Bakkali F, Averbeck S, Averbeck D, Idaomar M. Biological effects of essential oils—a review. Food Chem Toxicol. 2008;46:446–75. https://doi.org/10.1016/j.fct.2007.09.106.

Bang KH, Lee DW, Park HM, Rhee YH. Inhibition of fungal cell wall synthesizing enzymes by trans-cinnamaldehyde. Biosci Biotechnol Biochem. 2000;64:1061–3. https://doi.org/10.1271/bbb.64.1061.

Barchiesi F, Silvestri C, Arzeni D, Ganzetti G, Castelletti S, Simonetti O, et al. In vitro susceptibility of dermatophytes to conventional and alternative antifungal agents. Med Mycol. 2008;47:321–6. https://doi.org/10.1080/13693780802641920.

Barros FMC, Pippi B, Dresch RR, Dauber B, Luciano SC, Apel MA, et al. Antifungal and antichemotactic activities and quantification of phenolic compounds in lipophilic extracts of Hypericum spp. native to South Brazil. Ind Crop Prod. 2013;44:294–9. https://doi.org/10.1016/j.indcrop.2012.11.017.

Basak S, Guha P. A review on antifungal activity and mode of action of essential oils and their delivery as nano-sized oil droplets in food system. J Food Sci Technol. 2018;55:4701–10. https://doi.org/10.1007/s13197-018-3394-5.

Binns SE, Arnason JT, Baum BR. Phytochemical variation within populations of Echinacea angustifolia (asteraceae). Biochem Syst Ecol. 2002;30:837–54. https://doi.org/10.1021/jf011439t.

Borjihan H, Ogita A, Fujita K, Hirasawa E, Tanaka T. The vacuole-targeting fungicidal activity of amphotericin B against the pathogenic fungus Candida albicans and its enhancement by allicin. J Antibiot (Tokyo). 2009;62:691–7. https://doi.org/10.1038/ja.2009.103.

Burt S. Essential oils: their antibacterial properties and potential applications in foods-a review. Int J Food Microbiol. 2004;1:223–53. https://doi.org/10.1016/j.ijfoodmicro.2004.03.022.
Butts A, Palmer GE, Rogers PD. Antifungal adjuvants: preserving and extending the antifungal arsenal. Virulence. 2017;8:198–210. https://doi.org/10.1080/21505594.2016.1216283.
Copetta A, Bardi L, Bertolone E, Berta G. Fruit production and quality of tomato plants (Solanum lycopersicum L) are affected by green compost and arbuscular mycorrhizal fungi. Plant Biosyst. 2011;145:106–15. https://doi.org/10.1080/11263504.2010.539781.
Cordell GA. The alkaloids chemistry and biology, vol. 67. London: Academic Press; 2009.
Danielli LJ, Dos Reis M, Bianchini M, Camargo GS, Bordignon SAL, Guerreiro IK, et al. Antidermatophytic activity of volatile oil and nanoemulsion of Stenachaenium megapotamicum (Spreng.). Baker Ind Crops Prod. 2013;50:23–8. https://doi.org/10.1016/j.indcrop.2013.07.027.
Danielli LJ, Pippi B, Soares KD, Duarte JA, Maciel AJ, Machado MM, et al. Chemosensitization of filamentous fungi to antifungal agents using Nectandra Rol. ex Rottb. Species essential oils. Ind Crop Prod. 2017;102:7–15. https://doi.org/10.1016/j.indcrop.2017.03.013.
De Oliveira LFS, Fuentefria AM, Klein FS, Machado MM. Antifungal activity against Cryptococcus neoformans strains and genotoxicity assessment in human leukocyte cells of Euphorbia tirucalli L. Braz J Microbiol. 2014;45:1349–55. https://doi.org/10.1590/s1517-83822014000400027.
Dharmaratne HR, Piyasena KG, Tennakoon SB. A geranylated biphenyl derivative from Garcinia mangostana. Nat Prod Res. 2005;19:239–43. https://doi.org/10.1080/14786410410001710582.
Emile A, Waikedre J, Herrenknecht C, Fourneau C, Gantier JC, Hnawia E, et al. Bioassay-guided isolation of antifungal alkaloids from Melochia odorata. Phytother Res. 2007;21:398–400. https://doi.org/10.1002/ptr.2078.
Faria NC, Kim JH, Goncalves LA, Martins Mde L, Chan KL, Campbell BC. Enhanced activity of antifungal drugs using natural phenolics against yeast strains of Candida and Cryptococcus. Lett Appl Microbiol. 2011;52:506–13. https://doi.org/10.1111/j.1472-765X.2011.03032.x.
Favre-Godal Q, Dorsaz S, Queiroz EF, Conan C, Marcourt L, Wardojo BP, et al. Comprehensive approach for the detection of antifungal compounds using a susceptible strain of Candida albicans and confirmation of in vivo activity with the Galleria mellonella model. Phytochemistry. 2014;105:68–78. https://doi.org/10.1016/j.phytochem.2014.06.004.
Ferheen S, Ahmed E, Afza N, Malik A, Shah MR, Nawaz SA, et al. Haloxylines A and B, antifungal and cholinesterase inhibiting piperidine alkaloids from Haloxylon salicornicum. Chem Pharm Bull (Tokyo). 2005;53:570–2. https://doi.org/10.1248/cpb.53.570.
Fuentefria AM, Pippi B, Dalla Lana DF, Donato KK, de Andrade SF. Antifungals discovery: an insight into new strategies to combat antifungal resistance. Lett Appl Microbiol. 2018;66:2–13. https://doi.org/10.1111/lam.12820.
Guo N, Liu J, WuX BX, Meng R, Wang X, et al. Antifungal activity of thymol against clinical isolates of fluconazole-sensitive and -resistant Candida albicans. J Med Microbiol. 2009;58:1074–9. https://doi.org/10.1099/jmm.0.008052-0.
Guo N, Wu X, Yu L, Liu J, Meng R, Jin J, et al. In vitro and in vivo interactions between fluconazole and allicin against clinical isolates of fluconazole-resistant Candida albicans determined by alternative methods. FEMS Immunol Med Microbiol. 2010;58:193–201. https://doi.org/10.1111/j.1574-695X.2009.00620.x.
Hammer KA, Carson CF, Riley TV. Melaleuca alternifolia (tea tree) oil inhibits germ tube formation by Candida albicans. Med Mycol. 2000;38:354–61.
Han Y. Synergic anticandidal effect of epigallocatechin-O-gallate combined with amphotericin B in a murine model of disseminated candidiasis and its anticandidal mechanism. Biol Pharm Bull. 2007;30:1693–6. https://doi.org/10.1248/bpb.30.1693.
Hyldgaard M, Mygind T, Meyer RL. Essential oils in food preservation: mode of action, synergies and interactions with food matrix components. Front Microbiol. 2012;3:1–24. https://doi.org/10.3389/fmicb.2012.00012.
Iwazaki RS, Endo EH, Ueda-Nakamura T, Nakamura CV, Garcia LB, Filho BP. In vitro antifungal activity of the berberine and its synergism with fluconazole. Antonie Van Leeuwenhoek. 2010;97:201–5. https://doi.org/10.1007/s10482-009-9394-8.

Karpinski T. Essential oils of Lamiaceae family plants as antifungals. Biomol Ther. 2020;10:103. https://doi.org/10.3390/biom10010103.

Lohombo-Ekomba ML, Okusa PN, Penge O, Kabongo C, Choudhary MI, Kasende OE. Antibacterial, antifungal, antiplasmodial, and cytotoxic activities of Albertisia villosa. J Ethnopharmacol. 2004;93:331–5. https://doi.org/10.1016/j.jep.2004.04.006.

Lopes G, Pinto E, Salgueiro L. Natural products: an alternative to conventional therapy for dermatophytosis? Mycopathologia. 2017;182:143–67. https://doi.org/10.1007/s11046-016-0081-9.

Ma WG, Fuzzati N, Li QS, Yang CR, Stoeckli-Evans H, Hostettmann K. Polyphenols from Eriosema tuberosum. Phytochemistry. 1995;39:1049–61. https://doi.org/10.1016/0031-9422(95)00052-9.

Machado GDRM, Pippi B, Dalla Lana DF, Amaral AP, Teixeira ML, Souza KC, et al. Reversal of fluconazole resistance induced by a synergistic effect with Acca sellowiana in Candida glabrata strains. Pharm Biol. 2016;54:2410–9. https://doi.org/10.3109/13880209.2016.1158286.

Machumi F, Samoylenko V, Yenesew A, Derese S, Midiwo JO, Wiggers FT, et al. Antimicrobial and antiparasitic abietane diterpenoids from the roots of Clerodendrum eriophyllum. Nat Prod Commun. 2010;5:853–8.

Mahboubi M, HeidaryTabar R, Mahdizadeh E. The anti-dermatophyte activity of Zataria multiflora essential oils. J Mycol Med. 2017;27:232–7. https://doi.org/10.1016/j.mycmed.2017.03.001.

Mahboubi M, Kazempour N. The antifungal activity of Artemisia sieberi essential oil from different localities of Iran against dermatophyte fungi. J Mycol Med. 2015;25:e65–71. https://doi.org/10.1016/j.mycmed.2015.02.042.

Marcos-Arias C, Eraso E, Madariaga L, Quindós G. In vitro activities of natural products against oral Candida isolates from denture wearers. BMC Complement Altern Med. 2011;11:119–26. https://doi.org/10.1186/1472-6882-11-119.

Mello JCPD, Mentz LA, Petrovick PR. Farmacognosia: Da Planta ao Medicamento; Porto Alegre. Porto Alegre, Brazil: EFRGS; 2003.

Meng F, Zuo G, Hao X, Wang G, Xiao H, Zhang J, et al. Antifungal activity of the benzo[c]phenanthridine alkaloids from Chelidonium majus Linn against resistant clinical yeast isolates. J Ethnopharmacol. 2009;125:494–6. https://doi.org/10.1016/j.jep.2009.07.029.

Miron D, Battisti F, Silva FK, Lana AD, Pippi B, Casanova B, et al. Antifungal activity and mechanism of action of monoterpenes against dermatophytes and yeasts. Rev Bras Farmacogn. 2014;24:660–7. https://doi.org/10.1007/s11418-014-0884-2.

Moraes RC, Dalla Lana AJ, Kaiser S, Carvalho AR, De Oliveira LFS, Fuentefria AM, et al. Antifungal activity of Uncaria tomentosa (Willd.) D.C. against resistant non-albicans Candida isolates. Ind Crops Prod. 2015;69:7–14. https://doi.org/10.1016/j.indcrop.2015.01.033.

Naghdi BH, Soroushzadeh A, Rezazadeh SA, Sharifi M, Ghalavand A, Rezaei A. Evaluation of phytochemical and production potential of borage (Borago officinalis L.) during the growth cycle. J Med Plant. 2008;7:37–43.

Nazzaro F, Fratianni F, Martino L, Coppola R, Feo V. Effect of essential oils on pathogenic bacteria. Pharma (Basel). 2013;6:1451–74. https://doi.org/10.3390/ph6121451.

Negri M, Salci TP, Shinobu-Mesquita CS, Capoci IR, Svidzinski TI, Kioshima ES. Early state research on antifungal natural products. Molecules. 2014;19:2925–56. https://doi.org/10.3390/molecules19032925.

Nkunya MH, Makangara JJ, Jonker SA. Prenylindoles from Tanzanian Monodora and Isolona species. Nat Prod Res. 2004;18:253–8. https://doi.org/10.1080/14786410310001620529.

Orafidiya LO, Oyedele AO, Shittu AO, Elujoba AA. The formulation of an effective topical antibacterial product containing Ocimum gratissimum L. leaf essential oil. Int J Pharm. 2001;224:177–83. https://doi.org/10.1016/s0378-5173(01)00764-5.

Ostrosky-Zeichner L, Casadevall A, Galgiani JN, Odds FC, Rex JH. An insight into the antifungal pipeline: selected new molecules and beyond. Nat Rev Drug Discov. 2010;9:719–27. https://doi.org/10.1038/nrd3074.

Palmeira-de-Oliveira A, Salgueiro L, Palmeira-de-Oliveira R, Martinez-de-Oliveira J, Pina-Vaz C, Queiroz JA, et al. Anti-candida activity of essential oils. Mini Rev Med Chem. 2009;9:1292–305. https://doi.org/10.2174/138955709789878150.

Pauli A. Anticandidal low molecular compounds from higher plants with special reference to compounds from essential oils. Med Res Rev. 2006;26:223–68. https://doi.org/10.1002/med.20050.
Perry NB, Brennan NJ. Antimicrobial and cytotoxic phenolic glycoside esters from the New Zealand tree Toronia toru. J Nat Prod. 1997;60:623–6. https://doi.org/10.1021/np970013z.
Pessini GL, Holetz FB, Sanches NR, Cortez DAG, Dias Filho BP, Nakamura CV. Avaliação da atividade antibacteriana e antifúngica de extratos de plantas utilizados na medicina popular. Rev Bras Farmacogn. 2003;13:21–4. https://doi.org/10.1590/S0102-695X2003000300009.
Pierce CG, Srinivasan A, Uppuluri P, Ramasubramanian AK, Lopez-Ribot JL. Antifungal therapy with an emphasis on biofilms. Curr Opin Pharmacol. 2013;13:726–30. https://doi.org/10.1016/j.coph.2013.08.008.
Pinto E, Pina-Vaz C, Salgueiro L, Goncalves MJ, Costade-Oliveira S, Cavaleiro C, et al. Antifungal activity of the essential oil of Thymus pulegioides on Candida, Aspergillus and dermatophyte species. J Med Microbiol. 2006;55:1367–73. https://doi.org/10.1099/jmm.0.46443-0.
Pippi B, Lana AJ, Moraes RC, Güez CM, Machado M, de Oliveira LF, et al. In vitro evaluation of the acquisition of resistance, antifungal activity and synergism of Brazilian red propolis with antifungal drugs on Candida spp. J Appl Microbiol. 2015;118:839–50. https://doi.org/10.1111/jam.12746.
Rajeh MAB, Zuraini Z, Sasidharan S, Latha LY, Amutha S. Assessment of Euphorbia hirta L. leaf, flower, stem and root extracts for their antibacterial and antifungal activity and brine shrimp lethality. Molecules. 2010;15:6008–18. https://doi.org/10.3390/molecules15096008.
Raut JS, Karuppayil SM. A status review on the medicinal properties of essential oils. Ind Crop Prod. 2014;62:250–64. https://doi.org/10.1016/j.indcrop.2014.05.055.
Ribeiro DI, Alves MDS, Faria MGI, Svidzinski TIE, Nascimento IA, Ferreira FBP, et al. Determination of antifungal activity of essential oils of Curcuma longa L. (Zingiberaceae) and Achillea millefolium (Asteraceae) grown in the Northwest Paraná. Arq Ciênc Saúde UNIPAR. 2010;14:103–9.
Roemer T, Krysan DJ. Antifungal drug development: challenges, unmet clinical needs, and new approaches. Cold Spring Harb Perspect Med. 2014;4:1–14. https://doi.org/10.1101/cshperspect.a019703.
Rojas JJ, Ochoa VJ, Ocampo SA, Muñoz JF. Screening for antimicrobial activity of ten medicinal plants used in Colombian folkloric medicine: a possible alternative in the treatment of non-nosocomial infections. BMC Complement Altern Med. 2006;6:2. https://doi.org/10.1186/1472-6882-6-2.
Rubio OC, Cuellar Cuellar A, Rojas N, Castro HV, Rastrelli L, Aquino R. A polyisoprenylated benzophenone from Cuban propolis. J Nat Prod. 1999;62:1013–5. https://doi.org/10.1021/np980339n.
Schmidt B, Ribnicky DM, Poulev A, Logendra S, Cefalu WT, Raskin I. A natural history of botanical therapeutics. Metabolism. 2008;57:S3–9. https://doi.org/10.1016/j.metabol.2008.03.001.
Scorzoni L, Sangalli-Leite F, de Lacorte SJ, de Paula E, Silva AC, Costa-Orlandi CB, Fusco-Almeida AM, et al. Searching new antifungals: the use of in vitro and in vivo methods for evaluation of natural compounds. J Microbiol Methods. 2016;123:68–78. https://doi.org/10.1016/j.mimet.2016.02.005.
Sharma M, Manoharlal R, Negi AS, Prasad R. Synergistic anticandidal activity of pure polyphenol curcumin I in combination with azoles and polyenes generates reactive oxygen species leading to apoptosis. FEMS Yeast Res. 2010;10:570–8. https://doi.org/10.1111/j.1567-1364.2010.00637.x.
Shinobu-Mesquita CS, Bertoni TA, Guilhermetti E, Svidzinski TIEIE. Antifungal activity of the extract of Curcuma zedoaria (christm.) roscoe, zingiberaceae, against yeasts of the genus Candida isolated from the oral cavity of patients infected with the human immunodeficiency virus. Rev Bras Farmacogn. 2011;21:128–32. https://doi.org/10.1590/S0102-695X2011005000017.
Silva-Santos A, Antunes AMS, Bizzo HR, d'Avila LA. A participação da industria óleo-citrícola na balança comercial brasileira. Rev Bras Plant Med. 2006;8:8–13.

Singh U, Sarma B, Mishra P, Ray A. Antifungal activity of venenatine, an indole alkaloid isolated from Alstonia venenata. Folia Microbiol. 2000;45:173–6. https://doi.org/10.1007/BF02817419.

Soares LA, de Cássia Orlandi Sardi J, Gullo FP, de Souza Pitangui N, Scorzoni L, Leite FS, et al. Anti dermatophytic therapy prospects for the discovery of new drugs from natural products. Braz. J Microbiol. 2014;44:1035–41. https://doi.org/10.1590/S1517-83822014005000011.

Tariq S, Wani S, Rasool W, Shafi K, Bhat MA, Prabhakar A, et al. A comprehensive review of the antibacterial, antifungal and antiviral potential of essential oils and their chemical constituents against drug-resistant microbial pathogens. Microb Pathog. 2019;134:103580. https://doi.org/10.1016/j.micpath.2019.103580.

Tempone AG, Sartorelli P, Teixeira D, Prado FO, Calixto IARL, Lorenzi H, et al. Brazilian flora extracts as source of novel antileishmanial and antifungal compounds. Mem Inst Oswaldo Cruz. 2008;103:443–9. https://doi.org/10.1590/s0074-02762008000500006.

Vermerris W, Nicholson R. Families of phenolic compounds and means of classification. In: Phenolic compound biochemistry. Netherlands: Springer; 2008. p. 1–34.

Wilson RM, Danishefsky SJ. Small molecule natural products in the discovery of therapeutic agents: the synthesis connection. J Org Chem. 2006;71:8329–51. https://doi.org/10.1021/jo0610053.

Xiao C, Ji Q, Rajput ZI, Wei Q, Liu Y, Bao G. Antifungal efficacy of Phellodendron amurense ethanol extract against Trichophyton mentagrophytes in rabbits. Pak Vet J. 2014;34:219–23.

Zhao YP, Li JH, Yang ST, Fan J, Fu CX. Effects of postharvest processing and geographical source on phytochemical variation of Corydalis rhizome. Chin Herbal Med. 2013;5:151–7. https://doi.org/10.3969/j.issn.1674-6348.2013.02.013.

Chapter 17
Antifungal Vaccines: Current Status and Future Directions

Andrés Tirado-Sánchez, Denisse Vazquez-González, Brenda Sáenz-Dávila, and Alexandro Bonifaz

Contents

A. Tirado-Sánchez (✉)
Dermatology Department, Hospital General de México "Dr. Eduardo Liceaga", CDMX, Mexico City, Mexico

Clinical Medicine Department, Instituto Mexicano del Seguro Social, CDMX, Mexico City, Mexico

D. Vazquez-González
Dermatology Department, Hospital General de México "Dr. Eduardo Liceaga", CDMX, Mexico City, Mexico

B. Sáenz-Dávila
Clinical Medicine Department, Instituto Mexicano del Seguro Social, CDMX, Mexico City, Mexico

A. Bonifaz
Dermatology Department, Hospital General de México "Dr. Eduardo Liceaga", CDMX, Mexico City, Mexico

Mycology Laboratory, Hospital General de México "Dr. Eduardo Liceaga", CDMX, Mexico City, Mexico

M. G. Frías-De-León et al. (eds.), *The Impact of Climate Change on Fungal Diseases*, Fungal Biology, https://doi.org/10.1007/978-3-030-89664-5_17

Abbreviations

APC	Antigen-presenting cell
GM-CSF	Granulocyte-macrophage colony growth-stimulating factor
HSP90	Heat-shock protein 90
PAMP	Pathogen-associated molecular patterns
PRR	Pattern recognition receptors

17.1 Introduction

Currently, there are effective vaccines around the world against some viral and bacterial diseases that would otherwise represent important public health problems (Delany et al. 2014); the development of safe and effective vaccines has been a major barrier for other infectious agents including fungi, partly due to the lack of knowledge about the mechanisms involved in protective immunity (Zheng et al. 2018).

Some of the most severe fungal diseases involve endemic agents, restricted in certain geographical areas (Souza et al. 2019), and in many countries like Mexico, where preventive vaccination is not always mandatory, there are no reliable statistics or a cost-benefit analysis that justifies the need for vaccine development (Guadarrama-Orozco et al. 2015).

The renewed interest is mainly related to the changes in the epidemiological landscape of fungal diseases; some reports analyze the usefulness of preventing various systemic mycoses and the impact on their incidence, for example, vaccination of seronegative adults for coccidioidomycosis residents of the San Joaquin Valley in California, would prevent up to 11 deaths per year and also save three million dollars in healthcare expenses (Oliveira et al. 2021).

Vaccine development is required for severe fungal pathogens, including *Candida albicans*, *Cryptococcus neoformans*, *Aspergillus fumigatus*, and dimorphic fungi (Table 17.1). Many challenges confront fungal vaccine development, including different host risk factors and fungal pathogenesis (Kischkel et al. 2020). No single antigen can be used in a "Universal Fungal Vaccine"; rather, specifically tailored vaccines will be required for the major fungal pathogens (Hamad 2012).

In the case of invasive candidiasis, the mortality rate has remained stable (around 0.4 per 100,000 inhabitants in the United States) from 1997 to 2003 despite the development of new antifungal drugs effective against *Candida* spp.

Table 17.1 Main agent-producing primary mycosis and opportunistic fungi

Primary pathogens	Opportunistic pathogens
Coccidioides immitis/C. posadasii	*Candida* spp.
Histoplasma capsulatum	*Cryptococcus neoformans*
Blastomyces dermatitidis	*Aspergillus* spp.
Paracoccidioides brasiliensis	*Pneumocystis jirovecii*

In hematopoietic cell transplant recipients, the response to antifungal drugs as prophylaxis against invasive aspergillosis achieves 31%, and mortality persists in about 60% of cases (Kischkel et al. 2020).

Important advances have been made in the understanding of the immune mechanisms of the host against pathogenic fungi, their virulence mechanisms, genetic sequencing, as well as the recognition of antigenic molecules (carbohydrates and especially proteins) that stimulate a protective immunity, being suitable candidates for effective vaccines (Krylov and Nifantiev 2020).

One difficulty observed in the development of a fungal vaccine is that the patients who are most susceptible to opportunistic fungal infections are those least able to mount protective responses; as in cases of pulmonary blastomycosis and histoplasmosis, $CD4^+$ cells are important in vaccine immunity against the causal extracellular pathogen and the facultative intracellular pathogen, respectively (Iannitti et al. 2012). Another impediment relates to the limited number of licensed vaccine adjuvants. Candidate adjuvants that act on multiple innate and antigen-specific host defense pathways are likely to be the most effective in protecting against opportunistic fungal infections. The definition of adjuvants has mostly been restricted to those that stimulated antibody titers (e.g., *Pneumococcus*) or in the case of the Calmette-Guerin bacillus vaccine, delayed-type hypersensitivity responses. Recently, the concept of adjuvants has been expanded to include soluble mediators and antigenic carriers (e.g., endotoxins, Flt3 ligand, and heat-shock protein) that activate antigen-presenting cells and stimulate innate and cellular immunity (Portuondo et al. 2015).

17.2 Basics about Vaccination and Types of Immunization

The basic characteristics of an effective vaccine are the induction of specific immunity to a certain etiological agent, avoiding the invasion of the microorganism, and neutralizing its toxins. On the other hand, the protection provided must be durable, which depends on the effective stimulation of memory T and B lymphocytes (Da Silva et al. 2020; Nami et al. 2019).

17.2.1 Active Immunization

Active immunization is carried out by an antigen or immunogenic molecule that triggers a cellular immune response and the production of antibodies (Zheng et al. 2018).

Infections exclusively human, whose etiological agents are stable, will be more susceptible of being controlled by vaccination, but not those whose causative agents show a wide antigenic variation, an animal or environmental reservoirs for the disease, or highly infectious microorganisms; primary and opportunistic fungal pathogens mostly meet the latter characteristics, hence the difficulty of developing vaccines that achieve the eradication of the different mycoses; however, the antifungal vaccines developed to date do not seek disease eradication, but rather an effective immunological mechanism for disease control in specific populations (Lionakis and Levitz 2018).

The types of vaccines that generate active immunization by promoting various defensive mechanisms are those of live-attenuated or inactivated cells, while those of killed agents and those of protein subunits predominantly trigger CD8+ T cytotoxic responses; this type of immune response contributes to the defense against most pathogenic fungi, such as *Histoplasma capsulatum* (sl), *Paracoccidioides brasiliensis* (sl), *C. neoformans*, and *Pneumocystis jirovecii* (da Costa et al. 2015; Wuthrich et al. 2003).

17.2.2 Passive Immunization

It consists of the transfer of preformed specific antibodies (immune serum) against various antigens. It is characterized by being transient since it is only maintained for the life of the antibody, but it does not induce their autonomous production or a cellular immune response; furthermore, it does not induce specific memory in the event of subsequent pathogen exposure (García-Carnero et al. 2018). An example of this type of immunization is immune sera antibodies against snake venom and scorpion. Mucosal antibodies, certain monoclonal antibodies (human and murine), and some genetically engineered antibody fragments have been successfully used in certain mycoses; several of these sera can be administered to patients at risk before the period of immunosuppression; effectiveness is due to the longevity of IgG (weeks to months) even during periods of prolonged immunosuppression. Gigliotti et al. (2002) showed that the prophylaxis with monoclonal antibodies (IgM and IgG1) in a murine model is effective against *P. jirovecii*.

The use of recombinant human monoclonal antibodies is currently being tested, for example, against certain antigens such as HSP90 (heat-shock protein 90). A very interesting aspect about the use of antibodies has been observed by Nosanchuk et al. (2003), with systemic histoplasmosis, in which antibodies against a histone-like surface protein promote opsonization, phagocytosis, and destruction of intracellular yeasts, which is protective in previously immunized murine models, showing a reduction in the fungal load and the pulmonary inflammatory process. Table 17.2 summarizes the different types of antifungal vaccines, the proposed antigens, and the immune mechanism underlying their protective function.

17.3 The Immune Response against Fungi

Based on the interaction with the host, the major human disease-causing fungi can be divided into primary and opportunistic pathogens. The defensive response of the host against different microorganisms, morphologically, genetically, and

Table 17.2 Main types of vaccines proposed for active and passive immunization

Type of vaccine	Antigens	Immune mechanism triggered
Whole cells or cell extracts		
Aspergillosis	Inactivated conidia	Undefined
	Live-attenuated conidia	Undefined
Blastomycosis	BAD1 strain (live attenuated)	TH1 cell immunity
Candidosis	CA2 strain (live attenuated)	TH1 cell immunity
	Ribosomal cell fraction	Cellular and antibody-mediated immunity
	Inactivated complete cells	Undefined
Coccidioidomycosis	Inactivated spherules (endospores)	TH1 cell immunity
Histoplasmosis	Ribosomal vaccine	Undefined
DNA vaccines		
Coccidioidomycosis	More than one gene	Undefined
Paracoccidioidomycosis	Gen *gp43*	TH1 and TH2 cellular immunity
	Gen *rPb27*	Cellular and antibody-mediated immunity
	Hsp65	Undefined
Pneumocystosis	Gen *Kexin*	Cellular and antibody-mediated immunity
T cells and CPA vaccines		
Aspergillosis	Dendritic cells loaded with *Aspergillus* spp. antigens	TH1 cell immunity
Candidosis	Dendritic cells loaded with *Candida* spp. antigens	TH1 cell immunity
Vaccine type	Antigens	Triggered immune mechanism
Protein subunits and glycoconjugates		
Aspergillosis	Asp f 3	TH1 cell immunity
	β-1-3-glucan	Growth inhibitory antibodies
Candidosis	Agglutinin-like sequences	Cellular immunity
	Secreted aspartic proteinase 2 (Sap2)	Anti-Sap2 antibodies
	65 kDa mannoprotein	Adhesin-neutralizing antibodies
	β-1-3-glucan	Growth inhibitor and fungicidal antibodies
	β-1-2-Mannosides	Opsono-phagocytic antibodies (possibly adhesion blockers)
Coccidioidomycosis	Antigen 2	TH1 cell immunity
	β-1-3-glucosyltransferases	Undefined
	Chimeric polyprotein	Undefined
Cryptococcosis	Capsular polysaccharide	Various mechanisms
	Glucoxylomannan conjugate vaccine	Unknown, possibly from antibodies
Paracoccidioidomycosis	Glycoprotein 43 kDa	Cellular and antibody-mediated immunity

(continued)

Table 17.2 (continued)

Type of vaccine	Antigens	Immune mechanism triggered
Pneumocystosis	P55 protein (major surface glycoprotein)	Not defined, possibly by antibodies
	Kexin protease	Cellular and antibody-mediated immunity
Antibodies		
Candidosis	Anti-HSP90 peptide antibodies	Unknown
	Anti-β-1-3-glucan 2G8 antibodies	Growth inhibition
	C7 antibody (stress mannoprotein)	Fungicide
	Single chain anti-idiotypic antibodies (variable fragment)	Fungal antibodies
	Anti-mannan C6 antibodies	Opsono-phagocytic
	Anti-glucosyl antibodies	Fungicides
Cryptococcosis	Anti-glucuroxylomannan 18B7 antibodies (murine)	Opsono-phagocytic
	Anti-glucuroxylomannan IgG2 antibodies (human)	Opsono-phagocytic
Histoplasmosis	Antibodies against histone-like proteins	Undefined

biochemically, requires adequate coordination of its two main constituents: innate and acquired immunity, the latter involves two effector mechanisms: humoral (antibody) and cell-mediated immunity (Netea et al. 2015).

17.3.1 Innate Immunity

The innate immune system is the frontline of defense against any pathogen (primary or opportunistic) and also contributes to the development of antifungal vaccines, mainly concerning antigen-presenting cell (APC) interaction. Highly conserved antigenic molecules on the fungal surface, also called pathogen-associated molecular patterns (PAMP), detect and interact with pattern recognition receptors (PRR) of the APCs, releasing inflammatory mediators and the recruitment of innate effector cells (neutrophils, macrophages, and dendritic cells), for the processing and further antigen presentation (Pilecki and Moeller 2020). Neutrophils are the main innate effector cells against fungi, due to the phagocytic and microbicide characteristics through the production of oxygen-free radicals (Gazendam et al. 2016).

APCs, through PAMP-PRR interactions, antigen processing, and presentation to T lymphocytes, constitute the link with acquired immunity, promoting the activation of naive T cells and the proper differentiation into Th1 and Th2 cells (Sirisinha 2011).

Among the PRRs of APCs identified in fungal infections, toll-like receptors, especially type 2 and 4, dectin-1, the Fcγ receptor, and mannose receptors, have prominent roles in detecting cell wall PAMPs (Thind et al. 2015).

The complement system and other humoral factors, such as antifungal peptides, mannose-binding lectins, defensins, and collectins, also provide fundamental defense mechanisms through the opsonization of fungi (Sharma and Bisht 2020). Recognition of deposited complement particles on β-(1,6)-glucans of the fungus surface by complement receptor 3 leads to elimination of pathogens by phagocytic cells, a process called opsonophagocytosis (Chamilos and Carvalho 2020). Defensins, secreted by the epithelium and Paneth cells in the gut, and collectins are involved in opsonizing, and also in the induction of inflammatory responses, collaborating with Th17-derived cytokines (Shankar et al. 2018).

17.3.2 Acquired Immunity

Cellular immunity is the fundamental defense mechanism against fungal infections, through its two effector mechanisms: cytotoxicity (secretion of cytokines) and humoral (antibody-mediated response) (Polonelli et al. 2000). The effective response against fungi is regulated by Th1 lymphocytes and dendritic cells, as well as the increase in INF-γ, TNF-α, and IL-12 levels. INF-γ induces cell-mediated immunity through phagocytes and Th17 cells to release IL-17 and IL-22 cytokines, mediating the neutrophilic response and the release of antimicrobial peptides (defensins) to the site of infection. Finally, regulatory T cells release TGF-α and IL-10 to repress the elevated levels of inflammatory responses (De Luca et al. 2010).

Some pathogenic fungi including *H. capsulatum* can block the aforementioned signals by binding to the CD11b glycoprotein receptor, which suppresses IL-12 production by macrophages, resulting in a poor Th1 response, with inadequate activation, and thus increasing fungus survival (Marth and Kelsall 1997).

Th2 responses to fungi are usually harmful to the host, mainly due to the production of IL-4, IL-5, and IL-10, which favors disease progression, increases the production of IgE, and increases the presence of eosinophils (O'Dea et al. 2014). However, Th2 mechanisms help to regulate the cellular immune response, avoiding uncontrolled tissue damage. In normal subjects, there is a balance between cellular and humoral immunity; immunosuppressed patients often change the balance of both immune responses, developing defective Th1 and active Th2 responses, leading to disease spreading (Li et al. 2020).

17.4 Fungal Vaccines

Different fungal antigens have been studied for use as vaccines, and they are usually classified into three main groups including live-attenuated, recombinant, and conjugate vaccines (Fig. 17.1).

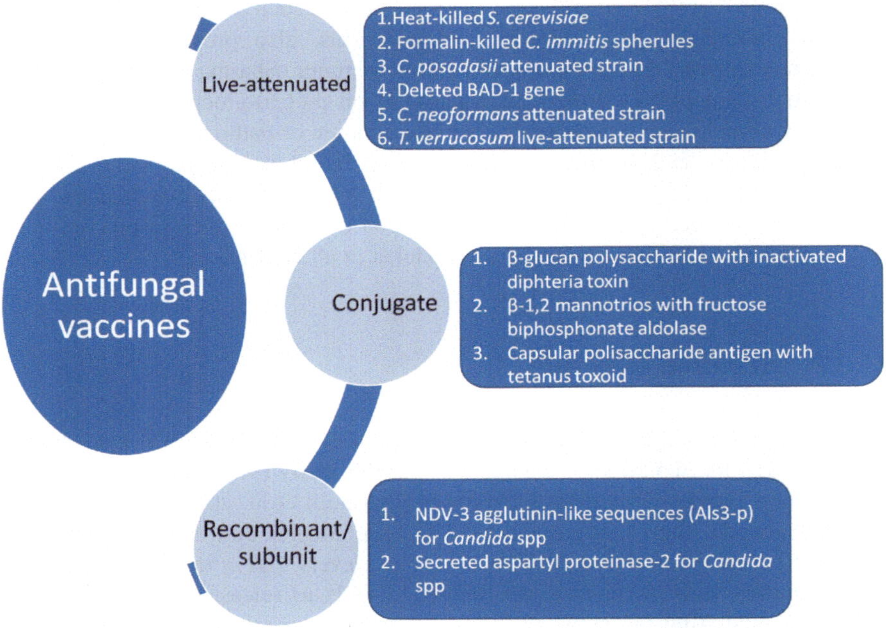

Fig. 17.1 Main categories of antifungal vaccines

17.4.1 Live-Attenuated or Inactivated Fungal Vaccines

According to the similarity of live-attenuated vaccines with infectious agents, they launch long-term and strong immune responses, efficient among immunocompetent patients. These vaccines will be applicable for endemic fungal pathogen prevention shortly in subjects with a healthy immune system living in endemic areas (Zheng et al. 2018).

Xue et al. (2009) used a genetically modified vaccine against coccidioidomycosis, where two chitinase genes (CTS2 and CTS3) have been modified to give way to an attenuated strain unable of endosporulation and, thus, which has no infectious potential but keeps its immunogenic properties, demonstrating that the triggered immune response was both Th1 and Th2 cell types with the respective production of cytokines, and developing well-defined granulomas on histopathological examination.

Other novel vaccines, including the heat-killed *Saccharomyces cerevisiae* yeast vaccine, have shown protective activity against virulent strains of *Coccidioides posadasii*, *C. albicans*, and *A. fumigatus* (Martinez et al. 2017). This type of vaccines is not for use in immunocompromised patients, and given the difficulties in their production, it is not possible to ensure consistency and standardization of their quality (Martinez et al. 2017); however, few studies in blastomycosis and experimental histoplasmosis suggest that the use of live-attenuated cell vaccines could be useful in immunocompetent individuals, and even in patients with specific $CD4^+$ failure, as evidenced by studies carried out with live-attenuated and modified strains of *Blastomyces dermatitidis*, which present loss of the BAD1 gene that encodes a surface protein (immunologic target and virulence determinant), which makes the vaccine highly protective in $CD4^+$ immunodeficient murine models (Szymczak and Deepe Jr. 2010). According to the results of these studies, the immune response induced by vaccination is carried out by $CD8^+$ cells, unlike immunocompetent patients, where the presence of $CD4^+$ cells is necessary for effective immune responses.

$CD8^+$ cells secrete IFN-γ, TNF-α, and granulocyte-macrophage colony growth-stimulating factor (GM-CSF), developing a protective immune response and long-lasting immune memory, which suggests that the behavior of $CD8^+$ T cells is different against viruses, bacteria, and fungi, as well as in immunocompetent and immunosuppressed patients (Liu et al. 2011). The main challenge in immunosuppressed patients is that live-attenuated vaccines do not induce another kind of disease (Pati et al. 2018), leading to poor prognosis.

17.4.2 Purified Antigen (Protein Subunit and Carbohydrate, Recombinant) Vaccines

Vaccines developed from protein and polysaccharide subunits are currently the most studied among fungal vaccines. They consist in the transfer and expression of a gene that encodes for an antigen that triggers the desired immune response. This antigen consists of one or more purified proteins (usually recombinant) and polysaccharides, often conjugated with highly immunogenic proteins such as toxoids, or an adjuvant (e.g. aluminum salts), increasing their immunogenic properties (Nami et al. 2019; Specht et al. 2017).

Some of these vaccines are based on peptide and polysaccharide antigens, commonly expressed in different fungal species, opening up the possibility of the so-called universal vaccine (Nami et al. 2019), where the same antigen protects against several pathogens, similar to the aforementioned *S. cerevisiae* vaccine.

Ito et al. (2006), in a study with sonicated hyphae (fragmented by ultrasound) of *A. fumigatus*, found antigenic molecules common to various filamentous fungi and yeast, such as the protein antigens Aspf3 (Jolink et al. 2015).

17.4.3 DNA Vaccines

DNA vaccines transfer DNA plasmids encoding one or more protein antigens, which stimulate $CD4^+$ and $CD8^+$ cellular immune responses through MHC II and I, respectively, with the cytokine production pathways and antibodies. One of the advantages of this type of immunization is the ability to control the orientation and magnitude of the immune response to certain genetic products, mainly due to the expression within the host's cells, allowing endogenous antigen processing and further presentation (Milan et al. 2004).

DNA vaccines are useful in immunocompromised patients with significant $CD4^+$ T cell deficiency, and it is possible to introduce additional DNA sequences (immunostimulatory) that encode cytokines and adjuvants of the immune response. Most studies with DNA vaccines have been carried out in diseases such as coccidioidomycosis and mainly paracoccidioidomycosis (Nami et al. 2019). One of the most widely studied antigens is a proline-rich protein, associated with a reduction in the fungal load in *C. posadasii* strains in studies with exposed mice (Herr et al. 2007).

The first fungal DNA vaccine was related to tinea due to *Trichophyton verrucosum* (Wawrzkiewicz and Wawrzkiewicz 1992). The exocellular protein gp43 from *P. brasiliensis* is the main antigen studied in vaccination trials against this disease and a good example of the use of DNA vaccines showing protection in mice susceptible to infection (Oliveira and Coelho 2017). Other antigens of *P. brasiliensis* have been fractionated and analyzed; two of the fractions obtained (F0 and FII) confer protection against paracoccidioidomycosis in mice, while another fraction (FIII) seems to exacerbate the disease, suggesting that the F0 and FII fractions contain candidate molecules for vaccine development (Diniz et al. 2004).

The most recently studied fraction is a 27 kDa (Pb27) protein of 219 amino acids that is part of the F0 fraction. Assays with cDNA fragments encoding this antigenic protein (rPb27) show that it is protective in vaccinated mice producing high levels of TNF-α, INF-γ, and IgG2b, moderate levels of IgG1, and low levels of IgG2a and IL-10, and developing a clinically limited disease (Coitinho et al. 2019). Similarly, genetic immunization studies are carried out with some other antigens including heat-shock protein 65 (DNA Hsp65 vaccine), with favorable results in both paracoccidioidomycosis and chromoblastomycosis due to *Fonsecaea pedrosoi*, and other non-fungal diseases such as tuberculosis and leishmaniasis (Siqueira et al. 2013).

17.4.4 Monoclonal Antibody Vaccines

Monoclonal antibody vaccines are currently used as prophylaxis or as adjuvants in immunotherapy with antifungal agents. The factors to take into account in the development of these antibodies are isotype affinity and specificity, whether as immunotherapy or prophylaxis. Among the most recent advances in the use of recombinant human monoclonal antibodies and their fragments against various systemic mycoses (Rodrigues et al. 2007) is an antibody called Mycograb, directed against HSP90, which has shown protection against systemic candidiasis (Cabezas et al. 2010). Other antibodies without the Fc region (crystallizable fragment) have antifungal efficacy, suggesting that they can act even in the absence of phagocytic effector cells or complement, which would be useful in these patients; thus, these antibodies act as opsonins and neutralize virulence (adhesins) and fungal growth (Nami et al. 2019). A histone-type surface antigen of *H. capsulatum* (H2B-like protein) has been identified against which specific antibodies are directed, increasing its opsonization, inhibition of cell growth, and destruction within macrophages, reducing fungal load and lung inflammation, and improving host survival (Nosanchuk et al. 2003).

In DBA/2 mice given *Candida* or *Aspergillus* intravenosuly, solubilized cell wall β-glucan was markedly decreased in a dose-dependent manner. Moreover, in deep mycoses, the anti-β-glucan antibody levels are usually decreased, and this change correlated with clinical symptoms and other parameters such as C-reactive protein. It is suggested that the anti-β-glucan antibody forms an antigen-antibody complex and participates in the immune response as a pathogenic recognizing fungi molecule (Ishibashi et al. 2005).

17.5 The Importance and Difficulties of Developing Antifungal Vaccines

Fungal diseases are rare in immunocompetent patients, whereas groups of immunocompromised individuals are often at risk of developing invasive fungal infections. The impact of the increase in the number of people affected by fungal diseases can be already seen, such as in the case of *C. albicans*, whose mortality rate can reach 60% when associated with invasive infection. Another example is *C. neoformans* yeasts, an environmental fungus causing the most common fungal meningoencephalitis worldwide in immunocompromised patients, accounting for more than 600,000 deaths per year; also, it has been found that other *Cryptococcus* spp., such as *C. gattii*, can also affect immunocompetent hosts (Desnos-Ollivier et al. 2015).

Species from the genus *Aspergillus* are associated with the second most common cause of nosocomial infection; the mortality rates for invasive aspergillosis can reach ≥50% in some cases, which is even more dramatic than candidiasis (Lestrade et al. 2019).

Most of these infections afflict patients with severe immunodeficiency. Current antifungal drugs have limitations, such as toxicity, availability, and the spectrum of activity, and may have major drug interactions. The development of resistance, when used for long periods, is also an important limitation. It is important to develop new strategies involving fungal vaccines to reduce the risk of death in these patients (Ulrich and Ebel 2020).

The goal of an efficient fungal vaccine is to generate immune responses that will lead to immunological memory and protection against recurrent exposure to fungi (as occur in certain endemic areas). In recent years, many vaccine candidates have been tested against some fungal pathogens, such as *C. albicans*, *Aspergillus* spp., *Cryptococcus* spp., *Blastomyces* spp., *P. brasiliensis*, and *Sporothrix* spp. (Medici and Del Poeta 2015); opportunistic pathogens represent a major challenge for vaccine development since the most affected group includes immunosuppressed patients that are ineligible for active immunization given their underlying immunological conditions (Cassone and Torosantucci 2006) (Fig. 17.2). However, several

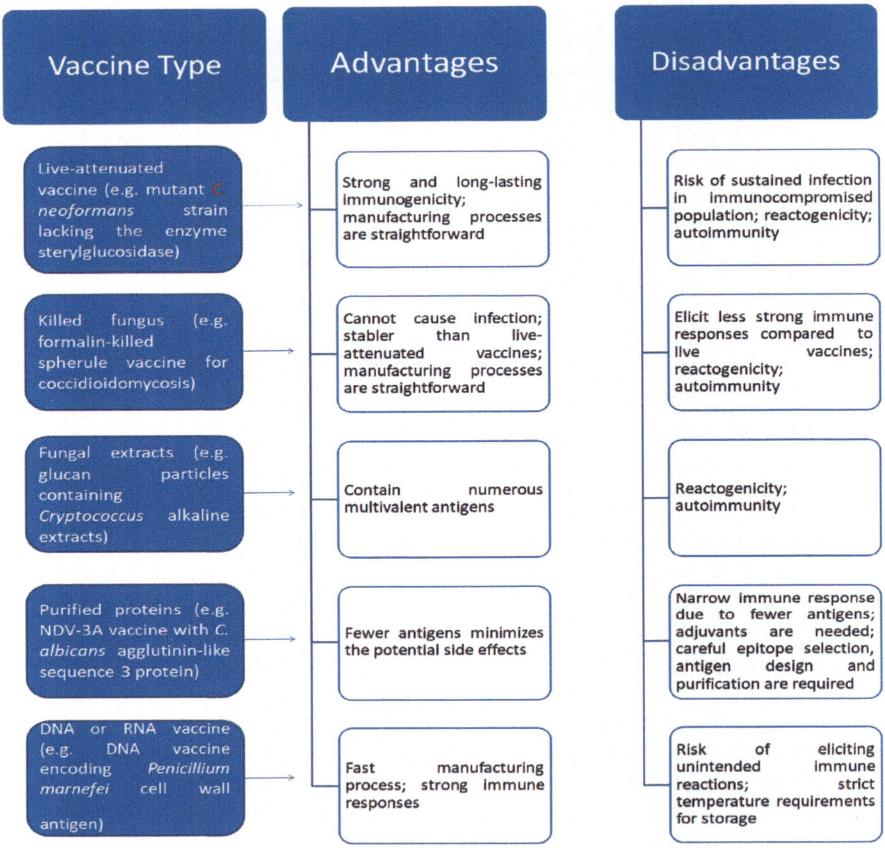

Fig. 17.2 Main advantages and disadvantages of the major antifungal vaccines

studies in murine models have shown that the immunization of mice against *Aspergillus* spp. is effective, before the use of immunosuppressive therapy by transplantation; protection has been presented during the period of neutropenia. It has also been used during treatment with potent immunosuppressant for chronic graft-versus-host disease in response to subsequent exposure to *Aspergillus* spp. conidia (Sahin et al. 2016).

Immunosuppressed patients present failures in various immunological mechanisms (cellular and humoral immunity, production of several cytokines and effector cells with immune functions), which makes vaccine design a challenge (Zubair et al. 2017); patients with HIV/AIDS who have a deficiency in $CD4^+$ Th1 lymphocytes do not have a good response in the subsequent production of INF-γ, TNF-α, and IL-12 cytokines, which are the main phagocytic cell stimulators. However, several studies indicate that certain fungal antigens can elicit immune responses of $CD8^+$ even in the absence of $CD4^+$ lymphocytes; therefore, in immunosuppressed patients, it is possible to induce the recruitment of

CD8+ cells by vaccination and stimulate effective immune protection (Nanjappa et al. 2012).

Such obstacles in the development of fungal vaccines have been gradually overcome in the last decade, and although to date there is no authorized fungal vaccine or in advanced phases of clinical trials for active immunization in humans, the main advances have been carried out in murine models with multiple candidate antigens to be used as vaccines. For more than 20 years, studies have been carried out on anti-dermatophyte vaccines for prophylaxis of bovine cattle, and there is a live-attenuated vaccine against *T. verrucosum* that is successfully used in countries like Norway, where systematic vaccination of cattle has made the disease practically inexistent. Unfortunately, the fungal vaccines used in veterinary medicine have manufacturing standards that do not match with the development of vaccines for human use (Nami et al. 2019). Only a prophylactic vaccine against *C. immitis*, made with formalin-inactivated spherules, has been tested in humans, with poor results, developing multiple side effects, mainly local reactions (edema and injection site pain), as well as insufficient protection against pulmonary coccidioidomycosis (Huppert et al. 1967). Moreover, additional difficulties posed by pathogens such as *Candida* spp., with its ability to evade the immune mechanisms of the host, make vaccine development complicated (Tso et al. 2018). The goal of developing a vaccine against *Candida* spp. is not to induce immunity against this fungus belonging to the fungal microbiota but to generate an immune response that controls the fungus and prevents severe symptoms, including systemic candidiasis and fungemia. Recently, two univalent vaccines against *Candida* (targeting a single specific antigen) have concluded the first phase of clinical trials (Martin-Cruz et al. 2021); however, some experts, such as Cassone (2013), suppose that these vaccines won't control invasive candidiasis since *Candida* has multiple virulence factors and easily evades the mechanism of action of a vaccine directed to a single antigen.

17.6 Conclusion

Currently, due to the knowledge about the immunopathogenic mechanisms of fungi and the technological advances in molecular biology, in addition to the new antifungal treatments, the development of fungal vaccines offers the possibility of addressing fungal infections from the immunotherapeutic point of view.

Fungal diseases are often endemic and may lead to poor prognosis, so the development of vaccines directed at these agents to prevent and treat these diseases is important and also challenging. One of the main factors that contribute to the increase in the frequency and severity of the disease includes cellular and humoral immune deficiency; vaccination alternatives need to be safe and effective in hosts with normal and deficient immune systems. Advances in proteomics and systems biology have facilitated the development of more vaccines and the characterization of their modifications, functions, and interactions. Advanced studies with fungal vaccines set the stage for the design of safe and effective preventive strategies.

References

Cabezas J, Albaina O, Montañez D, Sevilla MJ, Moragues MD, Pontón J. Potential of anti-Candida antibodies in immunoprophylaxis. Immunotherapy. 2010;2(2):171–83. https://doi.org/10.2217/imt.09.76.

Cassone A. Development of vaccines for Candida albicans: fighting a skilled transformer. Nat Rev Microbiol. 2013;11(12):884–91. https://doi.org/10.1038/nrmicro3156.

Cassone A, Torosantucci A. Opportunistic fungi and fungal infections: the challenge of a single, general antifungal vaccine. Expert Rev Vaccines. 2006;5(6):859–67. https://doi.org/10.1586/14760584.5.6.859.

Chamilos G, Carvalho A. Aspergillus fumigatus DHN-Melanin. Curr Top Microbiol Immunol. 2020;425:17–28. https://doi.org/10.1007/82_2020_205.

Coitinho JB, Costa MAF, Melo EM, Morais EA, de Andrade LGA, da Rocha AM, et al. Structural and immunological characterization of a new nucleotidyltransferase-like antigen from Paracoccidioides brasiliensis. Mol Immunol. 2019;112:151–62. https://doi.org/10.1016/j.molimm.2019.04.028.

da Costa TA, Di Gangi R, Martins P, Figuereido Longhini AL, Zanucoli F, Rodrigues de Oliveira AL, et al. Protection against Paracoccidioides brasiliensis infection in mice treated with modulated dendritic cells relies on inhibition of interleukin-10 production by CD8+ T cells. Immunology. 2015;146(3):486–95. https://doi.org/10.1111/imm.12526.

Da Silva LBR, Taborda LC, Nosanchuk JD. Advances in fungal peptide vaccines. J Fungi (Basel). 2020;6(3):119. https://doi.org/10.3390/jof6030119.

De Luca A, Zelante T, D'Angelo C, Zagarella S, Fallarino F, Spreca A, et al. IL-22 defines a novel immune pathway of antifungal resistance. Mucosal Immunol. 2010;3(4):361–73. https://doi.org/10.1038/mi.2010.22.

Delany I, Rappuoli R, De Gregorio E. Vaccines for the 21st century. EMBO Mol Med. 2014;6(6):708–20. https://doi.org/10.1002/emmm.201403876.

Desnos-Ollivier M, Patel S, Raoux-Barbot D, Heitman J, Dromer F, French Cryptococcosis Study Group. Cryptococcosis serotypes impact outcome and provide evidence of Cryptococcus neoformans speciation. MBio. 2015;6(3):e00311. https://doi.org/10.1128/mBio.00311-15.

Diniz SN, Reis BS, Goes TS, Zouain CS, Leite MF, Goes AM. Protective immunity induced in mice by F0 and FII antigens purified from Paracoccidioides brasiliensis. Vaccine. 2004;22(3–4):485–92. https://doi.org/10.1016/j.vaccine.2003.07.017.

García-Carnero LC, Pérez-García LA, Martínez-Álvarez JA, Reyes-Martínez JE, Mora-Montes HM. Current trends to control fungal pathogens: exploiting our knowledge in the host-pathogen interaction. Infect Drug Resist. 2018;11:903–13. https://doi.org/10.2147/IDR.S170337.

Gazendam RP, van de Geer A, Roos D, van den Berg TK, Kuijpers TW. How neutrophils kill fungi. Immunol Rev. 2016;273(1):299–311. https://doi.org/10.1111/imr.12454.

Gigliotti F, Haidaris CG, Wright TW, Harmsen AG. Passive intranasal monoclonal antibody prophylaxis against murine Pneumocystis carinii pneumonia. Infect Immun. 2002;70:1069–74. https://doi.org/10.1128/IAI.70.3.1069-1074.2002.

Guadarrama-Orozco JH, Vargas-López G, Viesca-Treviño C. Decisiones de los padres que no arriesgan la vida de sus hijos, pero que los exponen a daños serios: no a las vacunas [parental decisions that do not risk children's lives but place them at significant harm: foregoing vaccination]. Bol Med Hosp Infant Mex. 2015;72(5):353–7. https://doi.org/10.1016/j.bmhimx.2015.09.007.

Hamad M. Universal fungal vaccines: could there be light at the end of the tunnel? Hum Vaccin Immunother. 2012;8(12):1758–63. https://doi.org/10.4161/hv.21838.

Herr RA, Hung CY, Cole GT. Evaluation of two homologous proline-rich proteins of Coccidioides posadasii as candidate vaccines against coccidioidomycosis. Infect Immun. 2007;75(12):5777–87. https://doi.org/10.1128/IAI.00807-07.

Huppert M, Levine HB, Sun SH, Peterson ET. Resistance of vaccinated mice to typical and atypical strains of Coccidioides immitis. J Bacteriol. 1967;94(4):924–7. https://doi.org/10.1128/jb.94.4.924-927.1967.

Iannitti RG, Carvalho A, Romani L. From memory to antifungal vaccine design. Trends Immunol. 2012;33(9):467–74. https://doi.org/10.1016/j.it.2012.04.008.
Ishibashi K, Yoshida M, Nakabayashi I, Shinohara H, Miura NN, Adachi Y, et al. Role of anti-beta-glucan antibody in host defense against fungi. FEMS Immunol Med Microbiol. 2005;44(1):99–109. https://doi.org/10.1016/j.femsim.2004.12.012.
Ito JI, Lyons JM, Hong TB, Tamae D, Liu YK, Wilczynski SP, et al. Vaccinations with recombinant variants of Aspergillus fumigatus allergen Asp f 3 protect mice against invasive aspergillosis. Infect Immun. 2006;74(9):5075–84. https://doi.org/10.1128/IAI.00815-06.
Jolink H, de Boer R, Willems LN, van Dissel JT, Falkenburg JH, Heemskerk MH. T helper 2 response in allergic bronchopulmonary aspergillosis is not driven by specific Aspergillus antigens. Allergy. 2015;70(10):1336–9. https://doi.org/10.1111/all.12688.
Kischkel B, Rossi SA, Santos SR, Nosanchuk JD, Travassos LR, Taborda CP. Therapies and vaccines based on nanoparticles for the treatment of systemic fungal infections. Front Cell Infect Microbiol. 2020;10:463. https://doi.org/10.3389/fcimb.2020.00463.
Krylov VB, Nifantiev NE. Synthetic carbohydrate based anti-fungal vaccines. Drug Discov Today Technol. 2020;35-36:35–43. https://doi.org/10.1016/j.ddtec.2020.11.002.
Lestrade PP, Bentvelsen RG, Schauwvlieghe AFAD, Schalekamp S, van der Velden WJFM, Kuiper EJ, et al. Voriconazole resistance and mortality in invasive aspergillosis: a multicenter retrospective cohort study. Clin Infect Dis. 2019;68(9):1463–71. https://doi.org/10.1093/cid/ciy859.
Li LX, Hole CR, Rangel-Moreno J, Khader SA, Doering TL. Cryptococcus neoformans evades pulmonary immunity by modulating xylose precursor transport. Infect Immun. 2020;88(8):e00288–20. https://doi.org/10.1128/IAI.00288-20.
Lionakis MS, Levitz SM. Host control of fungal infections: lessons from basic studies and human cohorts. Annu Rev Immunol. 2018;36:157–91. https://doi.org/10.1146/annurev-immunol-042617-053318.
Liu M, Clemons KV, Bigos M, Medovarska I, Brummer E, Stevens DA. Immune responses induced by heat killed Saccharomyces cerevisiae: a vaccine against fungal infection. Vaccine. 2011;29(9):1745–53. https://doi.org/10.1016/j.vaccine.2010.12.119.
Marth T, Kelsall BL. Regulation of interleukin-12 by complement receptor 3 signaling. J Exp Med. 1997;185(11):1987–95. https://doi.org/10.1084/jem.185.11.1987.
Martin-Cruz L, Sevilla-Ortega C, Benito-Villalvilla C, Diez-Rivero CM, Sánchez-Ramón S, Subiza JL, et al. A combination of polybacterial MV140 and Candida albicans V132 as a potential novel trained immunity-based vaccine for genitourinary tract infections. Front Immunol. 2021;11:612269. https://doi.org/10.3389/fimmu.2020.612269.
Martinez M, Clemons KV, Stevens DA. Heat-killed yeast as a pan-fungal vaccine. Methods Mol Biol. 2017;1625:23–30. https://doi.org/10.1007/978-1-4939-7104-6_2.
Medici NP, Del Poeta M. New insights on the development of fungal vaccines: from immunity to recent challenges. Mem Inst Oswaldo Cruz. 2015;110(8):966–93. https://doi.org/10.1590/0074-02760150335.
Milan R, Alois R, Josef C, Jana B, Evzen W. Recombinant protein and DNA vaccines derived from hsp60 Trichophyton mentagrophytes control the clinical course of trichophytosis in bovine species and Guinea-pigs. Mycoses. 2004;47(9–10):407–17. https://doi.org/10.1111/j.1439-0507.2004.01028.x.
Nami S, Mohammadi R, Vakili M, Khezripour K, Mirzaei H, Morovati H. Fungal vaccines, mechanism of actions and immunology: a comprehensive review. Biomed Pharmacother. 2019;109:333–44. https://doi.org/10.1016/j.biopha.2018.10.075.
Nanjappa SG, Heninger E, Wüthrich M, Sullivan T, Klein B. Protective antifungal memory CD8(+) T cells are maintained in the absence of CD4(+) T cell help and cognate antigen in mice. J Clin Invest. 2012;122(3):987–99. https://doi.org/10.1172/JCI58762.
Netea MG, Joosten LA, van der Meer JW, Kullberg BJ, van de Veerdonk FL. Immune defence against Candida fungal infections. Nat Rev Immunol. 2015;15(10):630–42. https://doi.org/10.1038/nri3897.

Nosanchuk JD, Steenbergen JN, Shi L, Deepe GS, Casadevall A. Antibodies to a cell surface histone-like protein protect against Histoplasma capsulatum. J Clin Invest. 2003;112:1164–75. https://doi.org/10.1172/JCI19361.

O'Dea EM, Amarsaikhan N, Li H, Downey J, Steele E, Van Dyken SJ, et al. Eosinophils are recruited in response to chitin exposure and enhance Th2-mediated immune pathology in Aspergillus fumigatus infection. Infect Immun. 2014;82(8):3199–205. https://doi.org/10.1128/IAI.01990-14.

Oliveira AF, Coelho PSR. Yeast expressing Gp43 protein as a vaccine against Paracoccidioides brasiliensis infection. Methods Mol Biol. 2017;1625:213–24. https://doi.org/10.1007/978-1-4939-7104-6_14.

Oliveira LVN, Wang R, Specht CA, Levitz SM. Vaccines for human fungal diseases: close but still a long way to go. NPJ Vaccines. 2021;6(1):33. https://doi.org/10.1038/s41541-021-00294-8.

Pati R, Shevtsov M, Sonawane A. Nanoparticle vaccines against infectious diseases. Front Immunol. 2018;9:2224. https://doi.org/10.3389/fimmu.2018.02224.

Pilecki B, Moeller JB. Fungal recognition by mammalian fibrinogen-related proteins. Scand J Immunol. 2020;92(4):e12925. https://doi.org/10.1111/sji.12925.

Polonelli L, Casadevall A, Han Y, Bernardis F, Kirkland TN, Matthews RC, et al. The efficacy of acquired humoral and cellular immunity in the prevention and therapy of experimental fungal infections. Med Mycol. 2000;38(Suppl 1):281–92.

Portuondo DL, Ferreira LS, Urbaczek AC, Batista-Duharte A, Carlos IZ. Adjuvants and delivery systems for antifungal vaccines: current state and future developments. Med Mycol. 2015;53(1):69–89. https://doi.org/10.1093/mmy/myu045.

Rodrigues ML, Shi L, Barreto-Bergter E, Nimrichter L, Farias SE, Rodrigues EG, et al. Monoclonal antibody to fungal glucosylceramide protects mice against lethal Cryptococcus neoformans infection. Clin Vaccine Immunol. 2007;14(10):1372–6. https://doi.org/10.1128/CVI.00202-07.

Sahin U, Toprak SK, Atilla PA, Atilla E, Demirer T. An overview of infectious complications after allogeneic hematopoietic stem cell transplantation. J Infect Chemother. 2016;22(8):505–14. https://doi.org/10.1016/j.jiac.2016.05.006.

Shankar J, Cerqueira GC, Wortman JR, Clemons KV, Stevens DA. RNA-seq profile reveals Th-1 and Th-17-type of immune responses in mice infected systemically with Aspergillus fumigatus. Mycopathologia. 2018;183(4):645–58. https://doi.org/10.1007/s11046-018-0254-9.

Sharma D, Bisht GS. Recent updates on antifungal peptides. Mini Rev Med Chem. 2020;20(4):260–8. https://doi.org/10.2174/1389557519666190926112423.

Siqueira IM, Ribeiro AM, Nóbrega YK, Simon KS, Souza AC, Jerônimo MS, et al. DNA-hsp65 vaccine as therapeutic strategy to treat experimental chromoblastomycosis caused by Fonsecaea pedrosoi. Mycopathologia. 2013;175(5–6):463–75. https://doi.org/10.1007/s11046-012-9599-7.

Sirisinha S. Insight into the mechanisms regulating immune homeostasis in health and disease. Asian Pac J Allergy Immunol. 2011;29(1):1–14.

Souza IEL, Fernandes FF, Schiavoni MCL, Silva CL, Panunto-Castelo A. Therapeutic effect of DNA vaccine encoding the 60-kDa-heat shock protein from Paracoccidioides brasiliensis on experimental paracoccidioidomycosis in mice. Vaccine. 2019;37(37):5607–13. https://doi.org/10.1016/j.vaccine.2019.07.090.

Specht CA, Lee CK, Huang H, Hester MM, Liu J, Luckie BA, et al. Vaccination with recombinant Cryptococcus proteins in glucan particles protects mice against cryptococcosis in a manner dependent upon mouse strain amBio.01872-17. nd cryptococcal species. MBio. 2017;8(6):e01872–17. https://doi.org/10.1128/mBio.01872-17.

Szymczak WA, Deepe GS Jr. Antigen-presenting dendritic cells rescue CD4-depleted CCR2−/− mice from lethal Histoplasma capsulatum infection. Infect Immun. 2010;78(5):2125–37. https://doi.org/10.1128/IAI.00065-10.

Thind SK, Taborda CP, Nosanchuk JD. Dendritic cell interactions with Histoplasma and Paracoccidioides. Virulence. 2015;6(5):424–32. https://doi.org/10.4161/21505594.2014.965586.

Tso GHW, Reales-Calderon JA, Pavelka N. The elusive anti-Candida vaccine: lessons from the past and opportunities for the future. Front Immunol. 2018;9:897. https://doi.org/10.3389/fimmu.2018.00897.

Ulrich S, Ebel F. Monoclonal antibodies as tools to combat fungal infections. J Fungi (Basel). 2020;6(1):22. https://doi.org/10.3390/jof6010022.

Wawrzkiewicz K, Wawrzkiewicz J. An inactivated vaccine against ringworm. Comp Immunol Microbiol Infect Dis. 1992;15(1):31–40. https://doi.org/10.1016/0147-9571(92)90099-d.

Wuthrich M, Filutowicz HI, Warner T, Deepe GS Jr, Klein BS. Vaccine immunity to pathogenic fungi overcomes the requirement for CD4 help in exogenous antigen presentation to CD8+ T cells: implications for vaccine development in immune-deficient hosts. J Exp Med. 2003;197(11):1405–16. https://doi.org/10.1084/jem.20030109.

Xue J, Chen X, Selby D, Hung CY, Yu JJ, Cole GT. A genetically engineered live attenuated vaccine of Coccidioides posadasii protects BALB/c mice against coccidioidomycosis. Infect Immun. 2009;77(8):3196–208. https://doi.org/10.1128/IAI.00459-09.

Zheng Z, Diaz-Arévalo D, Guan H, Zeng M. Noninvasive vaccination against infectious diseases. Hum Vaccin Immunother. 2018;14(7):1717–33. https://doi.org/10.1080/21645515.2018.1461296.

Zubair S, Azhar A, Khan N, Ahmad E, Ajmal M, Owais M. Nanoparticle-based mycosis vaccine. Methods Mol Biol. 2017;1625:169–211. https://doi.org/10.1007/978-1-4939-7104-6_13.

The manufacturer's authorised representative in the EU is Springer Nature Customer Service Centre GmbH, Europaplatz 3, 69115 Heidelberg, Germany. If you have any concerns regarding our products, please contact ProductSafety@springernature.com

Printed and bound by CPI Group (UK) Ltd, Croydon, CR0 4YY
30/09/2026
02225496-0008